HVACR 101

AIR CONDITIONING CONTRACTORS OF AMERICA
PHCC EDUCATIONAL FOUNDATION
REFRIGERATION SERVICE ENGINEERS SOCIETY

Australia • Brazil • Japan • Korea • Mexico • Singapore • Spain • United Kingdom • United States

HVACR 101
Air Conditioning Contractors of America, Plumbing-Heating-Cooling-Contractors– National Association Educational Foundation, Refrigeration Service Engineers Society

Vice President, Technology and Trades Professional Business Unit: Gregory L. Clayton

Product Development Manager: Ed Francis

Development: Nobina Chakraborti

Product Manager: Vanessa L. Myers

Director of Marketing: Beth A. Lutz

Executive Marketing Manager: Taryn Zlatin

Marketing Manager: Marissa Maiella

Production Director: Carolyn Miller

Production Manager: Andrew Crouth

Content Project Manager: Kara A. DiCaterino

Art Director: Benjamin Gleeksman

Production Technology Analyst: Thomas Stover

© 2009 Air Conditioning Contractors of America, Plumbing-Heating-Cooling-Contractors–National Association Educational Foundation, Refrigeration Service Engineers Society

ALL RIGHTS RESERVED. No part of this work covered by the copyright herein may be reproduced, transmitted, stored, or used in any form or by any means graphic, electronic, or mechanical, including but not limited to photocopying, recording, scanning, digitizing, taping, Web distribution, information networks, or information storage and retrieval systems, except as permitted under Section 107 or 108 of the 1976 United States Copyright Act, without the prior written permission of the publisher.

For product information and technology assistance, contact us at
Professional Group Cengage Learning Customer & Sales Support, 1-800-354-9706

For permission to use material from this text or product, submit all requests online at **cengage.com/permissions**.

Further permissions questions can be e-mailed to
permissionrequest@cengage.com.

Library of Congress Control Number: 2008903934

ISBN-13: 978-1-4180-6663-5
ISBN-10: 1-4180-6663-X

Delmar
5 Maxwell Drive
Clifton Park, NY 12065-2919
USA

Cengage Learning is a leading provider of customized learning solutions with office locations around the globe, including Singapore, the United Kingdom, Australia, Mexico, Brazil and Japan. Locate your local office at: **international.cengage.com/region**

Cengage Learning products are represented in Canada by Nelson Education, Ltd.

For your lifelong learning solutions, visit **delmar.cengage.com**

Visit our corporate website at **cengage.com**.

Notice to the Reader
Publisher does not warrant or guarantee any of the products described herein or perform any independent analysis in connection with any of the product information contained herein. Publisher does not assume, and expressly disclaims, any obligation to obtain and include information other than that provided to it by the manufacturer. The reader is expressly warned to consider and adopt all safety precautions that might be indicated by the activities described herein and to avoid all potential hazards. By following the instructions contained herein, the reader willingly assumes all risks in connection with such instructions. The publisher makes no representations or warranties of any kind, including but not limited to, the warranties of fitness for particular purpose or merchantability, nor are any such representations implied with respect to the material set forth herein, and the publisher takes no responsibility with respect to such material. The publisher shall not be liable for any special, consequential, or exemplary damages resulting, in whole or part, from the readers' use of, or reliance upon, this material.

Printed in Canada
1 2 3 4 5 XX 10 09 08

Brief Contents

Preface xii
Acknowledgments xiii
ACCA–PHCC Educational Foundation–RSES Subject Matter Experts . xiv
About the Author xv

Chapter 1   Welcome to the World of HVACR 1
Chapter 2   Introduction to Systems and Major Components . . . 19
Chapter 3   Introduction to Basic Electricity 51
Chapter 4   Introduction to Thermodynamics 81
Chapter 5   Basic Refrigeration Cycle 109
Chapter 6   Scheduled Maintenance 139
Chapter 7   Systematic Problem Solving 157
Chapter 8   Basic Installation and Repair Methods 181
Chapter 9   Energy Efficient Installations 227
Chapter 10   Selling and Customer Service 253

Answer Key 273
Glossary 279
Index 321

Contents

Preface xii

Acknowledgments xiii

ACCA–PHCC Educational Foundation–RSES Subject Matter Experts. xiv

About the Author. xv

Chapter 1 Welcome to the World of HVACR 1
   Introduction 2
   What Is HVACR? 5
   Why This Industry Provides an Exciting and Diverse Career! 5
      Where Will I Be Working? 6
      What Do Technicians Do?. 6
      Role of the Technician 10
   Professional Certification 10
   Technicians' Responsibility for Record Keeping 12
   Career Path. 13
   The Short History of Air Conditioning 13

Chapter 2 Introduction to Systems and Major Components 19
   Introduction 20
   How Does Cooling Occur? 20
   The Refrigeration Cycle. 21
      Refrigerant 22
      Compressors 22
      Condensers 24
      Metering Device 24
      Evaporators. 26
   Distribution Systems: Ducts, Refrigerant Piping, and Hydronic Piping 28
      Air Ducts 28
      Duct Distribution Systems 30
   Refrigerant Piping. 32
      Refrigerant Piping Identification. 34

Hydronic Piping 35
Steam Heating 37
Gas Heating 37
   Controls Section 37
   Blower Section 37
   Burner and Heat Exchanger Section 37
   Venting Section 38
Oil Heating 40
   Fuel Supply 40
   Burner Section 40
   Heat Exchanger 40
   Venting Section 41
Electric Strip Heat 42
Heat Pumps 42
   Cooling Mode 43
   Heating Mode 44
   Defrost Mode 44
Geothermal 44
Hydrogen Fuel Cell 46
Solar Energy 47

Chapter 3 Introduction to Basic Electricity 51

Introduction 52
I Want to Speak in Electrical Terms 53
   Conductors and Wires 53
   Volts 54
   Amperes 56
   Resistance 56
   Power 57
   Complete Circuit 57
Ohm's Law 57
Calculating Power 60
Introduction to Meters 61
   Multimeter 61
   Selecting a Good Multimeter 64
Clamp-on Ammeter 64
Electrical Symbols 66
Schematic Diagrams 70
Conductors and Wires 72

Chapter 4 Introduction to Thermodynamics 81

Introduction 82
Definitions and Concepts 82
   What Is Heat? 82
   British Thermal Units, Sensible Heat, and Latent
      Heat 83

Heat Quantity and Intensity 86
Temperature Scales 86
Heat Flow 87
    Conduction 88
    Convection 89
    Radiation 90
Characteristics of Solids, Liquids, and Gases 90
    Motion of the Molecules 90
    Melting and Fusion 91
    Sublimation 91
    Evaporation 92
    Boiling Point 93
    Sensible Heat of Vapor 93
    More on Latent Heat 94
    Applying Sensible and Latent Heat 94
    Specific Heat 94
    Latent Heat of Fusion 95
    Latent Heat of Vaporization 96
    Transferring Heat 96
What Is Pressure? 97
    Gauge Pressure 97
    Absolute Pressure 97
What Is Matter? 100
What Is a Scientific Law? 100
    The Laws of Thermodynamics 102
    Boyle's Law of Volume and Pressure 102
    Charles's Law of Volume and Pressure 103
    Dalton's Law of Volume and Pressure 104
    What Happens When Temperature, Pressure, and
        Volume All Change? 104

Chapter 5 Basic Refrigeration Cycle 109

Introduction 110
Basic Refrigeration Cycle 111
Saturation, Superheat, and Subcooling 112
    Saturation 112
    Superheat 120
    Subcooling 120
Compressors 122
    Reciprocating Compressors 123
    Operation of the Reciprocating Compressor . . . 123
    Semi-hermetic Compressors 125
    Scroll Compressor 125
    Rotary Compressor 126
Condenser 129
    Types of Condensers 129
Metering Device 129

Types of Metering Devices 131
Evaporator 134

Chapter 6 Scheduled Maintenance 139

Introduction 140
Scheduled Maintenance Is Not Service 140
Why Do Scheduled Maintenance? 140
Scheduling Maintenance and Inspection Work 142
Improvement Opportunities 142
*What to Check in a Scheduled Maintenance and
  Inspection Program* 142
Customer Complaints 143
Outdoor and Indoor Conditions 143
Air Filters 143
Cleaning Coils 146
Check Pressures 149
System Charge 149
Wiring, Connections, and Component Inspection . . . 149
Emergency Drain Pan and Float Switch 150
Duct Air Leaks 151
Thermostat Operation 152
Lubricate and Clean Blowers and Fans 152
Measure Amperage and Voltage 152
Additional Comments 153
Go Beyond Expectations 154
Review the Report with the Customer 154
Maintenance Contracts 154
Tools, Equipment, and Supplies Requirements 155

Chapter 7 Systematic Problem Solving 157

Introduction 158
Communication Is Important 158
To Obtain the Symptoms, Listen to the Customer . . . 158
Possible Causes 159
Finding Opens in the Control Sequence 160
Is It a Mechanical, an Electrical, or an Airflow
  Problem? 161
Electromechanical Sequence 162
Required Tools, Instrumentation, and Supplies 164
Diagnosis 166
Productivity 166
Evaluation 167
Case Studies 167
Case Study One 167
Case Study Two 168
Case Study Three 169

Case Study Four 171
Case Study Five 172
Case Study Six 174

Chapter 8 Basic Installation and Repair Methods . . 181

Introduction 182
Typical Field Tools 182
Basic Installation and Repair Methods 182
    Types of Air Handlers 183
    Clearance 183
    Outdoor Unit Placement 184
    Air Handler Installation 186
    Basic Wiring 188
    Installing the Condensing Unit or Package Unit 189
    Installing the Indoor Section 190
Start-up and Final Checkout 191
Cutting, Swaging, Soldering, Brazing, and Flaring . . 191
    Cutting and Swaging 191
    Soldering 193
    Brazing 195
    Flaring 199
Fasteners 201
Codes 202
Manufacturers' Requirements 203
Tools and Instruments 203
    Manifold Gauge Set 203
    Valve Core Removal Tool 205
    Temperature Testers 205
    Digital Psychrometer 205
    Power Tools 205
    Vacuum Pumps 207
    Airflow Measuring Devices 207
    Digital Scale 209
    Volt, Ohm, and Milliammeter 210
    Hand Tools 210
    Refrigeration Wrench 210
    Micron Gauge 211
    Various Hand Tools 211
Piping Materials and Methods 218
    Types and Sizes of Copper Tubing 218
    Copper Fittings 220
    Refrigeration Trap Requirements 221
Insulation 222
    Refrigerant Pipe Insulation 222
    Ductwork Insulation 222
    Condensate Drain Line Insulation 222

Condensate Drain 223
Typical Installation Problems and Failures 224

Chapter 9 Energy Efficient Installations 227

Introduction 228
First Cost, Maintenance Cost, and Operating Cost. . .228
    Maintenance Cost. 229
    Operating Cost. 229
Watts, Kilowatts, and Cost of Energy. 229
Conversion and Comparison of Fuels. 230
Measures of Efficiency: EER, SEER, COP, and HSPF . .231
    Energy Efficiency Ratio 231
    Seasonal Energy Efficiency Ratio. 231
    Coefficient of Performance 231
    Heating Season Performance Factor 233
Human Comfort 233
Alternative Energy Systems. 234
    Solar Energy 234
    Wind Energy 237
    Geothermal Systems .237
    Future Fuel Cells 238
    Green Buildings 239
Effects of Maintenance on Efficiency. 239
Energy Code. 240
Government and Utility Incentives 240
Energy Star Program .241
    Energy Star for the Home 241
Energy-Efficient Installation: Best Practices 241
    Cooling and Heat Load Calculations. 242
    Equipment Sizing, Selection, and Efficiency 243
    Duct Sizing, Support, and Sealing 245
    Refrigerant Pipe Sizing, Support, and Insulation . . .245
    Thermostat Installation. 247
    Miscellaneous Components. 247
    Replacing the Condensing Unit. 248

Chapter 10 Selling and Customer Service 253

Introduction 254
Customer Satisfaction 255
Listen. 256
Understanding Added Value from the Customer's Perspective 257
Value Selling 258
    Comfort Value 258
    Energy Bill Value 259

Indoor Air Quality (IAQ) 260
Scheduled Maintenance Value. 260
Other Problems Relating to HVAC 261
Why Replace Versus Repair 261
The Environment 262
Value of Investment 262
Role of the Technician as a Salesperson. 263
Flat-Rate Pricing. 264
Value of Energy Savings. 265
Uniforms, Language, and Customer Interaction. . . 266
Language 266
Customer Interaction 266
Quality and Cleanliness 266
Professional Affiliation and Certification 267
Air Conditioning Contractors of America (ACCA) . . . 268
Plumbing-Heating-Cooling Contractors–National
  Association (PHCC) 268
Refrigeration Service Engineers' Society (RSES) 268
North American Technician Excellence (NATE). 269
HVAC Excellence 269
UA STAR. 269
Performance of a System as Perceived by the
  Customer. 270
Performance of a System: What Do I Say to
  Mrs. Smith? 270
Right System for the Right Building at the
  Right Price 270
Unit and Component Capacity. 271

Answer Key. 273

Glossary. 279

Index. 321

Preface

This is the first book in a four-year series intended for the training of HVACR technicians. Suitable for use in formal educational or professional settings or for any apprenticeship or training-related program, the series emphasizes a blend of conceptual material and real world applications.

This book was developed in partnership with the following organizations: Air Conditioning Contractors of America (ACCA), Plumbing-Heating-Cooling-Contractors—National Association Educational Foundation (PHCC Educational Foundation), and Refrigeration Service Engineers Society (RSES). These organizations appointed a team of subject matter experts composed of outstanding HVACR professionals and educators. This group met on several occasions to develop a curriculum that would reflect industry needs and an approach to which today's technicians would relate. Throughout the series, topics are revisited in more depth and with additional applications to coincide with students' increasing work experience in the field.

Manuscript and proofs were reviewed, improved, and approved by the subject matter experts. The writers were approved and encouraged to follow specified guidelines. Each writer has a rich background full of field and teaching experience.

Contractors and manufacturers have requested a training program germane to the requirements of the industry. This book and series respond to that need.

Acknowledgments

The publisher wishes to thank the following companies and individuals who provided illustrations and permission to reproduce them:

Bill Johnson
Bristol Compressors International, Inc.
Carrier Corporation
Dick Wirz, Refrigeration Training Services
Emerson Climate Technologies
Eugene Silberstein
Ideal Industries, Inc.
International Association of Plumbing and Mechanical Officials
John A. Tomczyk
Knauf Insulation GmbH
National Fire Protection Association
North American Technician Excellence
Refrigeration Technologies
Ritchie Engineering Co. - YELLOW JACKET Products Division
Sporlan Division, Parker Hannifin Corporation
Trane Inc.
William C. Whitman

ACCA–PHCC Educational Foundation–RSES Subject Matter Experts

Hugh Cole (RSES), Certification and Training Services, Lawrenceville, GA
Greg Goater (ACCA), Isaac Heating & Air Conditioning, Inc., Rochester, NY
Al Guzik (ACCA), Energy Management Specialists, Inc., Cleveland, OH
Ivan Maas (RSES), North Dakota State College of Science, Wahpeton, ND
Terry Miller (ACCA), Energy Management Specialists, Cleveland, OH
Tom Moore (PHCC Educational Foundation), Climate Control, Inc., Leesburg, IN
Nick Reggi (RSES), Humber College, Toronto, ON, Canada
Dick Shaw (ACCA), Technical Education Consultant & Standards Manager, ACCA, Hesperia, MI
Jamie Simpson (PHCC Educational Foundation), Schaal Heating & Cooling, Des Moines, IA

About the Author

Joe Moravek is the author of the *Blueprint and Plans for HVAC* textbook. He holds a master's degree in education and is currently the corporate training coordinator at Hunton Trane. He has a Class A air conditioning and refrigeration license for the State of Texas, holds a third-grade stationary engineers license with the City of Houston, and is certified by North American Technician Excellence (NATE). He has also served as the lead air conditioning instructor at Lee College. Joe has been an air conditioning inspector and technician with Houston for six years, as well as the owner of Mechanical Training Services since 1993. Joe currently resides in Houston, Texas.

CHAPTER 1

Welcome to the World of HVACR

LEARNING OBJECTIVES

The student will:
- Describe what a HVACR technician does on the job.
- List the various career paths available for a technician.
- List reasons it is important to maintain professional records.
- Describe which part of the HVACR industry sounds most interesting.

INTRODUCTION

Why is a heating, ventilation, air conditioning, and refrigeration career so interesting? The answer is that you can do almost anything you want. You can work in the residential, commercial, industrial, or transportation environment. If you like assembling or putting things together, there is a spot for you. Want to diagnose and correct problems like a doctor? There is a job for that. Would you like to be a quality controller? Inspectors do that. Regardless of specialty, every technician needs to meet several goals in order to be successful and progress in this profitable and rewarding career.

You will be working on residential equipment, Figure 1-1, and commercial and industrial equipment, Figure 1-2. Normally, you will not be working on both unless your company specializes in all types of equipment. Most companies work on air conditioning or refrigeration equipment and further subdivide the jobs into residential and commercial applications.

The first goal facing every potential technician is the desire to be in the industry. Second, educate yourself like you are doing now. Third, learn in the field. Learn from others in addition to applying what you have learned and what you will be learning in the classroom. Finally, continuing education is important. This trade is like many technical and professional careers in that there is always more to learn. That is why it is such an interesting career—you will never learn it all. Just like the medical field has specialists, the HVACR field has specialists. You will decide what specialty you want as your career.

Figure 1-1
Gas furnace is illustrated on the left (Courtesy of Carrier) and a residential condensing unit on the right. (Photo by Susan Brubaker) This is common equipment that the technician maintains and repairs.

Figure 1–2
Two large chillers. This equipment can be larger than a full-size car. It is maintained by an industrial technician. (Photos by Lynn Talbot)

This chapter is designed to expose you to the various career possibilities in our profession. We are reliant upon heating, ventilation, air conditioning, and refrigeration. Think of a world without professionals to design, install, and maintain this equipment. In the winter, people would be very uncomfortable without heat, and the hot summers would drain us of our energy without air conditioning to refresh us. Human productivity in hot and cold climates would be limited. Our food and many drugs would have a short shelf life, and you would need to go to market every day to purchase eggs, milk, and meat because they would need to be used before they spoiled. Temperature-sensitive drugs would need to be picked up and administered daily at the pharmacy. In effect, our whole lifestyle would be

Field Problem

The thermostat can help you do general troubleshooting of the air conditioning and heating system. The thermostat is a temperature-sensitive switch that controls the operation of the system, as in Figure 1–3. The system is generally divided into three major components:

1. Indoor section or blower
2. Outdoor section or condensing unit
3. Furnace section that works with and is part of the blower section

Let's review a quick system checkout. Turn the thermostat to the on position and switch it to the cool mode. Lower the temperature in the cooling mode. When you lower the thermostat below the room temperature, the indoor blower and outdoor section should begin to operate, although some systems have a time delay before the blower or indoor section begins to operate. If the indoor blower section or outdoor section does not operate after a short delay, a problem exists in one of these components. Check to see if the air is blowing from the grille, and then go outside and see if the outdoor section is operating. If neither component operates, the problem is common to both the blower and the condensing unit. It is not important that you troubleshoot the exact problem at this time. It is important that you can isolate the problem, thus making discovery of the problem easier.

The same sequence can be tried with the heating system. Place the thermostat in the on position and in the heat mode. Adjust the temperature setting above the room temperature, and the heating system should operate after a couple minutes. Depending on the type of heater, a time delay may allow the heating system to warm prior to starting the blower. You should be able to feel warm air blowing from the grilles after the time delay. You will need to go the furnace to see or listen for burner operation of a gas or oil furnace. Electric heat and heat pumps will only have the blower sound when operating. If only one part of the system does not operate, the problem has been isolated to that part, but if nothing operates, it means that the problem is common to the blower and the heating system. Checking the thermostat operation is a good first step in troubleshooting. Sometimes customers have the wrong setting, preventing the system from operating in the desired heating or cooling mode.

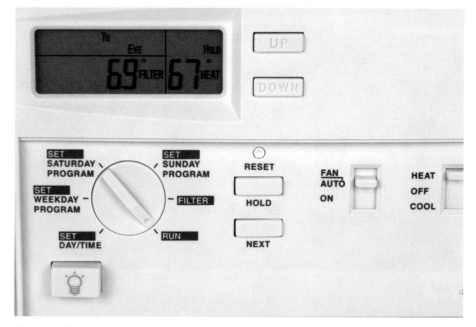

Figure 1–3
Digital thermostat with temperature reading. Heating, cooling, and fan settings are available below the readout. (Courtesy of iStockphoto)

changed. We would be spending more time just surviving and less time enjoying life. The comforts our profession offers are taken for granted, and that is OK. It is your job to make sure it stays that way.

Finally, this chapter will show that this is an exciting career. We will discuss the many job opportunities found in our profession. The chapter will also discuss the various types of certifications that are available to exhibit your credentials and knowledge. Information from the United States Bureau of Labor Statistics confirms job security and a variety of work environments.

WHAT IS HVACR?

The term "HVACR" is tossed around in our industry without everyone knowing what it means. *"HVACR" means heating, ventilation, air conditioning, and refrigeration.* The letter "R" was added recently when refrigeration was logically linked to this group. As you can see from the name, the field is diversified and provides many different opportunities.

WHY THIS INDUSTRY PROVIDES AN EXCITING AND DIVERSE CAREER!

The meaning of the acronym HVACR provides some insight into the wide range of exciting opportunities available. The starting job in this profession does not need to be and probably will not be the position you stay with throughout your career. For example, you can start as an installation helper and work your way into a lead installation technician. You can also start as a scheduled maintenance technician and eventually work your way into a service technician, servicing the very equipment you maintained years ago. You can start as a parts sales associate and end up as a technician replacing the parts you sold many years ago. You can also start in the field and transfer to job supervision, system design, or sales.

Like any job, the HVACR profession is what you make of it. You are not expected to stay in one position for the rest of your life. There are many options to choose. Here is just a sample of some of the options available to you:

- *Installation Technician*: An installation technician assembles and installs heating equipment, ventilation equipment, air conditioning equipment, and/or refrigeration equipment. The installation technician usually does not work with all of these components, instead specializing in one or two areas. For example, an installation technician may install residential heating and cooling equipment but not refrigeration equipment. The installation professions also include commercial kitchen hoods, piping, and cooling towers.
- *Service Technician*: The service technician troubleshoots and repairs problems. This profession is also divided into many different areas, such as residential, commercial, and industrial equipment servicing and includes heating, ventilation, air conditioning, and refrigeration. A service technician usually specializes in one or two of these areas. Cross-training to work on different equipment is very possible.
- *Owner/Operator*: With the proper training and a few years of experience, you can operate your own business. Most states require a few years of experience in order to qualify to take an exam to become a contractor; check your state's requirements to see what additional preparation you may need. You may be the only employee in your business, or you can expand and hire employees to meet your business goals. You are the boss.
- *Sales*: The sales opportunities are vast. You can sell parts and equipment to technicians and installers. You can be a representative for a company that specializes in a component line such as motors, and you may have a number of dealers to whom you supply equipment. Technical and sales skills merge well together. You can be involved with inside sales, assisting technicians as they secure components to repair equipment. If you like to travel and meet interesting people, you can be involved in outside sales, calling on contractors and engineers who need your company's parts and equipment.
- *Inspectors*: The experienced technician can work for a city, county, state, or federal inspection program. Installed equipment must be checked to ensure compliance with all codes, so inspectors are important in code enforcement and assuring safe equipment operation.
- *Supervisors and estimators*: Larger companies need supervisors to organize the work schedule and ensure compliance with maintenance and installation contracts. Estimators are used to determine the material necessary to complete installation jobs, including sizing equipment and duct work and listing

all the material required for successful completion of the job. Some estimators also develop bid proposals for jobs.

These are not the only jobs that are available, but the list covers the majority of the HVACR trades. You can also be an engineer, designer, educator, consultant, or researcher. These jobs require more formal education, but your field experience would enhance your abilities in those professions.

Working in HVACR is an exciting career because it offers so many options. Most technicians do not stay in the same HVACR track their whole careers. They move into closely related fields or obtain promotions within the same career track. This is the flexibility of our trade. The profession is always changing and posing new and interesting challenges.

Want to talk about job security? The HVACR industry is the most stable (and growing) industry for one big reason: your job cannot be outsourced to another country. You need to be here to diagnose the customer's problem and respond to the issues. A repair cannot be accomplished over the phone on a customer service hotline. Your work at the site is very important. Everyone and every building that people occupy needs our services.

Where Will I Be Working?

Again, the work environment is varied. You will find yourself working in buildings under construction; homes; commercial and industrial complexes; health care facilities; transportation, manufacturing, and educational complexes; military sites; refineries and chemical plants; and locations that you never realized had HVACR. Some technicians prefer one working environment over the others; only your actual work in the field will help you determine what your preference is.

Even though you will be working in and around buildings facing cold or hot extremes, your customers will let you know how delighted they are to be cool or warm again or how relieved they are to be able to keep the frozen products from thawing. You are working toward being the mechanical doctor who can keep the system healthy and fix the problem when it breaks and at the same time have a feeling of accomplishment.

What Do Technicians Do?

The focus of this series of books will be on the technician; therefore, this section will discuss what technicians do on a daily basis. The term "technician" as it relates to HVACR has a wide range of meanings. In this section, we will discuss the activities of a field service technician who diagnoses and repairs HVACR problems and an installation technician.

Heating, ventilation, and air conditioning systems control the temperature, humidity, and total air quality in residential, commercial, industrial, and other buildings. Refrigeration systems make it possible to store and transport food, medicine, and other perishable items. Heating, ventilation, air conditioning, and refrigeration service technicians and installers install, maintain, and repair such systems. Because heating, ventilation, air conditioning, and refrigeration systems often are referred to as HVACR systems, these workers are usually called HVACR technicians.

Heating, ventilation, air conditioning, and refrigeration systems consist of many mechanical, electrical, and electronic components, such as circuit boards, gas valves, motors, compressors, pumps, fans, refrigerants, ducts, pipes, thermostats, and switches, some of which are seen in Figure 1–4 through Figure 1–9. In central forced-air heating systems, for example, a furnace heats air that is distributed throughout the building via a system of metal, flexible, or fiberglass ducts. Service technicians must be able to maintain, diagnose, and correct problems throughout the entire system. To do this, they adjust system controls to recommended settings and test the performance of the entire system using special tools and test equipment.

Technicians often specialize in either installation or maintenance and repair, although they may be trained to do both. They also may specialize in doing heat-

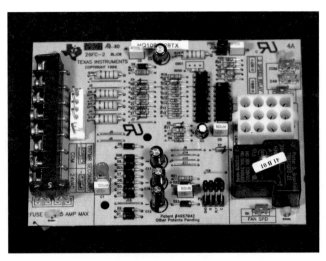

Figure 1-4
Solid-state circuit board that is used to control the heating, cooling, and blower operations. The solid-state board replaces many of the mechanical switches and relays commonly found on air conditioning systems. (Courtesy of Bill Johnson)

Figure 1-5
This is a gas valve. The gas valve and accessory kit control the flow of gas to the burners. (Photo by Susan Brubaker)

ing, ventilation, air conditioning, or refrigeration work. Some specialize in one type of system—for example, hydronics (water-based heating systems), chillers, or commercial refrigeration. Technicians also try to sell service contracts to their clients, which provide for regular maintenance of the heating and cooling systems and help to reduce the seasonal fluctuations of this type of work.

Installation technicians follow blueprints or other specifications to install oil, gas, electric, solid-fuel, and multiple-fuel heating systems and air conditioning systems. After putting the equipment in place, they install fuel and water supply lines, air ducts and vents, pumps, and other components. They may connect electrical wiring and controls and check the unit for proper operation. To ensure the proper functioning of the system, technicians often use sophisticated combustion test equipment, such as carbon dioxide testers, carbon monoxide testers, combustion analyzers, and oxygen testers.

After a furnace or air conditioning unit has been installed, service technicians often perform routine maintenance, diagnostic work, and repair work to keep the

Figure 1–6
A type of motor you will troubleshoot and replace. (Photo by Susan Brubaker)

Figure 1–7
Cutaway of scroll compressor. The motor is located at the bottom and the scrolls near the top of the compressor body. (Courtesy of Emerson Climate Technologies)

systems operating efficiently and safely as designed. They may adjust burners and blowers and check for leaks. If the system is not operating properly, they check the thermostat, gas pressure, controls, or other parts to diagnose and then correct the problem.

During the summer, when the heating system is not being used, heating equipment technicians do maintenance work, such as replacing filters, ducts, and other parts of the system that may accumulate dust and impurities during the operating season. During the winter, air conditioning technicians inspect the systems and do required maintenance, such as overhauling and rebuilding compressors.

Refrigeration technicians install, service, and/or repair industrial and commercial refrigeration systems and a variety of refrigeration equipment. They follow blueprints, design specifications, and manufacturers' instructions to install motors,

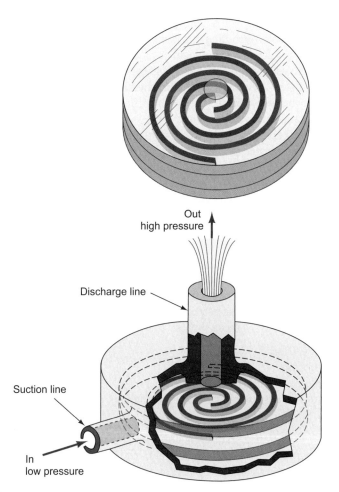

Figure 1-8
An illustration of the operation of a scroll compressor mechanism.

Figure 1-9
This is a thermostatic expansion valve. The valve opens and closes, maintaining the correct level of cold refrigerant in the cooling coil. (Courtesy of Sporlan Division, Parker Hannifin Corporation)

compressors, condensing units, evaporators, piping, and other components. They connect this equipment to the ductwork, refrigerant lines, and electrical power source. After making the connections, they charge the system with refrigerant, check it for proper operation, and program control systems.

Heating, ventilation, air conditioning, and refrigeration technicians are adept at using a variety of tools, including hammers, wrenches, metal snips, electric drills, pipe cutters and benders, measurement gauges, and acetylene torches, to work with refrigerant lines, refrigerants, and refrigerant pressure-temperature charts as shown in Figure 1–10. They use voltmeters, thermometers, pressure gauges, manometers, and other testing devices to check airflow, refrigerant pressure, electrical circuits, burners, and other components.

Other trade workers sometimes install or repair cooling and heating systems. For example, on a large air conditioning installation job, especially where workers are covered by union contracts, ductwork might be done by sheet metal workers and duct installers; electrical work by electricians; and installation of piping, condensers, and other components by pipe layers, plumbers, pipe fitters, and steamfitters. Home appliance repairers usually service room air conditioners and household refrigerators.

Role of the Technician

The function of the service technician is to solve the problem as quickly as possible and to the customer's satisfaction. This may mean making a temporary fix because a replacement part may not be readily available at the time of the repair. Technicians do scheduled maintenance, such as the following:

- Changing filters
- Cleaning coils
- Checking the charge
- Assuring the customer that the equipment is operating to manufacturers' standards

Scheduled maintenance is usually done by the technician who wants to gain experience. The purpose of starting a technician at this level is to give him an opportunity to become familiar with various pieces of equipment under conditions with less pressure than troubleshooting has.

Most service technicians are mobile and dispatched electronically, and they receive assignments throughout the day. The technician reports to the job site to conduct diagnosis, troubleshooting, and repair. The technician will work on the air conditioner, heating system, or refrigeration units, depending on the technician's area of specialty. When working on air conditioning or refrigeration systems, the technician is likely to work on refrigerant-side or electrical-side problems. Most problems are electrical. The technician may need to change the thermostat, blower motor, compressor, or some control component that is used to switch the equipment on and off. Changing the compressor will take the most time compared with changing other components.

If the problem is related to heating, it will more than likely be an electrical problem. The modern furnace has many more electrical components than furnaces manufactured twenty years ago. The technician may be diagnosing the blower motor, gas valve, circuit board, or a safety device. Gas furnaces have heat exchangers, and electric heaters have heat strips, both of which rarely fail as often as electrical components. Most electrical heating system components can be changed in less than an hour. In many cases, it takes longer to diagnose the problem than to replace the part.

PROFESSIONAL CERTIFICATION

When air conditioning and refrigeration technicians service equipment, they must use care to conserve, recover, and recycle chlorofluorocarbon (CFC), hydrochlorofluorocarbon (HCFC), hydrofluorocarbon (HFC), and other refrigerants used in air conditioning and refrigeration systems. The release of these refrigerants can be harmful to the environment. Technicians conserve the refrigerant by making sure that there are no leaks in the system, and they recover it by placing the refrigerant into proper cylinders. The refrigerant may then be recycled or reclaimed, as seen in Figure 1–11.

TEMPERATURE	REFRIGERANT						TEMPERATURE	REFRIGERANT						TEMPERATURE	REFRIGERANT					
°F	12	22	134a	502	404A	410A	°F	12	22	134a	502	404A	410A	°F	12	22	134a	502	404A	410A
−60	19.0	12.0		7.2	6.6	0.3	12	15.8	34.7	13.2	43.2	46.2	65.3	42	38.8	71.4	37.0	83.8	89.7	122.9
−55	17.3	9.2		3.8	3.1	2.6	13	16.4	35.7	13.8	44.3	47.4	66.8	43	39.8	73.0	38.0	85.4	91.5	125.2
−50	15.4	6.2		0.2	0.8	5.0	14	17.1	36.7	14.4	45.4	48.6	68.4	44	40.7	74.5	39.0	87.0	93.3	127.6
−45	13.3	2.7		1.9	2.5	7.8	15	17.7	37.7	15.1	46.5	49.8	70.0	45	41.7	76.0	40.1	88.7	95.1	130.0
−40	11.0	0.5	14.7	4.1	4.8	9.8	16	18.4	38.7	15.7	47.7	51.0	71.6	46	42.6	77.6	41.1	90.4	97.0	132.4
−35	8.4	2.6	12.4	6.5	7.4	14.2	17	19.0	39.8	16.4	48.8	52.3	73.2	47	43.6	79.2	42.2	92.1	98.8	134.9
−30	5.5	4.9	9.7	9.2	10.2	17.9	18	19.7	40.8	17.1	50.0	53.5	75.0	48	44.6	80.8	43.3	93.9	100.7	136.4
−25	2.3	7.4	6.8	12.1	13.3	21.9	19	20.4	41.9	17.7	51.2	54.8	76.7	49	45.7	82.4	44.4	95.6	102.6	139.9
−20	0.6	10.1	3.6	15.3	16.7	26.4	20	21.0	43.0	18.4	52.4	56.1	78.4	50	46.7	84.0	45.5	97.4	104.5	142.5
−18	1.3	11.3	2.2	16.7	18.2	28.2	21	21.7	44.1	19.2	53.7	57.4	80.1	55	52.0	92.6	51.3	106.6	114.6	156.0
−16	2.0	12.5	0.7	18.1	19.6	30.2	22	22.4	45.3	19.9	54.9	58.8	81.9	60	57.7	101.6	57.3	116.4	125.2	170.0
−14	2.8	13.8	0.3	19.5	21.1	32.2	23	23.2	46.4	20.6	56.2	60.1	83.7	65	63.8	111.2	64.1	126.7	136.5	185.0
−12	3.6	15.1	1.2	21.0	22.7	34.3	24	23.9	47.6	21.4	57.5	61.5	85.5	70	70.2	121.4	71.2	137.6	148.5	200.8
−10	4.5	16.5	2.0	22.6	24.3	36.4	25	24.6	48.8	22.0	58.8	62.9	87.3	75	77.0	132.2	78.7	149.1	161.1	217.6
−8	5.4	17.9	2.8	24.2	26.0	38.7	26	25.4	49.9	22.9	60.1	64.3	90.2	80	84.2	143.6	86.8	161.2	174.5	235.4
−6	6.3	19.3	3.7	25.8	27.8	40.9	27	26.1	51.2	23.7	61.5	65.8	91.1	85	91.8	155.7	95.3	174.0	188.6	254.2
−4	7.2	20.8	4.6	27.5	30.0	42.3	28	26.9	52.4	24.5	62.8	67.2	93.0	90	99.8	168.4	104.4	187.4	203.5	274.1
−2	8.2	22.4	5.5	29.3	31.4	45.8	29	27.7	53.6	25.3	64.2	68.7	95.0	95	108.2	181.8	114.0	201.4	219.2	295.0
0	9.2	24.0	6.5	31.1	33.3	48.3	30	28.4	54.9	26.1	65.6	70.2	97.0	100	117.2	195.9	124.2	216.2	235.7	317.1
1	9.7	24.8	7.0	32.0	34.3	49.6	31	29.2	56.2	26.9	67.0	71.7	99.0	105	126.6	210.8	135.0	231.7	253.1	340.3
2	10.2	25.6	7.5	32.9	35.3	50.9	32	30.1	57.5	27.8	68.4	73.2	101.0	110	136.4	226.4	146.4	247.9	271.4	364.8
3	10.7	26.4	8.0	33.9	36.4	52.3	33	30.9	58.8	28.7	69.9	74.8	103.1	115	146.8	242.7	158.5	264.9	290.6	390.5
4	11.2	27.3	8.6	34.9	37.4	53.6	34	31.7	60.1	29.5	71.3	76.4	105.1	120	157.6	259.9	171.2	282.7	310.7	417.4
5	11.8	28.2	9.1	35.8	38.4	55.0	35	32.6	61.5	30.4	72.8	78.0	107.3	125	169.1	277.9	184.6	301.4	331.8	445.8
6	12.3	29.1	9.7	36.8	39.5	56.4	36	33.4	62.8	31.3	74.3	79.6	108.4	130	181.0	296.8	198.7	320.8	354.0	475.4
7	12.9	30.0	10.2	37.9	40.6	57.8	37	34.3	64.2	32.2	75.8	81.2	111.6	135	193.5	316.6	213.5	341.2	377.1	506.5
8	13.5	30.9	10.8	38.9	41.7	59.3	38	35.2	65.6	33.2	77.4	82.9	113.8	140	206.6	337.2	229.1	362.6	401.4	539.1
9	14.0	31.8	11.4	39.9	42.8	60.7	39	36.1	67.1	34.1	79.0	84.6	116.0	145	220.3	358.9	245.5	385.9	426.8	573.2
10	14.6	32.8	11.9	41.0	43.9	62.2	40	37.0	68.5	35.1	80.5	86.3	118.3	150	234.6	381.5	262.7	408.4	453.3	608.9
11	15.2	33.7	12.5	42.1	45.0	63.7	41	37.9	70.0	36.0	82.1	88.0	120.5	155	249.5	405.1	280.7	432.9	479.8	616.2

VACUUM (in. Hg) – RED FIGURES
GAUGE PRESSURE (psig) – BOLD FIGURES

Figure 1–10
Tools and aids you will use servicing air conditioning and refrigeration equipment include hand tools, gauges, and special refrigerant charts like the one shown above. (Courtesy of Sporlan Division, Parker Hannifin Corporation)

Figure 1-11
Technician is recovering refrigerant from the air conditioning equipment and placing it in an approved recovery tank. Recovery tanks have a gray body and a yellow top. (Photo by Susan Brubaker)

Refrigerant handling certification is required by the federal government under the Clean Air Act passed by Congress in the late 1980s. This is also known as EPA certification because the Environmental Protection Agency is the governmental organization responsible for enforcing this act.

You will be required to obtain refrigerant handling certification prior to hooking up refrigerant gauges to an air conditioning or refrigeration system. The certification is divided into five categories:

- Core
- Type I
- Type II
- Type III
- Universal

The exam is divided into four 25-question exams, which are called Core, Type I, Type II, and Type III. If you pass all four exams with a 70% or better score, then you will be classified as a technician with Universal certification.

The Core exam has twenty-five multiple choice questions. It covers basic refrigerant handling and safety practices.
The Type I exam covers questions on small appliances such as window units and domestic refrigerators and freezers.
The Type II exam tests your knowledge of residential and commercial air conditioning and refrigeration equipment.
The Type III exam covers low-pressure chillers.

Many employers require industry-recognized certifications, and many times employers will help the technician train for and gain the certification.

TECHNICIANS' RESPONSIBILITY FOR RECORD KEEPING

As a first-year technician, you will be working with others who will supervise your activities. Once you learn a skill, you will be able to repeat it at other locations without supervision. When you begin to work independently, record keeping will be one of your responsibilities. As a professional, you must recognize the importance of record keeping. Record keeping is documenting what happened on each job that day. It will include the customer's name and address, perhaps a job number, a short description of the work completed, the hours spent on the job, and the parts and materials used to complete the task. Finally, the customer is asked to sign the form to ensure that the work is completed to the customer's satisfaction.

Record keeping is accomplished in many different ways, depending on the requirements of your employer. It may be in the form of an invoice that is presented to the customer when the service is completed, or it may be part of the time sheet that the technician submits to a supervisor each day. Few standards exist for record keeping, but it is important because it helps the company track the time and materials used on each job. This is especially helpful in some cases when the customer is billed at a later date, not on the day of service.

The use of laptop computers is rapidly replacing paper methods of documentation, but some employers ask their technicians to document their work on paper forms, in which case it is important to write or print clearly. Whether using a computer or paper document, this information is used to bill the customer and restock inventory, so it is important to document everything legibly. Forgotten items end up on the negative side of the ledger and take away from the company profit. A company needs to make profit in order to stay in business, expand its business opportunities, and give employees pay raises or improved benefits packages.

In summary, record keeping is very important for the survival of your company and yourself. It also gives the customer detailed records. Your attention to detail is required to meet these record keeping goals. Company invoices and records are legal documents that can be used in court as well as to help the company owner conduct analyses on profit and loss so that you can get paid and stay employed.

CAREER PATH

According to the United States Bureau of Labor Statistics, employment in the HVACR industry is projected to grow faster than the average career. This means a growth rate of 18% to 26% from 2004 to 2014. Job prospects are expected to be excellent, particularly for those with training from an accredited technical school or with formal apprenticeship training like you are now attending. Obtaining certification through one of several organizations is increasingly recommended by employers and may increase advancement opportunities.

Heating, ventilation, air conditioning, and refrigeration technicians and installers held about 270,000 jobs in 2004; almost half worked for plumbing, heating, and air conditioning contractors. The remainder was employed in a variety of industries throughout the country, reflecting a widespread dependence on climate-control systems. Some worked for fuel oil dealers, refrigeration and air conditioning service and repair shops, schools, and stores that sell heating and air conditioning systems. Local governments, the federal government, hospitals, office buildings, and other organizations that operate large air conditioning, refrigeration, or heating systems employed others. About 15% of technicians and installers were self-employed.

THE SHORT HISTORY OF AIR CONDITIONING

The term "air conditioning" refers to the cooling and heating of indoor air for human comfort. In a broader sense, the term can refer to any form of cooling, heating, or ventilation that modifies the condition of the air.

The concept of air conditioning is known to have been applied in ancient Rome, where aqueduct water was circulated through the walls of certain houses to cool them. Similar techniques in medieval Persia involved the use of cisterns and wind towers to cool buildings during the hot season. Cistern water evaporated, cooling the air in the building. This is similar to the air you feel when spraying water from a garden hose or the water spray from a mister. Modern air conditioning emerged from advances in chemistry during the nineteenth century. The first large-scale electric air conditioning system was invented and used in 1902 by Willis Haviland Carrier.

Carrier was a mechanical engineer who worked at the Buffalo Forge Company in Buffalo, New York. Companies carrying his name helped conquer the temperature-humidity relationship, merging theory with practical applications. Starting in 1902, he designed a spray-type temperature and humidity controlled system. His induction system for multiroom office buildings, hotels, apartments, and hospitals was just another of his air-related inventions. Many industry professionals and historians consider Carrier the father of air conditioning.

Prior to Carrier was Michael Faraday, who in 1820 discovered that compressing and liquefying ammonia could chill air when the liquefied ammonia was allowed to evaporate. In 1842, Florida physician John Gorrie used compressor technology to create ice, which he used to cool air for his patients in his hospital in Apalachicola, Florida. Gorrie was granted the first patent for an air conditioning-related device in 1851.

The first air conditioners and refrigerators used toxic or flammable gases like ammonia, methyl chloride, and propane, which could result in fatal accidents if they leaked. Thomas Midgley Jr. developed the first chlorofluorocarbon gas, called Freon, in 1928. Freon, manufactured by DuPont, was much safer for humans but was later found to be harmful to the atmosphere's ozone layer.

SUMMARY

Because of the increasing sophistication of heating, air conditioning, and refrigeration systems, employers prefer to hire people with technical school training or those who are attending or have completed an apprenticeship training program. Working and attending apprenticeship training is an easier way to learn and reinforce what is presented in the classroom.

Many secondary and postsecondary technical and trade schools, junior and community colleges, and the United States Armed Forces offer six-month to two-year programs in heating, air conditioning, and refrigeration. Students study theory, design, and equipment construction, as well as electricity. They also learn the basics of installation, maintenance, and repair in a classroom laboratory.

Apprenticeship programs frequently are run by joint committees representing local chapters of the Air Conditioning Contractors of America (ACCA), Plumbing-Heating-Cooling Contractors—National Association (PHCC), Refrigeration Service Engineers Society (RSES), and local chapters of the Sheet Metal and Air Conditioning Contractors' National Association (SMACNA) or the United Association of Journeymen and Apprentices of the Plumbing and Pipe Fitting Industry of the United States and Canada. Formal apprenticeship programs normally last three to five years and combine on-the-job training with classroom instruction. These programs also promote apprenticeship learning and working for wages.

Courtesy of ACCA

Courtesy of PHCC

Courtesy of RSES

Figure 1-12
A tech pressurizes the system with nitrogen and begins to look for a refrigerant leak. Nitrogen is used because it is environmentally friendly and an inert gas. Never use oxygen or compressed air. (Photo by Susan Brubaker)

Those who acquire their skills on the job usually begin by assisting experienced technicians, at first performing simple tasks such as carrying materials, insulating refrigerant lines, or cleaning furnaces. In time, they move on to more difficult tasks, such as cutting and soldering pipes and sheet metal and checking electrical and electronic circuits.

Advancement usually takes the form of higher wages. Some technicians, however, may actually change positions to that of a supervisor or service manager. Others may move into areas such as sales and marketing, and still others may become building superintendents, cost estimators, or, with the necessary certification, teachers. Those with sufficient money and managerial skills can open their own contracting business. The profession has unlimited potential and a variety of work environments. If you are looking for variety, a rewarding job in which you can really see and feel the difference you make, and the opportunity to always learn something interesting, this is the profession for you.

FIELD EXERCISES

1. Ask a supervisor how he got started in the HVACR business. Discuss how he changed his career roles from when he started to the position he holds today. Ask for advice on how to strengthen your plan for your own professional career.
2. Ask a coworker how he got started in the HVACR business. Discuss how he changed his career roles from when he started to the position he holds today. Ask him for advice on how to improve your professional career. The technician in Figure 1-12 learned to leak check a system from someone else, just as you learn new skills in the classroom and in the field.
3. Write down the five most important goals you want to work toward in our industry.

REVIEW QUESTIONS

1. What percentage of growth for HVACR employment is expected by the United States Bureau of Labor Statistics from 2004 to 2014?
2. What does the acronym HVACR mean?
3. Describe five different career opportunities found in our profession.
4. Discuss your goals in our profession. How do you expect to accomplish these goals?
5. What does a technician do at a job site?
6. What required certification will you need to obtain as a technician in order to legally install gauges on a refrigeration system?
7. Name five general locations where you might be working while doing HVACR work.
8. Why is record keeping a very important part of the technician's job?

CHAPTER 2
Introduction to Systems and Major Components

LEARNING OBJECTIVES

The student will:

- Draw the refrigeration cycle, showing the direction of refrigerant flow and labeling the four major components.
- Describe the function of the compressor.
- Describe the function of the condenser.
- Describe the function of the metering device.
- Describe the function of the evaporator.
- List three types of ducts used to distribute air in a duct system.
- List the types of tubing used for refrigerant lines and hydronic systems.
- Briefly describe three types of heating systems.
- Calculate the heat content of a heat strip.
- Describe how a hydrogen fuel cell operates.
- Name two different types of solar energy systems.

INTRODUCTION

The purpose of this chapter is to help you identify the major parts of an air conditioning system and give you a general idea how it operates. It is not important at this stage of your training that you know the detailed operation of the system components. The major components are the same whether it is an air conditioning or a refrigeration system. We will also discuss the various types of air and fluid distribution systems such as duct work, refrigerant piping used for cooling, and hot water piping used for heating.

This section will introduce four common types of heating systems: natural gas, oil, electric heat, and heat pumps. Other forms of heat are used in our industry, but these are the most common.

Finally, the chapter will discuss new technology like hydrogen fuel cells and solar energy systems.

You will learn many new terms in this chapter. It is important to be able to use the terms of the HVACR trade and to be able to professionally communicate with your customers, supervisors, and coworkers.

HOW DOES COOLING OCCUR?

Before we explore the refrigeration cycle, let us discover the different ways we can cool ourselves. Everyone enjoys a cold drink from the refrigerator. The temperature of the refrigerated drink is about 40°F. If you are hot, this cools you down even if you are outside in the summer heat. Additional cooling can be accomplished by pouring the drink over ice. Now the drink is about 32°F. That 8°F drop in temperature seems to make a significant difference in the chill factor of the drink. It is so cold that it becomes more difficult to drink quickly. You might even melt an ice cube in your mouth to cool yourself down some more. These are examples of heat transfer. Heat transfers from a higher temperature to a lower temperature. It transfers from warm to cool or hot to cold. The heat is leaving your body while warming up the cold fluid in your body. When this heat transfer takes place, your body temperature drops slightly and you become more comfortable. This is an example of heat transfer by conduction. *Conduction is the transfer of heat through a material*, in this case, our body tissue. We will see that the refrigeration cycle uses the same heat transfer principle.

Another form of cooling is observed when rubbing isopropyl alcohol on your skin, which gives you a cooling sensation. This example of heat transfer is called evaporation. The alcohol evaporates quickly, removing heat from the skin. The faster a liquid evaporates, the cooler the sensation. When you spray water, you feel

Field Problem

You have started assisting with scheduled maintenance contracts, which includes changing filters and cleaning coils. This is your first day on the job doing scheduled maintenance. About thirty minutes into the job, the lead technician is called away to a nearby high-priority no-cooling call. You are left to complete the job of cleaning the condenser coil. You have never completed this task and are not sure how to tackle the job. You think about it for a few minutes while gathering and connecting the water hose and coil cleaner sprayer. Examining the condenser, you notice that the air is pulled into the side, run through the coil, and discharged from the top of condensing unit. You conclude that the coil must be cleaned by applying a mild soap and flushing with water in the opposite direction of the air flow because flushing the coil in the direction of the air flow would drive the dirt deeper into the coil. Some of the dirt would be driven through the coil, but some would get trapped and become impacted in the fins, which would reduce the heat transfer of the condenser. You contact the lead technician and explain what you plan to do.

> **Tech Tip**
>
> It is very important to always be a professional on the job, on the road, and at the shop. You are the company in the eyes of the customer. Your behavior and actions are noted by the customer. Remember to keep your professional image at its peak. You never know who is watching or listening.

> **Tech Tip**
>
> What is refrigeration? *Refrigeration is the process of removing heat from one substance and transferring it to another substance.* It is also removing heat from one area and dumping it in an area where it has little or no effect.

cool air near the mist because some of heat from the air goes to evaporate the water droplets into a mist. Let's see how this is applied to air conditioning or refrigeration.

THE REFRIGERATION CYCLE

The refrigeration cycle includes four basic components:

- Compressor
- Condenser
- Metering device
- Evaporator

The refrigerant flows from the compressor to the condenser, to the metering device, and to the evaporator before returning to the compressor to start over again. Refrigerant is circulated through the cycle. Let's discover how a refrigerant works in conjunction with the refrigeration cycle.

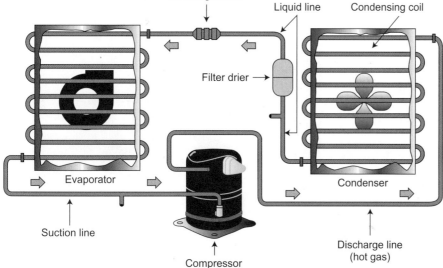

The Refrigeration Cycle

Courtesy of Carrier

Refrigerant

Water can be used to absorb heat or reject heat. Stepping into a cool shower will drop your body temperature quickly. Ice is used to chill food and prevent spoilage. Sliding into a hot tub in the winter warms your body. Water is also chilled and circulated in tubing and coils to cool an area of a building, or it can be heated and circulated in tubing embedded in a concrete floor to warm a space in the winter.

Refrigerant, like water, is a chemical fluid that can be used to absorb or reject heat. Even though refrigerant is not water, it has similar properties. For example, refrigerant and water both can exist as 100% liquid, 100% vapor, or a mixture of liquid and vapor. Refrigerant is like a heat sponge that can absorb and reject heat.

In summary, refrigerant is like a sponge that absorbs heat from the air or water passing through the walls of a metal heat exchanger, or what we call a coil. When heat is removed, the air or water temperature drops. The heat from the sponge is taken outside, where the heat is wrung out. The sponge is then once again ready to absorb heat from the air or water passing around or through it. See Figure 2–1.

Compressors

How do you get a fluid like air or liquid to move in a duct work or piping system? You can push or pull air with a fan or liquids with a pump. *A compressor is a vapor pump that compresses and circulates vapor to the condenser and returns it from the*

Dry ice freezes at −109.4°F

Ice melts at 32°F (0°C)

Figure 2–1
Heat transfers from warm to cold. Different substances have different freezing temperatures. For example, water will freeze at 32°F. Dry ice, which is solid carbon dioxide, freezes at −109.4°F. Anything warmer than these temperatures will transfer heat to the ice. The frozen water will transfer heat to the dry ice because the ice is warmer than the frozen carbon dioxide. (Courtesy of Photos.com)

evaporator to be recirculated. The compressor is sometimes called the heart of the system, like a heart pumping the lifeblood through the body. In this case, the refrigerant is the lifeblood of a refrigeration system. Common types of compressors are seen in Figure 2–2.

Cooling Capacities

The general classification of air conditioning compressors can be divided into the following categories:

1. Residential: $\frac{1}{2}$ to 5 tons
2. Light commercial: 6 to 25 tons
3. Industrial or heavy commercial: over 25 tons

A similar classification can be generally used for refrigeration compressors, except the term "horsepower" is used instead of "tons." In refrigeration, a unit of horsepower is usually equal to a little less than a ton of air conditioning.

1. Domestic refrigeration: $\frac{1}{6}$ to $\frac{1}{2}$ horsepower
2. Commercial refrigeration: usually over $\frac{1}{2}$ horsepower

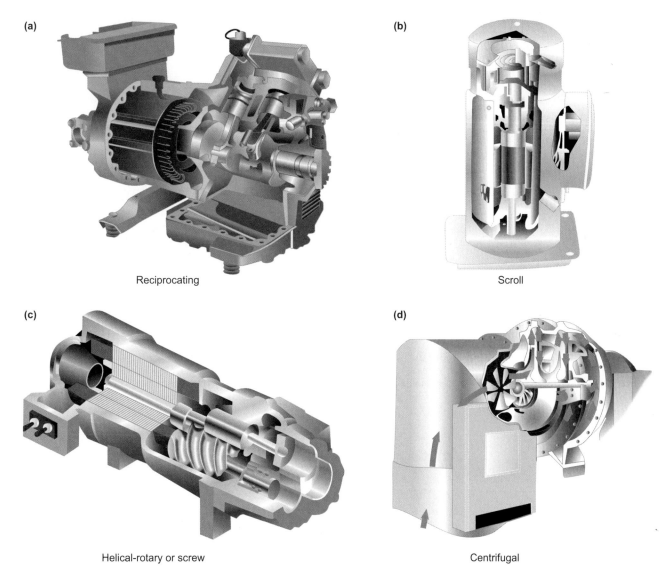

Figure 2–2
The reciprocating compressor uses a piston for compression. The scroll uses two rotating scrolls. The helical-rotary uses two screws. The centrifugal uses centrifugal force to create compression and a direction of flow. (Courtesy of Trane)

The break between the different levels of air conditioning and commercial refrigeration is not always clear. For example, there is an uncommon six-ton compressor that can be used for residential or commercial applications, and some light commercial units, around 25 tons, are large enough to be used in industrial applications.

Condensers

The condenser is the component that removes heat from the refrigerant. The condenser has air or water passing around the tubing and coil to remove heat that was absorbed by the refrigerant. The heat in the refrigerant comes from the BTUs absorbed from the air or water passing over it. The refrigerant also picks up some heat as it passes through the compressor. This is called the heat of compression, which generally includes the motor heat, heat from the compression of vapor, and heat generated by the friction of the compressor's moving parts. Three basic types of condensers are shown in Figure 2-3.

- Air-cooled condensers
- Water-cooled condensers
- Evaporative-cooled condensers

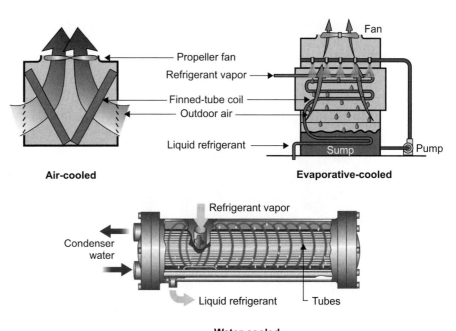

Figure 2-3
The purpose of the condenser is to remove heat from the refrigerant so that it can return and absorb more heat. The condenser uses air or water to remove heat from the refrigerant. The three types of condensers are air-cooled, water-cooled, and evaporative-cooled. (Courtesy of Trane)

Metering Device

A spray nozzle on a garden hose is an example of a metering device. The water flow is controlled by opening and closing the hose sprayer. Also, the length of the hose has an effect on the water flow and pressure. A longer hose or metering device will cause a greater pressure drop and a weaker refrigerant flow. If a larger amount of water is required, open the sprayer to allow maximum flow capacity, as shown in Figure 2-4. This could represent the need for maximum cooling on a hot day. Throttling the sprayer will reduce water flow when less is needed, as shown in Figure 2-5. This is similar to needing less cooling in the fall or spring when the outdoor temperature is lower. Water pressure is also a factor when considering water

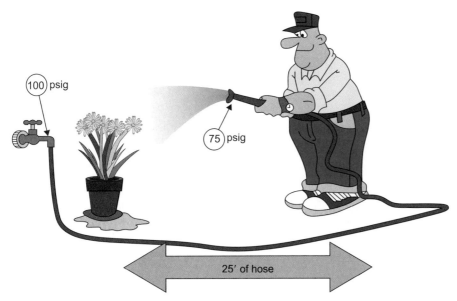

Figure 2-4
This is an example of a hand-operated metering device. The sprayer can be controlled by the operator. More water is supplied as the valve is opened. The pressure at the hose outlet is 75 psig. This is similar to a metering device with pressure dropping as the length of the hose increases. The pressure dropped from 100 psig to 75 psig through the hose and the spray nozzle. (Courtesy of Dick Wirz, Refrigeration Training Services)

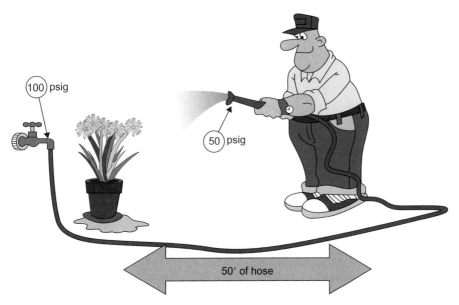

Figure 2-5
This is a different example of a hand-operated metering device with outlet pressure reduced to 50 psig. This is similar to a metering device throttling the refrigerant to reduce flow and capacity. The hose is longer than that in the prior example. A long hose creates more friction which will reduce the fluid flow and outlet pressure. Opening and closing the sprayer valve will also control the flow of fluid and change the pressure at the spray head. (Courtesy of Dick Wirz, Refrigeration Training Services)

flow. The greater the water pressure is, the greater the flow will be, as shown in Figure 2-6. Refrigerant pressure works the same way. The greater the refrigerant pressure is, the greater the refrigerant flow will be to the evaporator. More refrigerant flow equals more cooling or refrigeration capacity.

The metering device has many generic names. For example, you will hear it called a restrictive device, flow control, refrigeration control, and other terms.

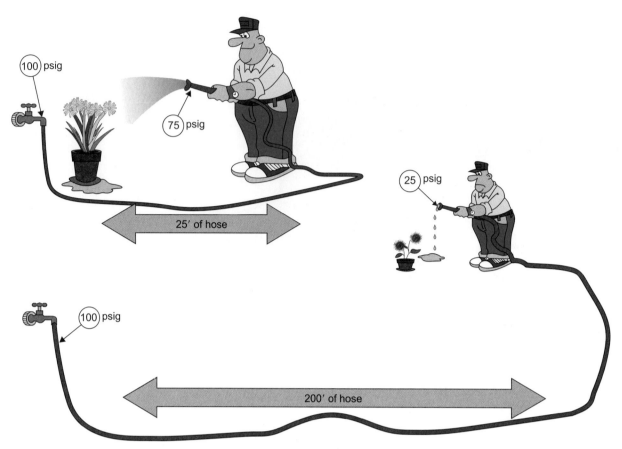

Figure 2–6
This is an example of a hand-operated metering device. The sprayer can be controlled by the operator. More water is supplied as the valve is opened. This is similar to a capillary tube with pressure dropping as the length of the hose increases. As the pressure drops, the flow decreases. (Courtesy of Dick Wirz, Refrigeration Training Services)

In summary, the metering device is found in the refrigeration cycle after the condenser coil. The purpose of the metering device is to reduce the pressure of the refrigerant. Just as you feel a cool spray when you throttle and reduce the pressure of a water sprayer, the metering device throttles and reduces the pressure of the liquid refrigerant. The reduction in pressure causes the temperature of the warm refrigerant from the condenser to evaporate to a cooler temperature. The refrigerant leaves the metering device as a lower pressure refrigerant that feeds the evaporator.

Evaporators

On a warm, dry day, as the moisture evaporates from your skin, you feel cooler. On a warm, humid day, the moisture does not evaporate as quickly, sweat collects on your skin, and you do not feel as comfortable. It takes heat to evaporate a liquid. Sweat evaporating from your skin creates cooling because heat causes the moisture to evaporate at the skin's surface. The heat it takes to evaporate the sweat is heat removed from your body. Another example of this is the rapid evaporation of alcohol on the skin.

How does this relate to the evaporator? The evaporator is similar in construction to a condenser coil, but smaller. The purpose of the evaporator is to remove heat from the air or water passing around it. The low-pressure liquid refrigerant evaporates inside the evaporator tubing, absorbing heat from the air or water passing around the tubing. The refrigerant changes are illustrated in Figure 2–7. The refrigerant leaving the evaporator will be a cool vapor. Remember, liquid has to absorb heat in order to evaporate. Under system pressures, refrigerant will evaporate at low temperatures.

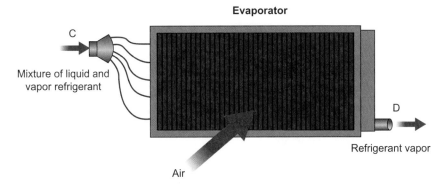

Figure 2-7
The evaporator receives a low-pressure, cold mixture of liquid and vapor refrigerant from the metering device. The cold refrigerant boils off as it removes heat from the air passing through it. The refrigerant leaves the coil as a cool vapor that returns to the compressor. (Courtesy of Trane)

The evaporator receives the cold, lower-pressure liquid refrigerant from the metering device. About 20% to 30% of what enters the evaporator is vapor because some of the liquid flashes off from a liquid to a vapor as it passes through the metering device. The remaining 70% to 80% is cold liquid that will do most of the cooling.

The lower pressure liquid refrigerant inside the evaporator coil absorbs heat from warmer air or water passing over it. The air temperature is dropped about 20°F as it passes through the coil. This is known as temperature difference and is abbreviated ΔT, which is said as "delta-T." The delta-T of a refrigerator or freezer coil will be lower. The delta-T of a water coil, normally called a chiller, is about 10°F. Figure 2–8 illustrates two different types of residential evaporator coils.

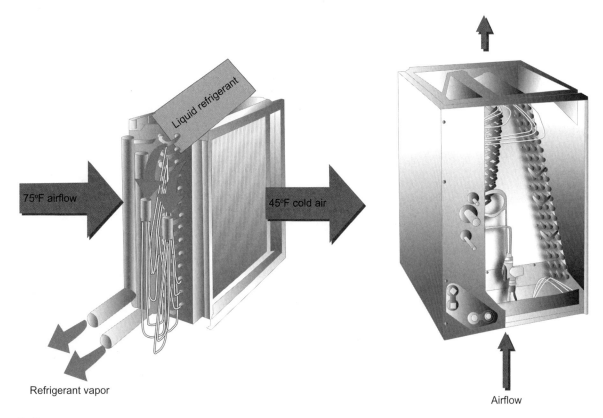

Figure 2-8
Slab coil on the left and "A" coil on the right. The coil on the left has two liquid lines and two suction lines and would be used in a commercial system. The temperature drop is 20°F. The coil on the right would be used on a vertical residential system. (Courtesy of Trane)

Liquid refrigerant evaporates inside the evaporator tubing, cooling the air or water passing around it. By the time the refrigerant leaves the evaporator outlet, it will have changed to vapor. The vapor returns to the compressor to be compressed and circulated in the refrigeration cycle again.

DISTRIBUTION SYSTEMS: DUCTS, REFRIGERANT PIPING, AND HYDRONIC PIPING

This section will introduce various types of distribution systems that you may see as an HVACR technician. This section will discuss air ducts, refrigerant piping, and hydronic piping.

Air Ducts

Air ducts are passageways for air to be supplied to the conditioned space and for air to be returned to the air handler to be heated or cooled again. They are also known as ducts or ductwork. The duct system must be sized properly to allow the proper airflow and air distribution throughout the structure.

The most common types of duct systems are sheet metal, flexible duct, and fiberglass duct board. This chapter will address these common duct materials. PVC, steel pipe, and fabric duct materials are also used in some air distribution applications.

Sheet Metal Ducts

Sheet-metal ducts, shown in Figure 2-9 and Figure 2-10, can be round, square, or rectangular. Metal ducts can be internally or externally insulated to reduce heat loss in the heating mode and heat gain in the cooling mode. Insulation also prevents moisture condensation on the duct surface in cooling. Foil faced insulation

Figure 2-9
Round sheet-metal duct. This is called metal snap-lock duct because it has a longitudinal split that is locked together when ready for installation. One advantage of this product is that the metal pieces can be placed inside of each other to save space when storing and when transporting to the installation site. (Photo by Joe Moravek)

Figure 2–10
Sheet-metal duct fabricated at the shop to be transported to the job for installation. This ductwork will need to be externally insulated if installed in an unconditioned space. (Photo by Susan Brubaker)

reflects heat away from the duct work. The thermal value of the duct insulation will be R-4, R-6, or R-8. The letter "R" means resistance to heat transfer. The higher the R-value is, the thicker the insulation, which transfers less heat. Sheet-metal duct is one of most expensive duct systems to install, but it is the strongest and most durable of the common duct materials. If insulated on the outside with an insulating duct wrap, the sheet-metal duct can be internally cleaned.

Square or rectangular sheet-metal ducts are partially or totally manufactured in a sheet-metal shop and transported to the job site. The insulation can be internally applied during fabrication, or duct wrap can be installed over the installed duct. Round ducts can be cut and assembled at the job site and will need exterior duct wrap insulation.

Flexible Duct

Flexible duct, sometimes called flexduct or factory-made duct, is constructed of an interior plastic liner that encloses a spring-like wire, as shown in Figure 2–11. The spiral wire gives the duct its support. On insulated flexible duct, the next layer of material is insulation. You will find R-4 duct insulation on older systems and higher R-values on newer installations where energy conservation is important.

The outer liner of the flexible duct is covered with a vapor barrier that prevents moisture from transferring through the outer skin into the insulation. Moisture in the insulation may condense into water and damage the insulation as well as cause water damage to the building structure. Foil barriers are more common on newer flexduct systems. The foil barrier also reflects heat away from the duct. The black vapor barrier is used for outside installations because it is more resistant to UV rays from the sun.

Fiberglass Duct Board

Fiberglass duct board is manufactured by compressing fiberglass into a rigid board with a foil vapor barrier glued to the surface, as Figure 2–12 shows. The insulation values are R-4, R-6, and R-8. R-4 is found on older duct installations, and R-6 and R-8 are found on newer installations where energy conservation is important.

Figure 2-11
Flexduct has an inner plastic liner that covers a spiral structural support. This is covered by the duct insulation and outer vapor barrier.

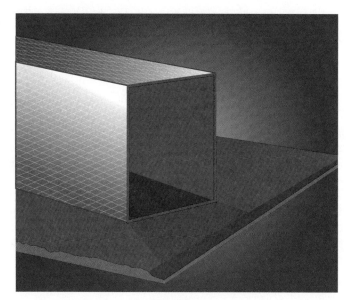

Figure 2-12
Fiberglass duct board manufactured from compressed fiberglass with a foil backing attached with an adhesive. (Courtesy of Johns Manville SuperDuct® RC™ Air Duct Board)

Fiberglass duct board is typically manufactured in 4-foot by 12-foot sheets that are 1 inch, $1\frac{1}{2}$ inch, or 2 inches thick. In most cases, the duct board is taken to the job site, where it is cut and assembled using hand tools or a duct board fabrication machine. The bottom part of Figure 2-13 illustrates the cutout of a sheet of duct board. The fiberglass must be grooved in order to fold the duct board into a square or rectangle.

Figure 2-14 shows a combination of the duct systems discussed in this section.

Duct Distribution Systems

Most duct systems can be classified as one of the following types:

- Radial duct system
- Extended plenum system
- Reducing extended plenum system
- Reduced trunk system
- Perimeter loop system

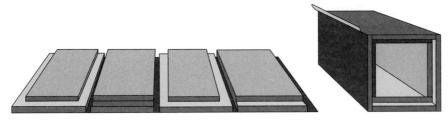

Figure 2-13
The duct board is fabricated on the site for immediate installation. The lower illustration shows how grooves are cut in the duct board to allow material to fold into the desired shape.
(Courtesy of Knauf Insulation GmbH)

Figure 2-14
This illustrates all three common types of duct systems. Starting at the top, you see sheet-metal duct with flexduct connecting the duct board.

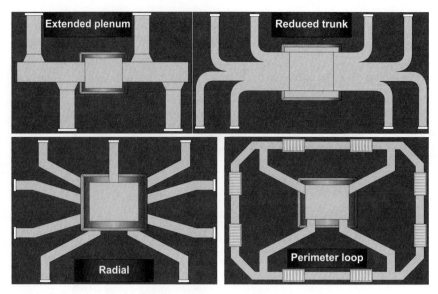

Figure 2–15
The four most common duct configurations. Extended plenum uses a long plenum with duct drops. The reduced trunk is similar to the extended plenum but reduces in size to maintain duct velocity. The radial duct design has duct radiating from one common plenum. The perimeter loop is a circular duct that maintains even pressure by four duct feeds from the main plenum. (Courtesy of Dick Wirz, Refrigeration Training Services)

Advantages and applications for each of these duct systems will be discussed in advanced training. At this time in your career, it is important to be able to identify the duct material and the duct designs, as in Figure 2–15.

REFRIGERANT PIPING

Copper air conditioning and refrigeration (ACR) tubing is used to distribute refrigerant through the refrigeration cycle. It differs from standard copper tubing in that it is charged with nitrogen and capped to prevent contamination. ACR tubing must be kept clean to prevent foreign objects from entering the refrigeration system. It is measured by the outside diameter, abbreviated OD. Water tubing is measured by the inside diameter, or ID. See Figure 2–16.

ACR tubing can be soft copper and sold in various lengths, as shown in Figure 2–17, or it can be hard-drawn copper sold in straight 10- or 20-foot sections, like those shown in Figure 2–18. The difference is that soft copper can be carefully bent to a 90-degree angle, reducing the use of elbows, as in Figure 2–19. In addition, soft copper can be bought in a 50-foot length, which reduces the number of couplings needed to assemble the tubing. Reducing the number of copper fittings reduces the number of potential leaks and the installation time. Hard-drawn copper is rigid, so it cannot be bent any more than a few degrees before it develops a kink, which

Tech Tip

Undersized duct systems reduce airflow, reduce comfort for the customer, reduce system capacity, increase operating noise, and increase the operating cost. Length of duct, elbows, turns, and dampers in the duct system will also reduce airflow. The duct system must be sized to compensate for these conditions. Second only to duct leaks, undersized duct systems account for the most significant loss in system efficiency.

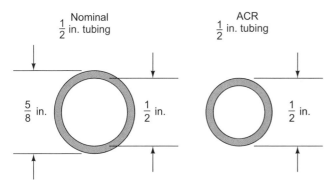

Figure 2-16
Copper tubing used in air conditioning is measured on the outside diameter. Copper tubing used in plumbing is measured on the inside diameter. In this example, $\frac{1}{2}"$ ACR tubing is less than $\frac{1}{2}"$ inside because the thickness of the walls needs to be subtracted.

Figure 2-17
Soft copper tubing comes in 50-foot rolls.

Figure 2-18
Straight hard-drawn copper comes in 20-foot lengths. It is nitrogen charged and capped on both ends. It is not bendable like soft copper tubing. (Photo by Joe Moravek)

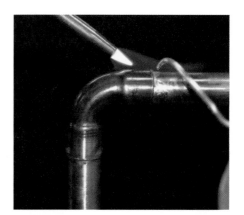

Figure 2-19
Copper elbow being soldered. (Courtesy of Bill Johnson)

Tech Tip

Following are some general guidelines for installing refrigerant tubing:
- Keep it simple.
- Minimize the number of joints.
- Use the smallest permitted size.
- Provide for oil return at minimum capacity.

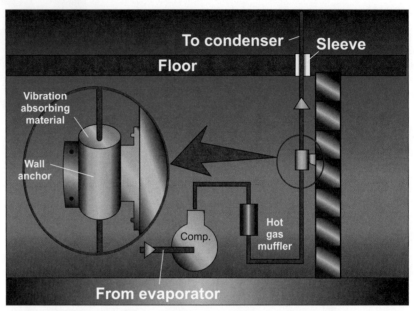

Figure 2-20
Pipe support is important. It must have horizontal and vertical support. The tubing must be free of vibrations that can weaken the copper and create noise in the structure. (Courtesy of Carrier Corporation)

restricts the refrigerant flow. However, it makes a better looking installation because it is difficult to perfectly straighten soft copper tubing.

Copper tubing is available in four standard wall thicknesses. Type K is a heavy duty wall construction. Type L is a bit thinner and is the most commonly used. Types M and DMV have the thinnest wall thickness and are generally not used for refrigeration systems.

General guidelines for the installation of refrigerant piping include the following:

- Provide adequate support, both on horizontal and vertical runs.
- Provide for expansion/contraction and vibration control (Figure 2-20).
- Provide proper insulation to prevent condensation and heat transfer.

Refrigerant Piping Identification

It is important to be able to identify the refrigerant tubing found in the refrigeration cycle. When you are on a job, you want to appear professional by knowing the correct terminology of the trade. Figure 2-21 shows the discharge line going from the compressor to the condenser. The liquid line connects the condenser and

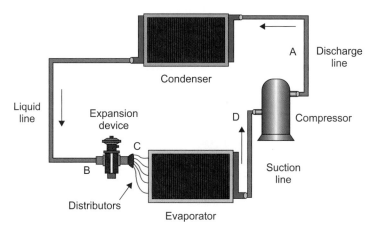

Basic Refrigeration System

Figure 2-21
Follow the basic cycle by tracing the refrigerant flow from the compressor to the condenser, to the expansion device, and as it returns to the evaporator. Pipe A is the discharge line. Pipe B is the liquid line. Pipes C are the distributor tubes, and pipe D is the suction line. (Courtesy of Trane)

metering device, and the distributor tube is between the metering device and the evaporator. Not all systems have distributor tubes. The suction line connects the evaporator to the suction side of the compressor.

HYDRONIC PIPING

When we refer to hydronics, we are talking about using hot water or steam circulated in tubing and pumped to a heat exchanger. Hydronics can also involve chilled water, but the term "hydronics" normally refers to heating and not cooling. We will limit this discussion to hydronic water heating.

According to the Uniform Mechanical Code, the maximum pressure design of a piping system for hydronic heating is 160 pounds per square inch gauge (psig), and the highest working temperature is 250°F. Higher temperature and pressure systems are designed for special heating applications that usually do not involve comfort heating. Most systems are low-pressure systems operating at less than 15 psi and below 180°F. The piping can be metal or a special plastic pipe. Check your local code for specific requirements.

A hydronic heat distribution system has two major advantages over duct heat:

1. It takes up less space than an air duct system, as shown in Figure 2-22.
2. It has less heat loss than an air duct system.

Figure 2-23 illustrates a simple hydronic system. Piping distributes heat to what is identified as a heat emitter. The most common use for hydronics is space heating.

The piping and pumping capacity must be designed and sized properly. Properly sized piping and pumps will develop a heating capacity of 10,000 BTUs per hour (BTUH) at a flow rate of 1 gallon per minute.

Water running through a pipe has to overcome the friction associated with piping walls, fittings, valves, controls, and heat exchangers. This friction causes a pressure drop in the system, called head. Circulators or pumps are sized to overcome the friction encountered in the piping.

Table 2-1 provides a rule of thumb for the maximum flow rate and heat-carrying capacity based on a 20°F temperature drop across a heat exchanger.

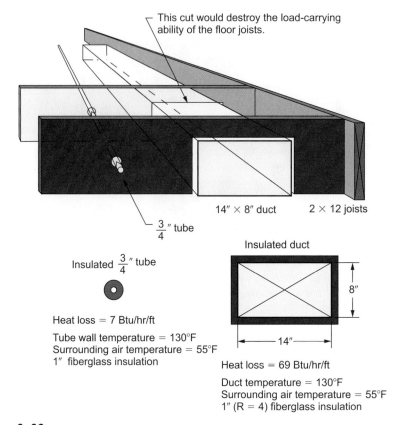

Figure 2-22
Hydronic heating has the advantage over duct heating in that it takes up less space to get heat to the needed area. There is less heat loss than with duct heat.

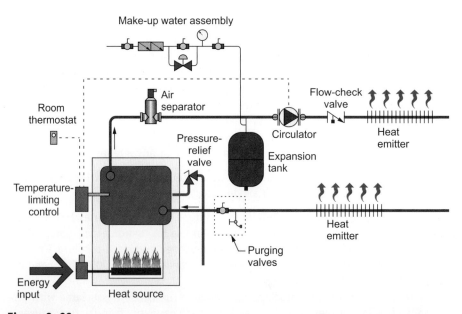

Figure 2-23
Hydronic heating has the advantage over duct heating in that it takes up less space to get heat to the needed area. There is less heat loss than with duct heat.

Table 2-1

Copper Pipe	Maximum Flow Rate	Heating Capacity
$\frac{1}{2}''$	$1\frac{1}{2}$ gpm	15,000 BTUH
$\frac{3}{4}''$	4 gpm	40,000 BTUH
$1''$	8 gpm	80,000 BTUH
$1\frac{1}{4}''$	14 gpm	140,000 BTUH

STEAM HEATING

Steam heat is sometimes used in residential, commercial, industrial, and institutional applications. Steam is derived from heating water under pressure until it is vaporized. Steam heat can be low or very high pressure. The steam flows through piping to a heat exchanger such as a baseboard heater or radiator. As you would expect, steam has more heat content than water. That is a major advantage of steam: more heat over the same heat exchanger surface and quicker heat than hot water provides.

GAS HEATING

Fuel gas heating systems can be natural gas, propane, or butane, but natural gas heating is the most common in urban areas. The gas furnace can be of an upflow, downflow, or horizontal air flow design, as shown in Figure 2-24. The gas furnace in Figure 2-25 is divided into four separate sections:

- Controls section
- Blower section
- Burner and heat exchanger section
- Venting section

Controls Section

The controls section has the electrical components found in the gas furnace. The new, high-efficiency furnaces have circuit boards that reduce the number of individual components found in the furnace. The modern furnace has numerous safety devices compared with the furnaces of a couple of decades ago. Most of the safety devices are thermal in nature (heat sensors) and will shut the furnace down to prevent overheating or a fire.

Blower Section

The blower section includes a motor, blower wheel, and blower housing. The belt drive blower motor usually is single speed and has pulleys and a belt. The direct drive blower motor usually has speed taps to operate on low, medium, or high speed. Blowers on high-end products use variable speed technology. The variable speed or electrical commutated motor varies the airflow output based on the control setup. The variable speed operation improves comfort and efficiency, making it easier to improve indoor air quality.

Burner and Heat Exchanger Section

The burners ignite inside the heat exchanger section. The heat exchanger section is a chamber that receives the heat from the burners through the middle of a hollow chamber or pipe. The outside of the heat exchanger has the air from inside the structure blowing across it. The byproducts from the burning of fuel gas travel

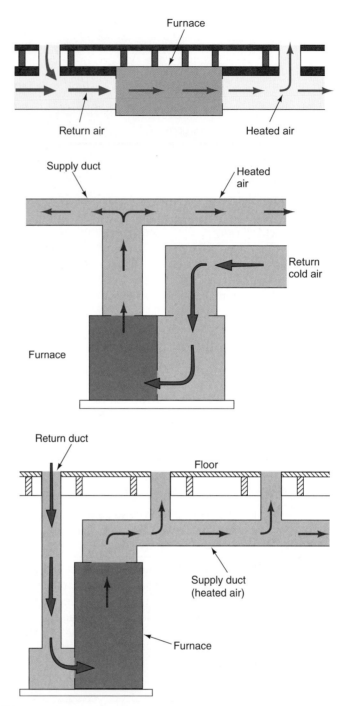

Figure 2-24
Top diagram is a horizontal application. The two lower furnaces are examples of upflow applications.

through the middle of the heat exchanger and cannot mix with the indoor air being heated. Figure 2-26 shows a cutaway view of this operation.

Venting Section

The venting section collects the byproducts of combustion (flue gases) and guides them to vent pipes that dispose of these potentially harmful gases. The byproducts of a complete combustion process include carbon dioxide, nitrogen, excess air, and

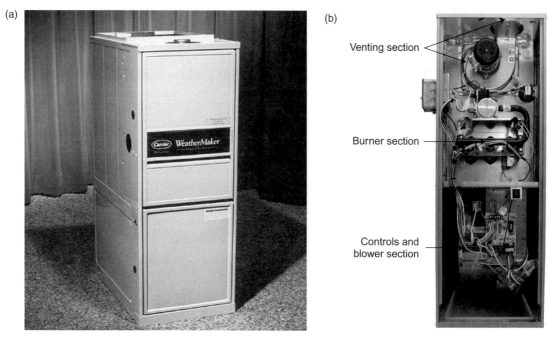

Figure 2-25
(a) Gas furnace with covers. (Courtesy of Carrier) (b) Major components of a gas furnace. (Photo by Susan Brubaker)

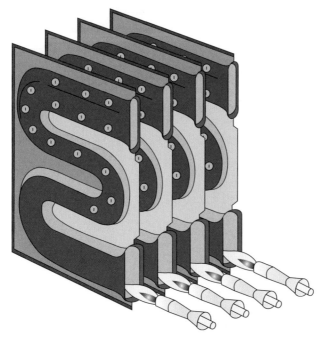

Figure 2-26
Burner and heat exchanger section of a gas furnace, showing the serpentine or S shape of the heat exchanger. The shape creates more heat contact time, which places more heat in the air passing around the heat exchanger.

moisture. Nitrogen is not used in the combustion process but passes through the heat exchanger because it is mixed with oxygen in the air.

The venting section includes a vent motor and vent pipe. The vent motor, commonly called the inducer draft motor, pulls the hot gases thought the heat exchanger and into the vent pipe. The metal vent pipe allows the light flue gases to rise and be safely removed from the furnace. Some very high-efficiency gas furnaces use a PVC vent pipe, which is possible because the gases are cooled to about 100°F.

Safety Tip

The products of incomplete combustion are odorless, tasteless, and undetectable by human senses. Carbon monoxide is formed when the gas and air mixture is out of balance or when the flame touches the metal heat exchanger. Carbon monoxide is not a normal by-product of combustion and is created only if there is a gas and air imbalance or the flame is cooled below its operating temperature.

OIL HEATING

Oil heating is similar to gas heating (see Figure 2-27). It uses light fuel oil and has a combustion chamber, a blower section, and a venting system. Most use #2 fuel oil, which is usually 15% hydrogen and 85% carbon. The heat content is approximately 139,400 BTUs per gallon. Like that for natural gas, the combustion process requires heat, fuel, and oxygen. Oil must be atomized into fine particles for proper combustion to occur.

Fuel Supply

The fuel supply includes a fuel tank that can be above or below ground and inside or outside the building. Upon receiving a call for heat, oil is pumped from the tank through an oil filter and to an oil burner assembly. The oil pump builds up high pressure that is forced through a small orifice called a nozzle.

Burner Section

The oil burner assembly, like the one shown in Figure 2-28, atomizes the oils and ignites the oil vapor, which shoots a flame into the combustion chamber.

Heat Exchanger

The heat exchanger heats up, and a blower circulates air around the outside of it, which is ducted into the space for heating. The combustion products do not come in contact with the air blowing across the heat exchanger.

Figure 2-27
Cutaway view of an oil tank. (Courtesy of Eugene Silberstein)

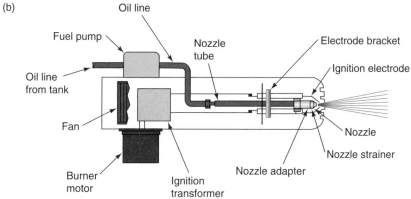

Figure 2-28
(a) Fuel oil gun that atomizes and ignites the fuel oil. (b) Cross-section of the gun assembly.

Venting Section

Like those in a gas furnace, the combustion products are collected and the hot gases rise out of a venting system. Oil furnace vent products can be much hotter than those from natural gas vents. A Type A vent pipe or chimney is necessary for venting oil combustion products.

Figure 2-29
Electric heat strips that produce 5 KW of heat, which is equivalent to 17,000 BTUH. The elements look similar to a toaster heating elements, only with larger wires. (Photo by Joe Moravek)

ELECTRIC STRIP HEAT

Electric strip heat is one of the simplest forms of heating. You have seen electric strip heat used in many forms in your life. It is used to toast bread. Portable electric heaters are used to temper cool air in a small space, and an electric range is used to heat or bake food. The use of electricity for human comfort operates on the same principle. High-resistance heating elements are used to heat the air in an air handler, as in Figure 2-29, which has a blower section that returns cool air from the space to be heated. The air is blown through supply ducts to heat the space.

Heat strips are rated in kilowatts (KW), each of which is equal to 1,000 watts. A common heat strip is rated at 5,000 watts or 5 KW. Watts can be converted to BTUs. One watt equals 3.4 BTUs of heat; therefore, one 5 KW heat strip will equal 17,000 BTUH, calculated as follows:

$$5{,}000 \text{ watts} \times 3.4 = 17{,}000 \text{ BTUH}$$

A heat system may require several 5 KW heat strips to provide enough heat for the conditioned space.

HEAT PUMPS

A heat pump is an air conditioner that can cool or heat. Because of a special reversing valve, the heat pump can convert the condenser into the evaporator and the evaporator into the condenser. The flow of refrigerant in the evaporator and condenser switches directions in the heat mode, while the flow of refrigerant in the compressor does not change direction. This can be confusing when talking about a heat pump problem with a supervisor or another technician. To reduce this confusion, the terms "outdoor coil" and "indoor coil" are used instead of "condenser" and "evaporator."

There are three types of heat pumps:

1. Air-to-air heat pumps
2. Liquid-to-air heat pumps
3. Liquid-to-liquid heat pumps

The most common type is the air-to-air heat pump. Liquid-to-air heat pumps are called geothermal heat pumps and are discussed in the next major section. The liquid-to-liquid heat pump is another type of geothermal heat pump that is used to chill or heat water.

Cooling Mode

The cooling mode is shown in the upper illustration in Figure 2–30. The cooling cycle is similar to a regular air conditioning cycle. The refrigerant is discharged into the lower side of the reversing value. The valve guides the refrigerant to the outdoor coil where heat is rejected from the refrigerant, and the refrigerant vapor is converted to liquid and goes to the metering device. The pressure and temperature drop as the liquid passes through the metering device. Cold liquid refrigerant enters the indoor coil and is used to cool and remove moisture from the air passing over the coil. The liquid vaporizes in the indoor coil as it absorbs heat from the air passing through it. The refrigerant returns through the reversing valve and the accumulator and then goes to the suction side of the compressor to start the process all over. The purpose of the accumulator is to collect any liquid refrigerant that might be returning to the compressor, allowing the liquid to vaporize prior to returning to the compressor. Liquid returning to the accumulator is more common in the heating mode.

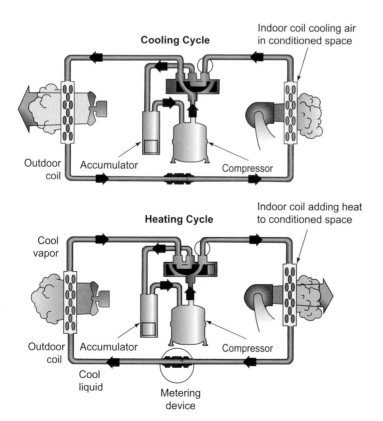

Figure 2–30
The upper picture is the cooling cycle. The lower picture is the heating cycle. The defrost cycle is similar to cooling cycle except the outdoor fan is turned off and auxiliary heat is supplied to the space.

Heating Mode

In the heating mode, shown in the bottom half of Figure 2-30, the reversing valve shifts position so that the outdoor coil adsorbs heat from the cold outdoor air. The hot discharge vapor leaves the compressor and goes to the reversing valve, which directs the refrigerant to the indoor coil. The hot indoor coil heats up the air that passes through it, the refrigerant condenses to a liquid as the indoor air passes over the indoor coil, and the liquid goes through the metering device, where it is converted to a cold, low-pressure liquid and vapor mixture. This cold mixture goes to the outdoor coil to absorb heat from the air. Heat transfers from warm to cold, so the refrigerant is colder than the outdoor air, allowing the heat to transfer. The heat-laden refrigerant is directed back through the reversing valve and the accumulator and then back to the suction side of the compressor to be recirculated.

Defrost Mode

Air-to-air heat pumps require a defrost mode. The defrost mode occurs in the heating cycle. In the heating mode, the outdoor coil can develop a coating of frost because it operates at about 15°F below the ambient temperature. If the heat pump runs long enough in the heating mode, it may also develop a layer of frost that reduces heat transfer into the outdoor coil. The frost will be removed in the defrost mode. This mode turns off the outdoor fan and, for a short time, switches the system to the cooling mode, in which the outdoor coil gets warm and melts the frost. When the frost is cleared from the outdoor coil, the heat pump switches back into the heat mode. Auxiliary heat, usually from electric heat strips, is used to temper the indoor air while the system is in the defrost mode.

GEOTHERMAL

A geothermal heat pump is a refrigeration system with a water-to-refrigerant heat exchanger. In the cooling mode, water is circulated through the water-cooled condenser to pick up heat from the refrigerant. The heat-laden water is pumped to a closed loop water source like a well, lake, or river, as seen in Figure 2-31. Most geothermal heat pumps use a coaxial heat exchanger, also known as a pipe-in-a-pipe or tube-in-a-tube heat exchanger. The water is circulated through the center of the pipe, and the refrigerant circulates through the outer wall.

In the heating mode, the refrigerant flow is reversed because it is a heat pump. Heat is absorbed from the fluid circulating through the coaxial coil. Figure 2-32 demonstrates how this heat is rejected in the indoor coil, which is used to heat the area.

There are two types of geothermal heat pumps:

1. Liquid-to-air heat pumps are the most common used to heat or cool air.
2. Liquid-to-liquid heat pumps are used to cool or heat water.

Geothermal systems have a closed-fluid loop or an open-fluid circulation loop and are referred to as closed loop or open loop. A single closed-loop system is shown in the left side of Figure 2-33. In most cases, multiple loops are buried more than 100 feet deep in a horizontal trench or vertical holes, as shown in the right side of Figure 2-33. An open loop is used to pump water from a well and discard the water back into another well some distance away. The water can be discharged into a waterway, as shown in Figure 2-34, but this is usually not recommended because it may deplete the aquifer being used by the well.

Geothermal heat pumps have very high efficiency ratings in the cooling and heating modes. A correctly designed system will outlast a conventional air source heat pump. However, geothermal heat pumps have limited use because of the high

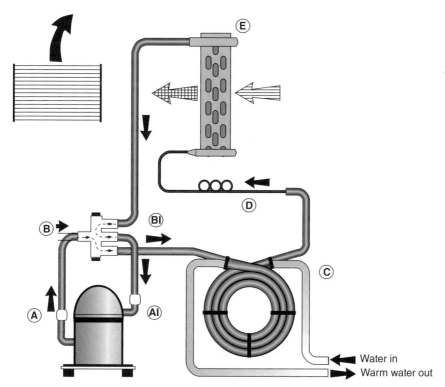

Figure 2–31
Cooling mode for a geothermal heat pump. The coaxial water-cooled condenser rejects heat from the refrigerant into the water circulated through it. The warm water is pumped to a lake, well or other large water source. (Courtesy of ClimateMasters, Inc.)

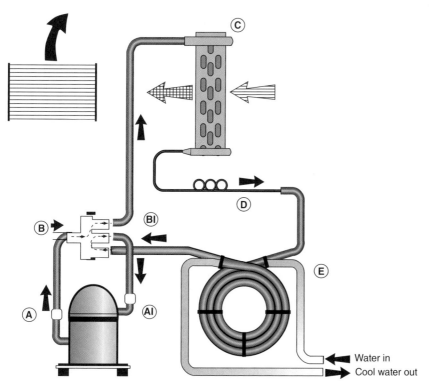

Figure 2–32
Heating mode for a geothermal heat pump. (Courtesy of ClimateMasters, Inc.)

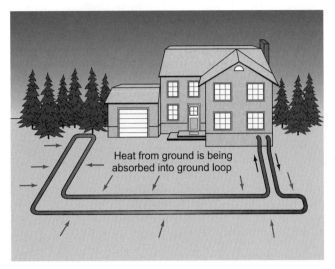
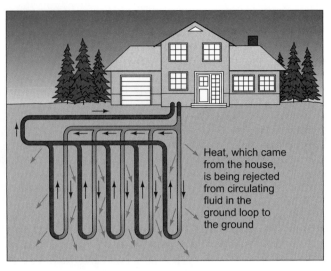

Figure 2-33
Ground closed-loop heat exchangers. The design on the left is a horizontal loop. The installation on the right is a vertical loop.

Figure 2-34
Open-loop system that uses a water well to remove or add heat to the refrigerant. The pond is used to collect the pumped water. The water can be pumped down to a different water level to keep the local aquifers charged with water.

initial installation cost of drilling or trenching. In some areas of the country, water drilling is limited or not allowed.

HYDROGEN FUEL CELL

A hydrogen fuel cell is an electrochemical energy conversion device. It produces electricity from external supplies of fuel (on the anode side) and oxidant (on the cathode side), as demonstrated in Figure 2-35. These react in the presence of an

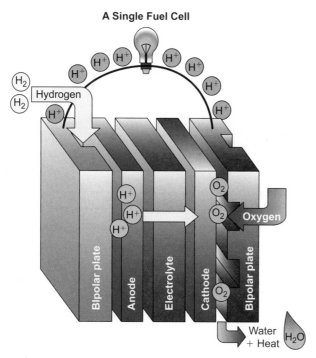

Figure 2-35
Hydrogen fuel cell as a power source.

electrolyte. Generally, the reactants flow in and reaction products flow out while the electrolyte remains in the cell. Fuel cells can operate virtually continuously as long as the necessary flows are maintained.

Fuel cells differ from batteries in that they consume reactants, which must be replenished, while batteries store electrical energy chemically in a closed system. Additionally, while the electrodes within a battery react and change as a battery is charged or discharged, a fuel cell's electrodes are catalytic and relatively stable.

Many combinations of fuel and oxidant are possible. A hydrogen cell uses hydrogen as fuel and oxygen as oxidant. Other fuels include hydrocarbons and alcohols, and other oxidants include air, chlorine, and chlorine dioxide.

SOLAR ENERGY

Solar energy has a variety of meanings and applications. For example, there are passive and active solar energy systems. A passive solar design can be the one built into new home plans. A rectangular house with a solar energy system should be built with the length facing north and south and the width facing east and west because these exposures have the greatest heat gain in the summer. Minimizing this exposure reduces the size of the cooling equipment needed and the amount of heat gain in the summer. With this design, the winter heat gain is also optimized because the best winter heat gain is on the south-facing wall. This is classified as a passive solar design because nothing has to be done after the house is built in order to obtain benefits from the sun.

Active solar energy systems include photovoltaic systems and water-heating applications. Photovoltaic systems convert sunlight to direct current (DC) voltage, as shown in Figure 2-36. The DC voltage can be used to power a direct current appliance or converted to household alternating current (AC).

Figure 2-36
An array of photovoltaic (PV) collectors. (Courtesy of iStockphoto)

Figure 2-37
Flat plate solar collectors on a tile roof of a new house to be used for water heating. (Courtesy of iStockphoto)

Water-heating applications circulate water or an antifreeze solution through solar collectors on the roof of the building. The hot water is collected in a large storage tank. If the water heating collector uses antifreeze solution, it will need to have a heat exchanger to prevent antifreeze contact with potable (drinkable) water, as shown in Figure 2-37.

SUMMARY

This chapter is an overview of the major components and systems you will find in our industry. These items will be discussed in detail in later chapters. The important thing at this time is to gain general information about HVACR systems and learn the language of the trade. Every industry has it special language, and ours is no exception.

It is important to be able to identify equipment and systems by their correct names. First, it helps you seem professional to your customer, coworkers, and supervisor. Second, it reduces confusion when communicating a problem. For example, you may call your supervisor and state that you are working on a heating problem. If you identify the heater as an electric furnace when it is really a heat pump, this will cause confusion when seeking advice on how to solve the problem. The solutions offered to you will not repair a heat pump if you are actually dealing with an electric heating system. Another example is mixing up the terms "evaporator" and "condenser." They are both coils, but the evaporator absorbs heat and the condenser rejects heat—a big difference!

Finally, you learned about different types of duct systems and refrigerant piping. The new language is important in communicating your professionalism. When you go to a doctor, she speaks with medical terms that seem to be a foreign language. If a customer listens to your conversation with a coworker, she may think the same thing. It is not a foreign language; it is the language of the HVACR professional. As you are learning the new terminology, ask for the correct name of each component. You will learn that some components have more than one name. Learn them all.

FIELD EXERCISES

1. Follow the refrigeration cycle on a split system and a package unit. Identify the major components and label the refrigerant lines.
2. Locate a heating system. Determine the fuel type and how the blower section operates.
3. Identify two different types of duct systems on a job.

REVIEW QUESTIONS

1. Draw the refrigeration cycle by connecting the refrigerant lines in Figure 2–38.
2. Label the refrigerant lines in Figure 2–38.
3. Which component in the refrigeration cycle rejects heat?
4. Which component in the refrigeration cycle absorbs heat?
5. Which components change the pressure in the refrigeration cycle?
6. What are the major components of the gas furnace?
7. What is hydronic heating?
8. Draw the refrigeration cycle for the heat pump in the cooling mode.
9. Which component in the heat pump heats the indoor air?
10. What are two sources of heat for a geothermal heat pump?
11. What is the difference between an open-loop and a closed-loop geothermal heat pump system?
12. Give two examples of active solar energy applications.
13. What is the difference between ACR tubing and water piping?
14. What are three common duct materials?
15. How many BTUH are produced by electric heat strips that equal 10 KW?

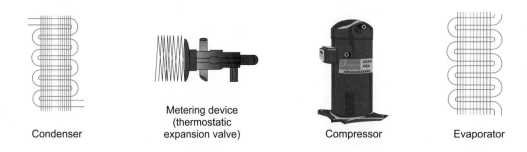

Figure 2-38
Draw the refrigeration cycle by connecting the components. (Courtesy of Emerson Climate Technologies)

CHAPTER 3

Introduction to Basic Electricity

LEARNING OBJECTIVES

The student will:
- Define "volts," "amps," "ohms," and "watts."
- Explain how to measure AC and DC voltage with a multimeter.
- Explain how to measure amperage with a clamp-on ammeter.
- Use Ohm's law to calculate volts, amps, and ohms.
- State four safety procedures to follow when using a ladder.
- Calculate watts using the power formula.
- Describe how a magnetic field is used to operate electrical components.
- Draw and identify twenty common electrical symbols.
- Draw a simple electrical schematic.
- Describe ten safety precautions to take when working around electricity.
- List the factors that determine maximum current draw of a conductor.

INTRODUCTION

When you go into a dark room, you turn on the light. What if the light does not operate? What is the problem? Your first thought might be that the light bulb is burned out. You change the light bulb, but it still does not light. What can be the problem? You think of the possibilities—it may be the circuit breaker, the light switch, or a wiring problem. Finally, you decide to try a second new light bulb. Bingo! The second light works. The previous new bulb must have been defective. This is an example of a simple electrical problem that has many possible solutions.

You need to understand the operation of electricity in order to arrive at a solution to a problem you encounter. You understood that the light did not work and tried the easiest, most common solutions until you were able to solve the lighting problem. If the easy solutions do not solve the problem, you will need to measure voltage with a meter to locate the exact problem.

In order to complete a simple lighting troubleshooting problem, you needed a basic understanding of the terms and the operation of the light circuit. You needed to know the terms "light," "light switch," "circuit breaker," and "wiring." You did not know exactly how these components operated, but you did need a general understanding of the operation or you would still be in the dark. In other words, you need to know that the bulb needs to be good and a switch on, or "closed," to bring electricity to the light. If that did not work, you needed to understand that a fuse or circuit breaker might be blown, or "open," and in need of replacement or resetting. Troubleshooting begins with an understanding of the circuit.

Understanding electricity is one of the most important aspects of being a good technician. One reason is that most problems are electrical, and you need to be able to understand how the electrical circuit operates to troubleshoot and solve the problem. An even more important reason for learning about electricity is your own safety. Working around HVACR equipment means working near electrically energized components. Understand that electricity in this equipment can injure or kill you. Even if equipment is off and not operating, it can still have power and be dangerous. This chapter will point out safety recommendations when working around electricity. It will introduce how to use multimeters so that you can check voltage prior to sticking your hands into the equipment.

This chapter will also discuss how to calculate voltage, amperage, resistance, and watts. In order to use a map, you need to know what the symbols on the map mean. For example, a blue squiggly line is a river, and ++++ lines may indicate a railroad track. Similarly, electrical circuits have symbols and maps also. You will learn the symbols that will help you read electrical diagrams, which are the maps to finding an electrical problem. Finally, the chapter will discuss wires, known as conductors, and wire size. You see large wires going to your car battery. These are the largest wires in a vehicle. The wires going to the lights, radio, and computer are much smaller. These conductors are sized to match the amperage requirements of the vehicle and battery. Undersized wiring in your car will overheat, burn out, or possibly cause a fire. The same holds true for HVACR system wiring. The incorrect size may cause component or wire damage and cause a fire.

This is an important chapter in your quest to become a good technician. At the end of the chapter, it is normal to be a little overwhelmed with the new material you have learned. The electrical topics will be covered again in future years of training. It is not something that you will totally grasp the first time through. Reread and review the material until it begins to sink in. Ask questions in the field when you see someone working on equipment. You will understand electricity as you continue to gain years of exposure and experience in our interesting profession. Learning about circuit electricity spills over into other professions like vehicle wiring, appliance troubleshooting, and building lighting circuits.

Even with your limited but growing knowledge of the HVACR system, you can try to locate some of the obvious problems. What the new technician did were the

Safety Tip

What is PPE? The term is commonly used in the commercial and industrial section of the HVACR trade. *PPE means "personal protective equipment."* For technicians to have the greatest protection, they should wear the following pieces of PPE:

- Eye and face protection
- Metal-free, rubber-soled, insulated safety shoes
- Leather gauntlet gloves over rubber insulated gloves when working around high voltage
- Non-conductive clothing—no metal buckles or buttons and no synthetic fibers

Field Problem

As a first-year technician, you are asked to meet an experienced coworker at a convenience store, and you are the first to arrive. Your partner calls to say he is running late from his current job, but he wants you to start by talking to the store owner and finding out more about the air conditioning problem. You have not been to a job by yourself, but you realize this is important because this is a long-time customer who has been waiting several hours for service.

You go into the store and introduce yourself to the owner. She is upset that it took so long for someone to respond on a hot July day. The front and back doors of the store are open to allow some ventilation into the hot store, but the store seems warmer than the outside temperature. After looking around, you notice the lighting and the refrigeration equipment rejecting heat inside the store. In addition to the air conditioning not working, maybe these internal heat generators are the reason it is warmer inside than outside.

You start checking out the problem with the minimum knowledge you have at this time in your career; you know the customer wants to see some action. If you do not find anything obvious, you can call your partner to ask him what to do next. You ask the customer about the problem and any prior problems. There have been no prior problems, and the equipment had its scheduled maintenance in May. You ask for the thermostat location. The store owner shows you the thermostat in a hallway toward the back of the small store. You verify that the thermostat is set for cooling and with a set point low enough to operate the system. The thermostat fan switch is set in the automatic blower position. Nothing seems to be working. You turn the thermostat to the fan on position, but the blower does not operate.

Next, you decide to investigate the equipment. The system is a package unit and located on the roof. The package unit is quiet. You remove a panel and find that the blower is off, and you do not hear any compressor or indoor fan sound. What are the possible things that could cause a total system shutdown?

You have checked the thermostat, even resetting it to ensure good thermostat switch contact. You ask the owner for the location of the circuit breaker. The owner says that she checked the breakers and that they are OK. Because you have nothing else to check at this time, you courteously ask to see the breaker panel. As the owner stated, all the breakers look set and in the closed position, completing voltage to all circuits. You turn off a 40-amp double breaker that is labeled "A/C unit," and you notice that it seemed to turn off with less resistance than other breakers you have turned off in the past. You reset the breaker and hear the air conditioner begin to operate. About that time your partner arrives and does not find an obvious problem. The lead technician reports this to the customer and discusses what might have caused this problem. The customer said she had a very brief power outage in the morning and that the air conditioning equipment did not come back on after that outage. The lead technician explains that might have been the problem because returning power may not be at the correct voltage level, causing higher amperage draw and tripping the breaker. Breakers are amperage sensitive. The technician recommends that a time delay be installed to prevent future problems like power short-cycling from the utility company or someone playing with the thermostat. The owner agrees, and the safety device is installed and tested.

first steps to troubleshooting: see what is working or not working and check the thermostat and related common components. Sometimes the problems are simple and sometimes complex. It is a matter of experience.

I WANT TO SPEAK IN ELECTRICAL TERMS

Before venturing into the world of electricity, we need to learn new terms that relate to this topic. Some of these terms are used interchangeably and sometimes incorrectly. Understanding the correct definitions will help you communicate a problem more professionally and understand what another technician or supervisor is conveying to you. Studying electricity is not a one-time process; it is something that should be repeated because it plays such an important part of your daily behavior in troubleshooting electrical problems.

Conductors and Wires

The term "conductor" means a wire or pathway for electricity to follow. Think of a conductor as a road for your car. In order for you to get to and from your destination, you will need a complete pathway, a two-way street. The conductor is composed of a metal molecule, which is composed of subatomic parts called neutrons,

> **Safety Tip**
>
> Shock occurs when part of your body comes in contact with electricity. Damaging effects can vary from minor tingling to burns, internal organ damage, and even death. Arc blast occurs when a current jumps from the electrical circuit to another conductor or ground. The intense heat can cause equipment fire or skin burns. This can also occur when meter probes are placed too close to high voltage.

protons, and electrons. When the molecule contains the same number of neutrons, protons, and electrons, it is considered neutral and a poor conductor of electricity. When the protons outnumber the neutrons and electrons, the molecule is positively charge and again not a good conductor of electricity. A good conductor of electricity has more electrons than protons, as shown in Figure 3-1. The negatively charged molecule contains one or more extra electrons, which are required for a good conductor of electricity. Electricity or current flows with the transfer of electrons between the molecules. In Figure 3-1, the circles with the dashes (-) represent the electrons, shown in the two outer molecule orbits. Excess electrons are found when there are more electrons than protons. It is important to know that there are conductors, like copper, that are friends to the flow of current. There are insulators, like glass and rubber, that stop current flow because there are no extra electrons in the outer orbit. Finally, there is resistance material, like carbon or nickel-chromium wire, that allows a reduced flow of electrons. In our profession, we will be involved with all three materials: conductors (and semiconductors), insulators, and resistors.

Volts

The concept of voltage can be described as similar to manually pushing a car down the street. Your pushing force, or voltage, is the pressure to move the car. The car is the current flow or movement down a pathway, the street. *A volt is defined as a unit of electrical pressure that pushes electrons through a wire and a circuit.* The higher the voltage is, the greater the electrical pressure. Without adequate electrical pressure, current will not flow in a circuit. A volt is also called voltage, potential difference, and electromotive force and can be abbreviated as EMF, V, or E. These are all terms that can be used interchangeably to relate to the word "volts." There is direct current (DC) and alternating current (AC) voltage. Let's discuss the difference between the two.

Direct Current

Direct current (DC) flows only in one direction and does not reverse its direction of flow, as shown in Figure 3-2. DC can be positive or negative. Even though it is referred to as direct current, it relates to voltages also. The best example of DC is a battery: a battery has a chemically charged material that allows the free flow of electrons from the positive side, through the load, and back to the negative side. DC is also used in electronic and solid-state devices, and it is found in furnaces and condensers that have circuit boards.

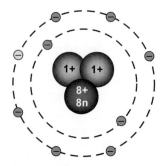

10 protons and 11 electrons

Figure 3-1
The negatively charged molecule contains extra electrons that are required for a good conductor of electricity. Electricity flows by the transfer of electrons between the molecules. The circles with the dashes (-) represent the electrons. The two outer molecule orbits represent electrons. Excess electrons are found when there are more electrons than protons. (Courtesy of Dick Wirz, Refrigeration Training Services)

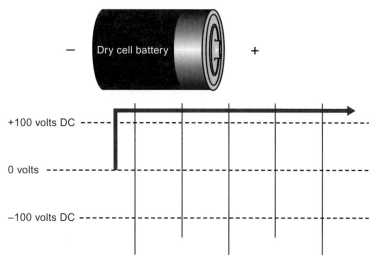

Figure 3-2
Direct current (DC) can be negative or positive polarity. DC voltage quickly reaches its peak voltage and stays at that point until turned off or the electron transfer diminishes, as is the case in a battery. (Courtesy of Dick Wirz, Refrigeration Training Services)

Alternating Current

Alternating current is current that reverses its direction of flow in a positive and negative direction 60 times a second. One positive and negative event is called one cycle or one hertz. This is known as frequency. Most of the power generated in the United States is developed at a frequency of 60 hertz. Some countries generate power at a 50 hertz frequency. A sine wave, or one complete cycle, is pictured in Figure 3-3. This complete cycle includes a positive and a negative wave form. Equipment designed for a specific hertz level can be damaged if operated at a different frequency. Figure 3-3 shows a single conductor (wire) in the magnetic field of a generator. The magnetic field varies between the north and south poles. As the magnetic field varies, so does the voltage induced into the conductor. When the north pole magnet is inducing a magnetic field in the conductor, it creates a positive voltage. When the south pole magnet is inducing a magnetic field, it creates a negative voltage. To create alternating current, the magnetic field is varied between the north and south poles. These alterations create the change in voltage

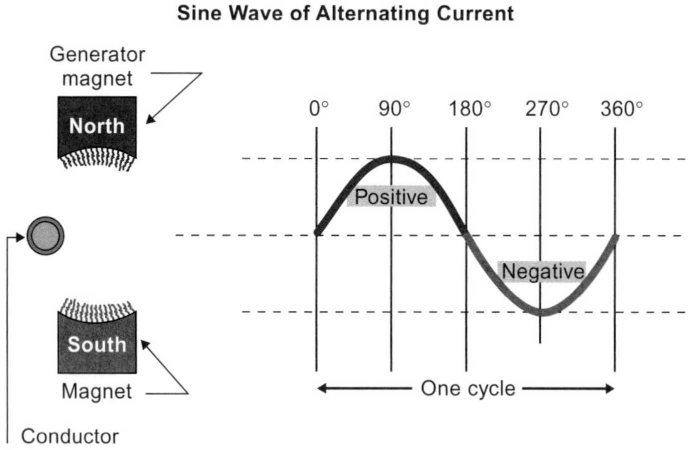

Figure 3-3
This is the sine wave of a 60-cycle, single-phase generator. The magnetic flow in the generator changes the positive and negative polarity in the conductor. The conductor is rotated 60 cycles per second through the magnetic field to create positive and negative changes. The conductor supplies voltage to transmission lines that is distributed to residential, commercial, and industrial customers. (Courtesy of Dick Wirz, Refrigeration Training Services)

as indicated in this drawing. This is considered single-phase voltage, which is used in residential applications and some very small commercial buildings. Most commercial and industrial applications use three-phase generated power.

Three-phase, AC-generated power is shown in Figure 3-4. Like single-phase power, there are two opposite magnetic poles. There are also three conductors that are evenly placed and rotated in the magnetic field. If you consider the placement of the conductors as being inside a circle (360°), then the conductors are placed 120° from one another. This creates a three-phase power supply that is 120° out of phase with each of the other phases. It is like taking three single-phase power supplies and starting them at one-third intervals, or 120° apart. In reviewing Figure 3-4, you will notice that the overlapping development of three-phase voltage partially fills the valley in the positive and negative parts of the cycle. Compared to the open valleys of single-phase generation, the partial filling of the valleys of the three-phase operation improves the efficiency and starting torque of motors.

Amperes

An ampere is similar to the number of cars on the road. Let's think of a car as an electron. The greater the number of cars (electrons) passing a given point is, the greater the amperage. *An ampere is the number of electrons flowing in a circuit*. The higher the amperage is, the greater the electron flow. The ampere is also known as the intensity of electron flow. To be more specific, an ampere is defined as 1 coulomb flowing past a point in an electrical circuit in 1 second. A coulomb is equal to 6.25×10^{18} electrons flowing through a circuit in 1 second. That is a lot of electrons moving in the wire. Ampere has alternative names such as amperage, current flow, amps, and the abbreviations A or I. Amperage creates a magnetic field around a wire. The greater the amperage is, the greater the magnetic field developed around the wire carrying the current, as shown in Figure 3-5. The clamp-on ammeter measures the intensity of the magnetic field and will be discussed in the meter section.

Resistance

Resistance is like pushing the car uphill. The uphill climb reduces the speed of your progress as well as the number of cars passing a specific point on the road. *Electrical resistance is the opposition to current flow*. As the resistance gets higher, the current flow gets lower. Alternative names for resistance are ohms, a load, the abbreviation R, and the Greek letter omega (Ω). Large diameter wires and short wires have lower resistance, and higher temperatures increase the resistance of

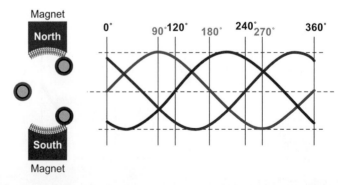

Figure 3-4
This is the sine wave of three-phase current flow. Three-phase power is developed by rotating three conductors, evenly spaced, through a magnetic field. The peaks and valleys in three-phase power are partially filled, making this power source more efficient and creating better motor starting torque than single-phase power sources. Windings that are 120° out of phase give three-phase motors their high starting torque. (Courtesy of Dick Wirz, Refrigeration Training Services)

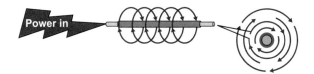

Figure 3-5
The flow of electrons or current in a wire creates a magnetic field around the wire. This magnetic field is called the line of flux. The greater the current flow is, the greater the magnetism or lines of flux. (Courtesy of Dick Wirz, Refrigeration Training Services)

the wire. Materials have various levels of conduction. For example, copper wire is a good, inexpensive conductor of electricity because of free electrons in the material. These free electrons can easily be exchanged, thus creating a current flow. Materials that have resistance to electron flow are known as insulators. Common insulators are glass, plastic, and rubber. They have few free electrons, thus reducing electron flow.

Power

Power is a measure of work being done. Power is calculated by multiplying the volts and amps in an operating circuit. Power is also stated as watts or wattage and abbreviated P or W. Most electric utilities measure and record power usage to bill their customers. Utility electric meters on customers' houses and buildings measure the wattage used. The utility company bills in this manner because pressure (volts) is required to push the electrons into the building and a high rate of electron flow is needed to operate the components in the building. The electric company must create enough pressure and provide enough electrons to satisfy its customers, some of whom are several hundred miles from the generating site.

Safety Tip

Do not directly touch a person receiving an electric shock. You could experience the same electrical shock by directly contacting that person. The human body is a conductor of electricity because it is primarily liquid. If you see a person being shocked, quickly turn the power off, push the person away with insulated material such as a piece of wood, or pull the person's clothes with an insulated tool.

Complete Circuit

A complete or closed circuit has an unbroken path for electrons to flow through a load. The components of a complete circuit are the power supply, the load, and interconnecting wires, as shown in Figure 3-6. This is a very simple complete circuit with no switches or other elements found in the air conditioning trade. Air conditioning circuits are more complex.

A circuit is incomplete or open when there is a break in the circuit, like the one in Figure 3-7. The break can be a break in a wire, an open motor winding, or an open switch.

OHM'S LAW

Many years ago, a man named Georg Ohm discovered the relationship between electrical pressure, current flow, and resistance. He found that the pressure drop,

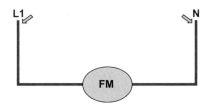

Figure 3-6
A complete circuit includes a power supply, a load, and interconnecting wires. This is a very simple circuit with voltage from L1 and neutral supplying power to a fan motor (FM). Additional controls like switches will be installed to control this fan motor. (Courtesy of Dick Wirz, Refrigeration Training Services)

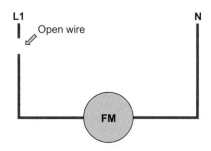

Figure 3-7
Open circuit due to a break in the power supply. This is also known as an incomplete circuit. An open circuit can be designed into the circuit, such as a switch or thermostat. (Courtesy of Dick Wirz, Refrigeration Training Services)

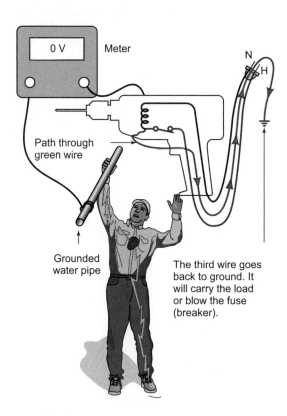

Figure 3-8
A grounded drill protects the technician if there is a short to the case. The electricity goes to ground rather than shocking the operator. Most new power tools are double insulated to prevent this danger. To check for a grounded tool, measure from the metal tool case to ground. A voltage reading to ground means the tool is dangerous.

> **Safety Tip**
>
> Figure 3-8 shows the importance of grounding a tool or piece of equipment. Without the ground, the electricity has a path through the operator. This can kill!

measured in volts, was equal to the current flow, measured in amperes, multiplied by the resistance. Resistance is measured in ohms, named in honor of the man who discovered it.

The electrical symbols for this formula are as follows:

- E for voltage
- I for amperage
- R for resistance

Using these symbols, the relationship can be expressed as a simple formula:

$$E = I \times R, \text{ or } E = IR$$

This formula is known as Ohm's law. It is a very fundamental electrical concept. It means that voltage is the product of current times the resistance.

Ohm's law can be expressed two other ways:

- When you are calculating current flow, the current flow equals the electrical pressure divided by the resistance: $I = E/R$.
- When you are calculating resistance in a circuit, the resistance equals the electrical pressure divided by the current: $R = E/I$.

For example, if a current of 5 amperes flows in a length of wire and the resistance of the wire is 10 ohms, then the applied voltage from one end of the wire to the other will be $E = I \times R$, or 50 volts = 5 amps × 10 ohms.

If another wire has a resistance of 8 ohms and an applied voltage of 80 volts, then current flows through it can be calculated by $I = E/R$, or 10 amperes = 80 volts/8 ohms.

In still another wire, we know the applied voltage is 110 volts when 10 amperes of current flows. So the resistance is $R = E/I$, or 11 Ω = 110 V/10 A.

Figure 3-9 provides an easy reference for the different ways of expressing Ohm's law. As shown in the upper left diagram, draw a circle and write an E in the upper portion and an I and an R in the lower portion of the circle. Dissect the circle with a horizontal line and a radius line splitting the bottom half of the circle into two equal parts. With your finger, cover up what you are trying to solve. If any one of the three letters is covered, representing the unknown quantity, then the two uncovered letters show how to obtain the answer. Thus, E, when covered, reveals the formula $E = I \times R$; and I, when covered, reveals $I = E/R$. When you cover R, the diagram gives you $R = E/I$.

In the Field

Unfortunately, the air conditioning trade uses several terms that mean the same thing. This is true in the electrical segment of the industry also. For example, voltage is known as E, V, volts, potential difference, electromotive force, or EMF. Current uses the terms I, A, amps, amperage, and current flow. Resistance has fewer synonyms, including R, ohms, and the Greek letter Ω.

All of these variations can be a little confusing when someone is trying to learn a new concept. To help you get accustomed to the new electrical language, these terms will be used interchangeably throughout the electrical chapter.

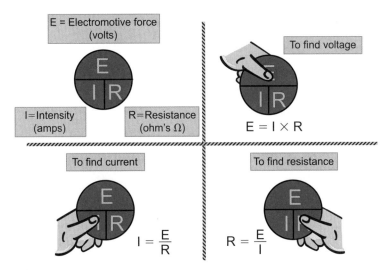

Figure 3-9
The Ohm's law disk can be used to calculate volts, amps, or ohms. Cover what you are calculating and the formula is exposed. (Courtesy of Dick Wirz, Refrigeration Training Services)

Safety Tip

Figure 3-10 shows a technician being shocked by an ungrounded air handler. It is important that every piece of equipment be properly grounded. If an energized wire or a component shorts to the case, it will be a danger to anyone touching it. If the technician is touching ground, like a pipe, the danger of receiving a fatal shock becomes greater.

Safety Tip

Ladders are on every construction site. Although they are a familiar sight, they are not always used properly and safely. Falls from ladders can be deadly and can occur when working from a ladder, climbing up or down a ladder, or when the ladder tips or slides. Any time you use a ladder, you must recognize the hazards involved and learn the proper ways to select, set up, use, and maintain ladders. Figure 3-11 offers other safety considerations.

Safety Tip

Before using a ladder, inspect it for any faults or defects. Look for broken or missing rungs, split rails, corroded components, and splinters. Never use a defective ladder. Mark and label it so that no one will use it. Learn about podium ladders, an alternative to stepladders used in construction. They provide a small platform to work from and come equipped with guardrails.

Ohm's law is accurate only with resistive loads like electric heat strips used for space heating. It is not accurate with inductive loads like motors or transformers. An inductive load is a device that uses magnetism to achieve an electrical and a mechanical result. An example of an inductive load is a motor that rotates due to a changing magnetic field in the motor windings. The magnetic field in the windings causes the motor to rotate by the attraction and repulsion in the magnetic field. It is similar to placing two magnets together. The magnets will attract each other or repel each other. The rotating magnetic field pushes and pulls the rotary and attached motor shaft to create motion.

CALCULATING POWER

Power is the rate of using energy or doing work. It is measured in watts and horsepower. In direct current, the formula for calculating power is $W = E \times I$. Thus, when 1 ampere of current flow is under pressure of 1 volt, energy is being consumed at the rate of 1 watt. A 6-volt battery connected to a small lamp causes 5 amperes of current to flow, producing power at the rate of 6 volts $\times$ 5 amps = 30 watts.

The electric power formula, mentioned above, also applies to single-phase alternating current (AC) circuits in which the load is pure resistance. Pure resistance is a circuit that only has resistance. It will not have inductive reactance (magnetic resistance), as found in a transformer or motor, or capacitive reactance (capacitor plate resistance), as found in a capacitor. An example of a pure resistance component is an electric heater. For example, an electric heater connected to 120 AC volts draws 5 amps of current. Therefore, its power is $P = I \times E$, or 600 watts = 5A $\times$ 120 V.

Another example is a small heater connected to 120 volts that draws 12 amps, so its power consumption is 1440 watts = 120 V $\times$ 12 A.

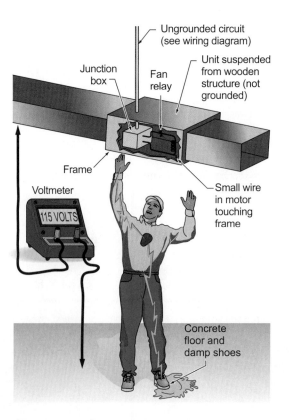

Figure 3-10
An ungrounded piece of equipment can be dangerous. If a component in the equipment shorts to ground, it can become dangerous to anyone touching the metal case while another part of his body touches ground.

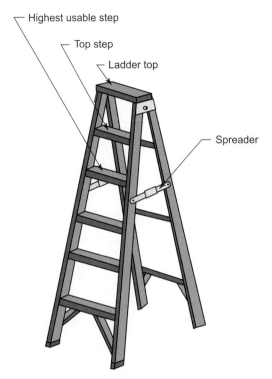

Figure 3-11
Ladder safety includes not exceeding the posted weight limits, which are listed on a label on the side of the ladder. Do not use the ladder top or top step. A-frame stepladders should only be used with the spreader bars down and locked.

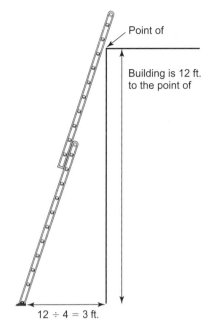

Figure 3-12
The safe use of an extension ladder is important to the user. The ladder must be at an angle to the building. In this example, the building is 12 feet tall. Divide the height of the building by 4 to obtain the minimal distance the ladder should be from the wall. The ladder should be tied off near the point of support.

Because 1000 watts equals a kilowatt or KW, the power of the previous heater can also be expressed as 1.44 KW, calculated by KW = watts/1000, so 1.44 KW = 1440 W/1000.

INTRODUCTION TO METERS

How do you know if voltage is applied to a circuit? If a light, motor, or radio is working properly, then you can assume that the correct voltage is applied, but just because a light, motor, or a radio is not working does not mean that voltage is not present. These items can be burned out or open even if voltage is applied. Is it the item that is defective, or is the power supply not available to operate these items? A defective component may have power applied. You need to know this prior to touching exposed wiring, which can be deadly. This is why we need to understand how to use meters to check voltage, amperage, and resistance of wires and components in a circuit. Meters are used for our safety as well as for troubleshooting the circuit or device that is not functioning.

In order to conduct electrical troubleshooting, the technician needs to know how to operate a variety of electrical instruments. The common meters used in our profession are the digital and analog multimeters. Digital and analog meters are shown in Figure 3-13 and Figure 3-14. We will generally discuss how to use these instruments, but it is important to read the meter's instructions so that you use the meter properly and safely.

Multimeter

A multimeter is an instrument used to measure voltage, amperage, and resistance. Sometimes this meter is called a VOM, which stands for volt-ohm-milliammeter.

Safety Tip

Selecting the right ladder for the task involves a variety of factors. For instance, remember that metal ladders and wet wooden ladders conduct electricity. In most cases, fiberglass ladders are recommended. Choose the right ladder length for the job with the recommendations provided in Figure 3-12. You do not want to lean your body off to one side and cause the ladder to become unstable. Make sure your weight, which includes your tools and parts, does not exceed the maximum capacity of the ladder.

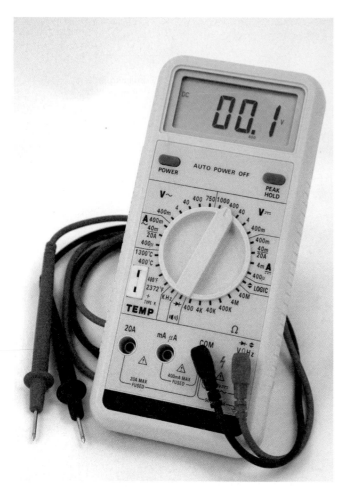

Figure 3-13
The digital multimeter has a direct readout. (Courtesy of iStockphoto)

Figure 3-14
Analog meters with a meter movement. The meter leads are inserted on the bottom side of the meter. (Courtesy of Photos.com)

> **Safety Tip**
>
> The best advice regarding the use of meters can be found in the literature provided with the instrument, which says "Be sure to read, understand, and follow the manufacturer's instructions."

Make sure that you have the meter set on the correct scale of volts, ohms, or amps before taking measurements. Taking a measurement on the wrong scale can result in an incorrect reading or no reading at all. In some cases, being on the wrong scale will damage the meter and possibly hurt the user.

There are analog meters and digital meters. The analog meter uses a needle meter movement, and the digital meter uses an electronic read out. The analog meter requires battery power only for the resistance measurements, but the digital meter requires battery power to display all options.

According to instrument historian Steve Solomon, in the 1920s it was a man whose last name was Simpson who put together three meters—a voltmeter, an ohmmeter, and a milliammeter—into what is known as the VOM. Before this VOM became available, technicians had to have each type of meter, which were sold separately. Over the years, the combined meter became known as a volt-ohm-multimeter. Eventually the word "volt" and "ohm" would be dropped from VOM, and the meter became known simply as a multimeter.

Measuring Voltage

Analog meters and many digital meters require that the technician select a voltage type and voltage scale. The user must select AC or DC voltage. Next, he must select the range of voltage to be measured, which can be 1–10 volts, 1–150 volts, 1–500 volts, or 1–1,000 volts. Always select the highest voltage range so as not to damage the meter with excessive voltage. The voltage range can be reduced if the reading is too low. Some digital meters have an autoranging feature. When using this feature, place the meter on the correct voltage type and the meter will measure voltage within its maximum range. Autoranging has one voltage, current, or resistance setting. There is no need to switch between various settings.

Measuring Resistance

When measuring resistance or ohms on an analog meter, the meter must select the range of resistance. The range of resistance can be any of the following:

- R × 1
- R × 10
- R × 100
- R × 1,000
- R × 10,000

The capital letter "R" refers to resistance, which is multiplied by the number that follows it. For example, if the meter is set to the R × 10 range, the meter reading will be multiplied by 10. If the meter reads 5, then the resistance will be 50 Ω (5 × 10 = 50 Ω). Digital meters will give a direct readout, while some digital models give a readout but require multiplication as indicated by a small "× 10," "× 100," or "1K" on the screen.

When using the ohm feature of a meter, it is important that the power source be removed. Measuring resistance with the power applied will blow the meter fuse or damage the meter beyond repair. It is also necessary to disconnect the component from the circuit. Measuring the resistance of a connected component in the circuit may give a false result because the meter may measure resistance through another component connected in series with it. Remember to disconnect power and remove the component for accurate resistance testing.

Some meters have a sound continuity option. If there is a complete circuit, the meter will signal with a sound. This is good if you are simply checking for a complete circuit. This option does not register a resistance reading but gives a beeping or warning sound. This option is also limited to a resistance of around 50 Ω. Many users of this sound option do not realize that the resistance level is this low. If they check a component that has high resistance, the meter will not signal with a sound, leading them to condemn the component thinking it is open when in reality it has a higher resistance that will not register with the sound continuity feature.

Measuring Amperage

A multimeter is an in-line amperage measuring device. In other words, the multimeter needs to be in series or in line with the amperage being checked. The circuit being measured must be deenergized and the meter placed in series with the load being checked. The circuit must be reenergized to read amperage, which creates a safety concern for the technician. This is why clamp-on meters are more commonly used on line voltage circuits. Most multimeter amperage measurements are less than 1 amp, usually in the milliamp or microamp range. A milliamp is 1 one-thousandth of an amp and is expressed as 0.001. A microamp is 1 one-millionth of an amp and is expressed as 0.000001. A common use for the microamp range is

---

Safety Tip

The National Electrical Code designates low voltage as 50 volts or less, line voltage as 51 to 600 volts, and high voltage as over 600 volts. Extreme caution should be exercised whenever working with voltages above 50 volts. Always follow applicable safety standards.

Safety Tip

When working with electric meters around live power circuits, keep in mind the following:

1. Replace the original fuse with a correct fuse.
2. Never bypass fuses—the voltage and amperage rating of the fuse are important to protect the circuit.
3. Always use the correct test tool for the job and place the meter on the correct setting.
4. Always use a properly rated multimeter.
5. When working on a live circuit, as is sometimes required to find the problem, try to keep one hand free.
6. Always use proper lockout/tag out procedures.
7. Routinely inspect your test leads.

Tech Tip

To use a multimeter safely and accurately, the technician must measure a known, live voltage source prior to using the meter for troubleshooting. Even a known voltage source may be turned off. If the multimeter does not measure a voltage, it does not mean that the circuit is dead. The meter could be defective, or the meter lead could be open or disconnected from the meter socket. For your safety, it is very important to double check. Remember, a meter can become defective at any time in the troubleshooting process.

measuring the current flow of a flame detector of a gas heating system. When the flame sensor "sees" a flame, it generates about 10 microamps.

Many multimeters can measure up to 10 amps. A multimeter is not commonly used to measure amperage above 1 amp because the circuit has to be deenergized and the meter hooked into the circuit. The next section will discuss an easier way to measure amperage above 1 amp.

Selecting a Good Multimeter

Select a multimeter based on features; price should be the least important part of your decision. The most important decision making aspect should be safety and meter features. An unsafe meter can put your safety in jeopardy. Here are some recommendations:

1. Is the meter fuse protected? Is the fuse voltage rating equal to or higher than the highest voltage measured by the meter? Amperage rating should be correct.
2. Verify that your instrument has been tested and certified by two or more independent testing laboratories such as Underwriters Laboratories (UL) in the United States, Canadian Standards Association (CSA) in Canada, and TUV Rheinland in Europe.
3. Look for 600 volt or 1,000 volt, CAT III or 600 volt, or CAT IV rating on the front of the meter and a double-insulated symbol on the back.
4. Verify that the ohms circuit is protected to the same level as the voltage test circuit.
5. Check for a broken case, worn test leads, or a faded digital display.
6. Does the meter have the features or options that are required for your job? A minimum recommendation is that the meter can measure voltage up to 600 volts, resistance up to 2 megaohms, and current measuring capabilities up to 10 amps. Having accessories like temperature measurement options and a capacitor checker can be an important part of your selection decision. Does it measure DC microamps?

CLAMP-ON AMMETER

Why is it important to measure amperage flow in a circuit? Excessive current flow is an indication of a problem. It could be the component that is defective, or it could be a short circuit to ground. Measuring amperage also confirms how hard a component or system is working. A nonoperating component or system will not draw amperage.

Excessive current flow will cause the overcurrent protection to open. If it were not for a correctly sized fuse or circuit breaker, the wire or component would overheat and burn up, possibly causing a fire in the equipment. How do you measure current flow in a circuit? Some multimeters have the capability to measure up to 10 amps, which has limited uses. The multimeter needs to be in series with the load it is measuring. This means that the system needs to be shut down and the meter inserted in series and turned on to take the measurement. In order to remove the meter, the reverse action is required. As you can see, this is very time consuming for a simple current measurement. The advent of the clamp-on ammeter changed all that.

A clamp-on ammeter, as shown in Figure 3–15, is used to measure amperage in a wire. The clamp-on ammeter jaws are clamped around a conductor and measure current flow in that conductor by measuring the electromagnetic field around it.

Clamp-on meters have optional functions such as the ability to read voltage and resistance. Most of these functions have limited ranges, and their accuracy may not be as good as on the multimeter. Many of the resistance readings are limited to 500 or 1,000 ohms. Troubleshooting some components requires a meter reading in the thousands of ohms to check coil continuity or resistance to ground. These are good options to have, but they should not be used to replace a full multimeter.

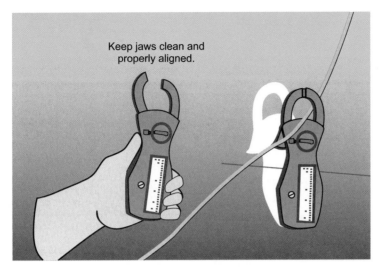

Figure 3-15
The clamp-on ammeter opens a set of jaws to take the reading of the magnetic field in one wire. The greater the magnetic field is, the higher the current flow. (Courtesy of Carrier Corporation)

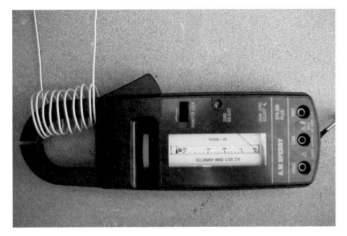

Figure 3-16
Ten wraps of wire around one jaw of the clamp-on ammeter will increase the meter reading by 10 times. (Courtesy of Eugene Silberstein)

Clamp-on meters come in analog or digital readouts. The digital readout tends to be more accurate because the readout is direct and there is less possibility of reading error. The analog meter has scales that need to be selected, and the user may misinterpret the reading. Digital clamp-on meters can also be purchased that read AC and DC current and measure peak amperage draw. The digital clamp-on meter does require battery power, while the analog meter does not need a battery. The analog receives its power from the induced magnetism from the wire it is measuring.

Finally, the clamp-on ammeter can be used to measure amperage below 1 amp. Measuring low amperage is done by wrapping the jaws with the wire that requires the amperage measurement. Each wrap increases the induced magnetic field into the jaws of the clamp-on meter. A common wrap is ten times around one the jaws, making the reading on the meter ten times higher than the actual current flowing in the wire. For example, if the technician wraps the ammeter clamp ten times, hooks it to the fan relay coil circuit, and reads 6 amps, the actual current flow will be 0.6 amps as calculated below and shown in Figure 3-16 and Figure 3-17.

$$\frac{6}{10} = 0.6 \text{ amps}$$

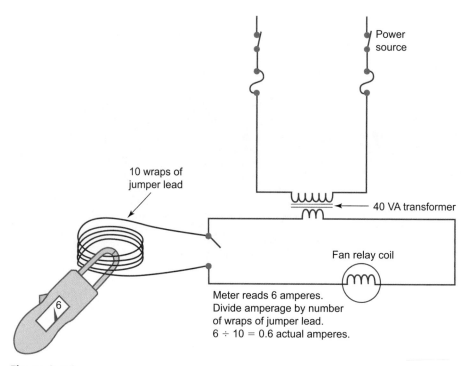

Figure 3-17
The clamp-on ammeter measures 6 amps, but the actual reading is 0.6 amps because the jaws are wrapped 10 times with wire in the circuit being measured.

ELECTRICAL SYMBOLS

Similar to a road map, electrical schematics and diagrams use symbols to represent electrical components in a circuit. They are signs that help you navigate through the electrical map that we call a wiring diagram. Wiring diagrams or schematics help you find your way through the electrical circuit. Without electrical symbols, the electrical circuit would be difficult to understand and troubleshoot. Not having a wiring diagram it is like taking a short trip without a map. You may eventually get to your destination, but a lot of time is wasted trying to find it.

Figure 3-18 illustrates some common symbols used in our profession. We will also discuss some other common air conditioning symbols not listed on this page and show their common images. Review and become familiar with these symbols because it will help you find your way quickly through a wiring diagram.

Figure 3-19 illustrates the symbols for a contactor. The contactor, as pictured in Figure 3-20, is an electromechanical switch. When power is applied to the coil, magnetism causes the contacts to close, completing a path for electricity to the equipment that is being controlled. There are single-pole, two-pole, three-pole, and four-pole contactors. A 240 volt single-pole contactor always has a bar that allows continuous power on one leg of the device. The technician must keep this in mind because power is still supplied to the equipment even if the component it is controlling is not operating.

Figure 3-21 shows a switching relay that is similar to a contactor. The difference between a contactor and a relay is that the switching relay handles lower amperage, usually 15 amps or less. The relay may have one or more sets of normally closed or normally open contacts, but the contactor has only normally open contacts. The symbol for the relay is similar to that for a contactor except for the closed contacts, which are expressed in Figure 3-22.

Figure 3-23 is a compressor terminal. The symbol in Figure 3-24 shows the diagram of a single-phase, split-phase compressor motor. There are also split-phase fan and pump motors that use the same symbol. "Split-phase" means that the motor has a start and run winding that splits the phase for better starting torque. This

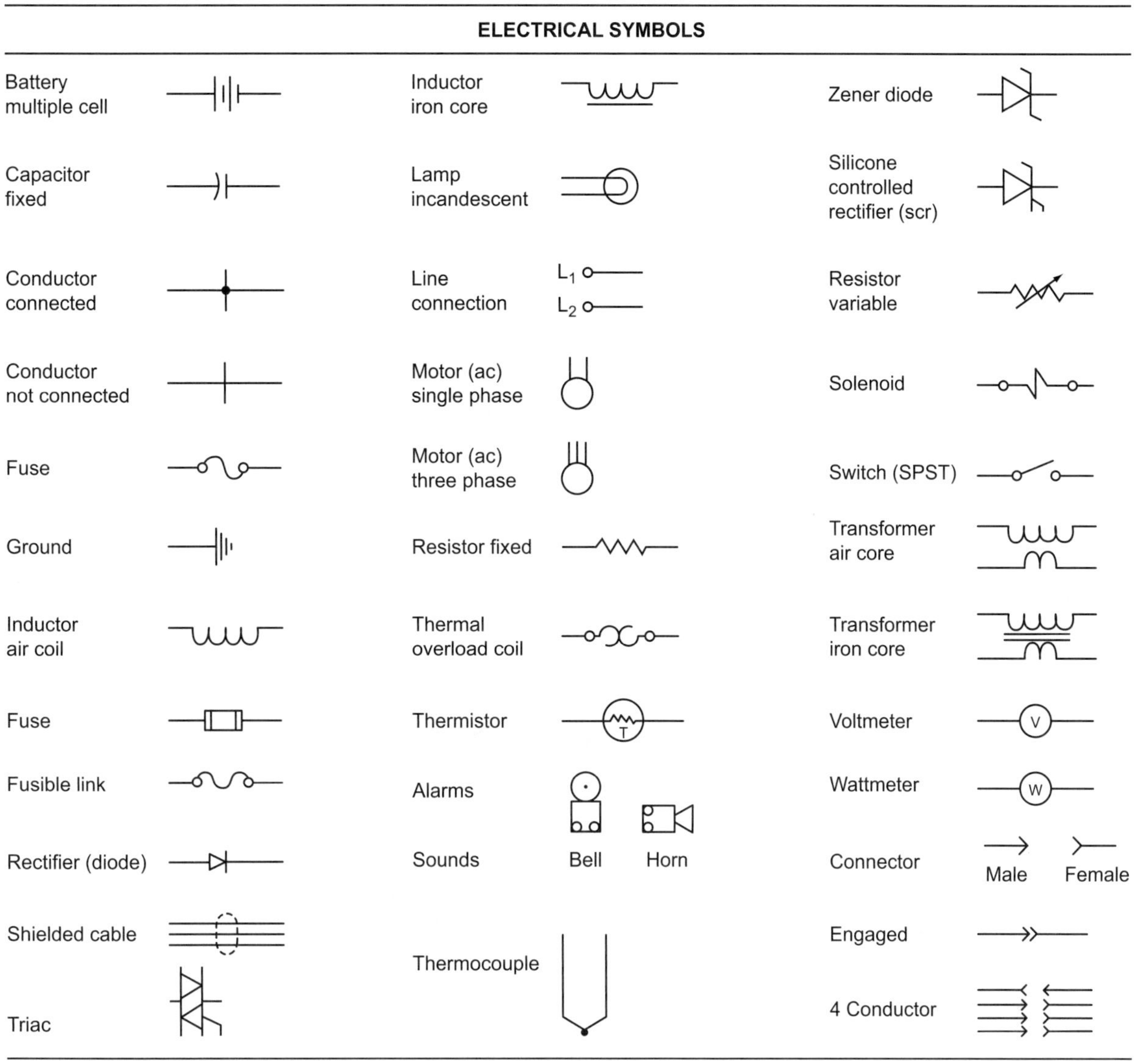

Electrical Symbols 67

ELECTRICAL SYMBOLS

Battery multiple cell	—|||—	Inductor iron core		Zener diode	
Capacitor fixed	—)|—	Lamp incandescent		Silicone controlled rectifier (scr)	
Conductor connected	—•—	Line connection	L_1 o— L_2 o—	Resistor variable	
Conductor not connected	—|—	Motor (ac) single phase		Solenoid	
Fuse		Motor (ac) three phase		Switch (SPST)	
Ground		Resistor fixed	—\/\/\/—	Transformer air core	
Inductor air coil		Thermal overload coil		Transformer iron core	
Fuse		Thermistor		Voltmeter	—(V)—
Fusible link		Alarms		Wattmeter	—(W)—
Rectifier (diode)	—▷|—	Sounds	Bell Horn	Connector	Male Female
Shielded cable		Thermocouple		Engaged	—»—
Triac				4 Conductor	

Figure 3–18
Electrical Symbols Chart

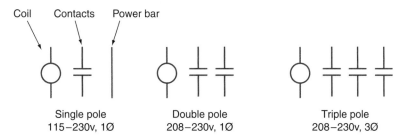

Coil   Contacts   Power bar

Single pole
115–230v, 1Ø

Double pole
208–230v, 1Ø

Triple pole
208–230v, 3Ø

Figure 3–19
Symbols for contactors. Some triple-pole contactors are rated for high voltage. The coil voltage to the contacts can be 24, 120, or 240 volts. The coil voltage can be determined by looking on the contactor data plate. (Courtesy of Dick Wirz, Refrigeration Training Services)

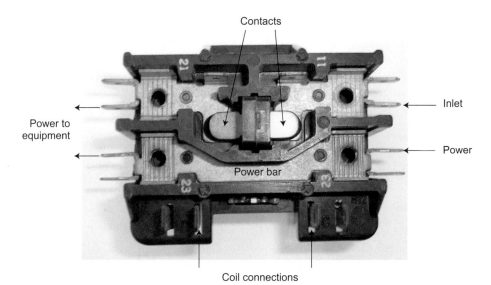

Figure 3-20
This is a single-pole contactor, which is an electromechanical switch. The two moveable contacts are near the middle of the contactor. The connections for the coils are on the lower portion of the contactor. When the coil is energized, it closes the contacts and allows for power to go to the equipment it is controlling. (Photo by Joe Moravek)

Figure 3-21
Two views of a switching relay. The top view shows the numbered terminals with raised symbols indicating whether they were normally open (NO) or normally closed (NC). The image below is a side view. The far left and right terminals on the side view are connected to the black coil wire. (Photos by Susan Brubaker)

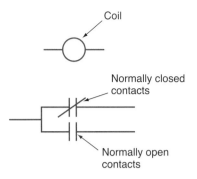

Figure 3-22
This relay has one normally open contact and one normally closed contact. Some relay designs have two of the outputs tied together, as shown in the left side of this symbol. This is classified as a single-pole, double-throw switch because the poles change positions when they are energized. This is written as SPDT.

Figure 3-23
Terminal of a common compressor. The common start and run terminals need to be wired correctly for the compressor to operate. Miswiring can damage the motor windings. (Photo by Susan Brubaker)

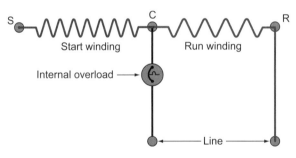

Figure 3-24
This is a symbol for a split-phase motor with an internal overload. If used with a run capacitor, it is called a permanent split capacitor. This can be a symbol for a single-phase compressor motor, fan motor, or pump motor. The internal overload is in series in with the start and run winding. (Courtesy of Dick Wirz, Refrigeration Training Services)

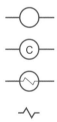

Figure 3-25
Four different symbols for a magnetic coil used to operate a contactor, relay, or other electromechanical switch. In this case, there is no standard when designers draft an electrical diagram using one of these coil symbols.

will be discussed in more detail when we discuss motors. Notice that Figure 3-24 does not indicate the motor application, only that it is a single-phase motor with a start and run winding.

By this time, you should have the idea that symbols are the language of a wiring diagram. Most symbols are universal, and only one symbol represents a component. An exception is the magnetic coil that operates contactors, relays, and other electromechanical switches. Figure 3-25 shows several different symbols for the coil. Now let us apply a little of this information to simple wiring diagrams, sometimes called schematics.

SCHEMATIC DIAGRAMS

Schematic diagrams are wiring maps that help us understand the operation of an HVACR circuit. In order to have a complete circuit, you need a power supply, interconnecting wires, and a load. Any additional components added are to the convenience of the circuit. For example, adding a switch to a circuit is a convenience because you will not need to disconnect the circuit by unplugging it.

Previously, we saw a simple circuit in Figure 3-6, and a slightly more advanced circuit is located in Figure 3-26. As soon as power is supplied to the circuit in Figure 3-26, it will begin to operate. Figure 3-26 is useful in our explanation that relates to designing diagrams. The power at the top of the diagram is labeled (phase conductor) HOT L1 on the left and (grounded conductor) NEUTRAL on the right. Following the circuit down the HOT L1 wire, we find a switch, a fuse, and a thermostat control switch that controls a light bulb. This could be some type of simple heating circuit because the light bulb will operate when the switch and thermostat are closed. The thermostat is designed to close on a drop in temperature.

Figure 3-27 is more complex than the previous diagram, but it is a diagram that you would see on a functioning system. This schematic represents a simple heating and cooling system with continued fan operation. Let's review the operation. The power supplied is listed as L1 and L2. This means the power is most likely 240 volts. When the disconnect switch on the upper left side is closed, the evaporator fan begins to operate. This motor is abbreviated as "Evap Mtr." The fan runs continuously when the power is applied. If the heating thermostat closes, it will complete the circuit to the heat strip, and heating will begin. As the space temperature reaches the set temperature, the thermostat contacts will open and the heat strips will be deenergized or go to the off position.

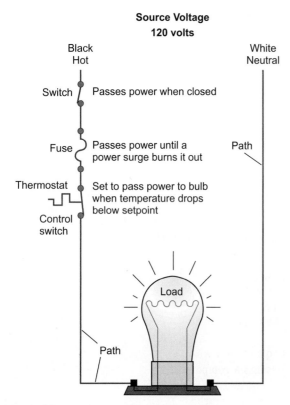

Figure 3-26
Simple-series electrical circuit that is complete when the switch, fuse, and thermostat are closed to bring voltage to the light.

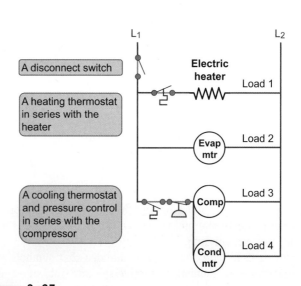

Figure 3-27
This simple series-parallel circuit can operate a heating or cooling circuit. There are more components in a standard functional HVAC system. This is used to illustrate the next level in diagram development. (Courtesy of Dick Wirz, Refrigeration Training Services)

If the cooling thermostat and pressure switch are closed, the compressor and condenser fan begins to operate. The system will cease the cooling operation when the air temperature is cooled to the thermostat set point, at which time the thermostat opens.

The final schematic diagram, Figure 3-28, is a function wiring design that you would find on a package unit. This schematic does not have a heating circuit. Notice the legend on the right side of the diagram. The purpose of the legend is to identify the symbols used in the diagram. It is an aid to the technician in tracing through the diagram.

Starting from the top of the diagram, you have the compressor, condenser fan, indoor fan, contactor coil circuit, transformer, thermostat, indoor fan relay coil, and control relay coil. The section above the transformer winding is the high-voltage section, and the section below the transformer is the low-voltage section. The low-voltage section is also called the control-voltage or 24-volt section.

Reading and understanding schematic diagrams are musts if you want to be able to understand the electrical/mechanical sequence and troubleshoot electrical problems. Through a process of elimination, you can isolate the section that is not working and possibly find the problem. If you can isolate the problem, it will be helpful should you need to call for assistance. One of the fundamentals of

In the Field

The words "schematic," "wiring diagram," and "ladder diagram" are used interchangeably in our industry, but there are differences in these terms. Schematic diagrams show their components in the electrical sequence without regard for physical location and are good to use for troubleshooting. Wiring diagrams show components mounted in their general locations with connecting wires and are used to hook up components or pieces of equipment. A ladder diagram shows components connected across the high- and low-voltage sources. The rungs of the ladder represent the voltage source, and the wires and components represent the ladder steps.

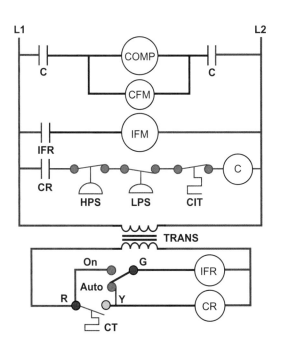

Figure 3-28
This is a functional diagram, a package unit schematic with legend. (Courtesy of Dick Wirz, Refrigeration Training Services)

Table 3–2 Table 310.16 Allowable Ampacities of Insulated Conductors Rated 0 Through 2000 Volts, 60°C Through 90°C (140°F Through 194°F), Not More Than Three Current-Carrying Conductors in Raceway, Cable, or Earth (Directly Buried), Based on Ambient Temperature of 30°C (86°F). (Reprinted with permission from the National Electrical Code® NFPA 70-2008, copyright © 2007 NFPA. This reprinted material is not the complete and official position of the NFPA on the referenced subject, which is represented only by the standard of its entirety.)

Size AWG or kcmil	Temperature Rating of Conductor [See Table 310.13(A).]						Size AWG or kcmil
	60°C (140°F)	75°C (167°F)	90°C (194°F)	60°C (140°F)	75°C (167°F)	90°C (194°F)	
	Types TW, UF	Types RHW, THHW, THW, THWN, XHHW, USE, ZW	Types TBS, SA, SIS, FEP, FEPB, MI, RHH, RHW-2, THHN, THHW, THW-2, THWN-2, USE-2, XHH, XHHW, XHHW-2, ZW-2	Types TW, UF	Types RHW, THHW, THW, THWN, XHHW, USE	Types TBS, SA, SIS, THHN, THHW, THW-2, THWN-2, RHH, RHW-2, USE-2, XHH, XHHW, XHHW-2, ZW-2	
	COPPER			ALUMINUM OR COPPER-CLAD ALUMINUM			
18	—	—	14	—	—	—	—
16	—	—	18	—	—	—	—
14*	20	20	25	—	—	—	—
12*	25	25	30	20	20	25	12*
10*	30	35	40	25	30	35	10*
8	40	50	55	30	40	45	8
6	55	65	75	40	50	60	6
4	70	85	95	55	65	75	4
3	85	100	110	65	75	85	3
2	95	115	130	75	90	100	2
1	110	130	150	85	100	115	1
1/0	125	150	170	100	120	135	1/0
2/0	145	175	195	115	135	150	2/0
3/0	165	200	225	130	155	175	3/0
4/0	195	230	260	150	180	205	4/0
250	215	255	290	170	205	230	250
300	240	285	320	190	230	255	300
350	260	310	350	210	250	280	350
400	280	335	380	225	270	305	400
500	320	380	430	260	310	350	500
600	355	420	475	285	340	385	600
700	385	460	520	310	375	420	700

Conductors and Wires 75

Temperature Rating of Conductor [See Table 310.13(A).]

Size AWG or kcmil	60°C (140°F) Types TW, UF	75°C (167°F) Types RHW, THHW, THW, THWN, XHHW, USE, ZW	90°C (194°F) Types TBS, SA, SIS, FEP, FEPB, MI, RHH, RHW-2, THHN, THHW, THW-2, THWN-2, USE-2, XHH, XHHW, XHHW-2, ZW-2	60°C (140°F) Types TW, UF	75°C (167°F) Types RHW, THHW, THW, THWN, XHHW, USE	90°C (194°F) Types TBS, SA, SIS, THHN, THHW, THW-2, THWN-2, RHH, RHW-2, USE-2, XHH, XHHW, XHHW-2, ZW-2	Size AWG or kcmil
	COPPER			ALUMINUM OR COPPER-CLAD ALUMINUM			
750	400	475	535	320	385	435	750
800	410	490	555	330	395	450	800
900	435	520	585	355	425	480	900
1000	455	545	615	375	445	500	1000
1250	495	590	665	405	485	545	1250
1500	520	625	705	435	520	585	1500
1750	545	650	735	455	545	615	1750
2000	560	665	750	470	560	630	2000

CORRECTION FACTORS

For Ambient temperatures other than 30°C (86°F), multiply the allowable ampacities shown above by the appropriate factor shown below.

Ambient Temp.(°C)							Ambient Temp. (°F)
21-25	1.08	1.05	1.04	1.08	1.05	1.04	70-77
26-30	1.00	1.00	1.00	1.00	1.00	1.00	78-86
31-35	0.91	0.94	0.96	0.91	0.94	0.96	87-95
36-40	0.82	0.88	0.91	0.82	0.88	0.91	96-104
41-45	0.71	0.82	0.87	0.71	0.82	0.87	105-113
46-50	0.58	0.75	0.82	0.58	0.75	0.82	114-122
51-55	0.41	0.67	0.76	0.41	0.67	0.76	123-131
56-60	—	0.58	0.71	—	0.58	0.71	132-140
61-70	—	0.33	0.58	—	0.33	0.58	141-158
71-80	—	—	0.41	—	—	0.41	159-176

*See 240.4(D).

SUMMARY

If you were not already familiar with electricity, this chapter may be a little overwhelming. Many new ideas and concepts presented here will need time to digest. This is normal. The way to overcome a feeling of not understanding electricity is to reread the parts you are having difficulty understanding. When you are in the field, ask the more experienced technicians to explain the operation of a piece of equipment. This is best done by using the schematic diagram as your road map and pointing to the parts of the unit as you trace the path of current flow. At first, try to trace the diagram yourself and identify as many components as you can. Many jobs do not allow time for this on-site self-training, but if a moment is available, use it for learning. The electrical part of our training seems to be the most difficult, but with time and practice you will improve.

The most important parts of the chapter are learning to use your test instruments and reading schematics. One of the best ways to learn to use your meter is to read the directions that come with it. Go through all the functions of the meter as listed in the instructions. Use your meter on the job even if it is to repeat what a senior technician has done. It is important that you buy a good-quality instrument for accurate readings and safe operation. A quality multimeter should be able to read volts, ohms, milliamps, and microamps and be able to check capacitors. The clamp-on ammeter, referred to as an Amprobe, is primarily used to measure amperage. Some clamp-on meters have additional features such as being able to read volts and ohms. In most models, the voltage and resistance range is very limited. This can lead to misdiagnosis because the technician may not realize that the resistance range stops at 100 Ω and the coil being measured has a resistance of 4,000 Ω.

Finally, you should memorize the symbols used in wiring diagrams. When you use these symbols on a regular basis, they will become part of the electrical language and you will not need to look at a legend or reference. It is only a matter of time before you will be mastering the electrical road maps. There are some diagram standards in our profession and many deviations. You will need to be exposed to those differences. Companies may have different diagram-drafting departments for commercial and residential schematics. However, they are all essentially the same. Some manufacturers use different symbols and names than what you see in the field. This is especially true in larger commercial and industrial pieces of equipment.

FIELD EXERCISES

1. Read the instructions for your multimeter. Notice all the features that are available. Practice using each feature in the instructions.
2. Compare two multimeter features and benefits.
3. Read the instructions for your clamp-on ammeter. Notice all the features that are available. Use each feature in the instructions.
4. Compare two clamp-on ammeter features and benefits.
5. Using your clamp-on ammeter, measure the amperage of the R terminal in the thermostat circuit when in the heating or cooling mode.
6. Compare two heating or cooling units that are made by two different manufacturers. Notice how they are similar and how they are different.

REVIEW QUESTIONS

1. Define "volt."
2. Define "ampere."
3. Define "ohm."
4. Define "watt."
5. How many volts will be measured in an electric heat strip if the resistance of a heater is 25 Ω and it measures 10 amperes?
6. How many amps are measured in a heating circuit that has 240 volts applied with a resistance of 10 Ω?
7. How many ohms of resistance will be in a heat strip that draws 20 amps at 240 volts?
8. How many watts will a set of electric heat strips draw if the total measured amperage is 45 A and they have an applied voltage of 250 V?
9. Convert the answer to question 8 to KW.
10. What are two precautions you must observe when measuring resistance?
11. What is the difference between an analog meter and a digital meter?
12. Name five safety features that you must require when purchasing a multimeter.
13. Draw and label twenty electrical symbols used in the HVACR industry.
14. What are the components of a complete circuit?
15. Draw an air conditioning circuit that includes the compressor, condenser, indoor fan, thermostat, and control circuit. Label the components in abbreviated form.
16. What factors determine the resistance of a conductor?
17. How many amps will a #12 AWG-TW (and UF) copper conductor handle in an air conditioning circuit?
18. How many amps will a #10 AWG-THHN copper conductor handle in a refrigeration circuit?
19. How do you measure low amperage using a clamp-on ammeter?
20. Name three types of wiring methods used in HVACR. What is the construction of each type?

CHAPTER 4

Introduction to Thermodynamics

LEARNING OBJECTIVES

The student will:
- Troubleshoot/solve a problem using Boyle's law.
- Troubleshoot/solve a problem using Charles's law.
- Troubleshoot/solve a problem using Dalton's law.
- Describe heat.
- Describe sensible heat.
- Describe latent heat.
- Describe specific heat.
- Discuss the three methods of heat transfer in substance.
- Describe the three states of a substance.
- Define "pressure."
- Discuss the difference between gauge pressure and absolute pressure.
- Discuss temperature conversions.

INTRODUCTION

You are asked to troubleshoot an air conditioning system without tools or instruments. You know that the duct system should be supplying cold air and the outdoor section should be rejecting heat. The system is operating but not cooling very well. The air coming out of the registers is slightly cool, and the air rejected by the condenser seems to be a few degrees warmer than the air going into the condenser. Not understanding thermodynamics is like not having tools and instruments to troubleshoot a system.

Thermodynamics will teach you the importance of knowing how heat is transferred, how to measure heat transfer and system pressure, and how temperature affects the operation of everything inside and outside the HVACR system. The term "thermodynamics" is derived from the words "thermo" (heat transfer) and "dynamics" (energy in motion; in our case mechanical energy in action). This may sound scientific, but we are only going to apply the practical parts of thermodynamics, the parts that we can we use to understand HVACR equipment. A good foundation of heat transfer is important when trying to understand system problems. We will be studying heat, pressure, and volume changes, all of which are important aspects of the way HVACR systems operate.

DEFINITIONS AND CONCEPTS

This section will discuss definitions and concepts important in thermodynamics. You are developing your understanding of the language of the HVACR trade, and we need to communicate at the same level before we can proceed in this chapter. It is important that you understand the terminology before we learn about heat transfer.

What Is Heat?

We talk about heat or a lack of heat in our daily lives. Heat is a form of energy that relates to the movement of molecules in a substance (a solid, liquid, or gas).

The principles of heat transfer are as follows:

1. Heat energy cannot be destroyed.
2. Heat always flows from a higher-temperature substance to a lower-temperature substance.
3. Heat can be transferred from one substance to another.
4. Heat exists in all substances down to a temperature of –463 degrees Fahrenheit.

Field Problem

You have just started working as a technician assistant for an air conditioning company and have enrolled in a course that will reinforce your work experiences. It is a Sunday in the middle of summer when your neighbor knocks on your door to ask for air conditioning advice. You had told her that you started working for an air conditioning company a month ago. She is very concerned because her air conditioning system went out yesterday and she could not get a service appointment until Monday morning. Her concern is increased because of her elderly mother's asthma problem and how it could worsen in the sweltering house. She says her deceased husband had some refrigerant and other air conditioning tools in the garage and is wondering if you could help her get the house cool until a service company can come out Monday morning. You find out that it is your company she has the appointment with.

Trying to do your neighbor a favor without violating the company policy on "moonlighting," you look in her garage for the air conditioning tools she mentioned. You call your supervisor, who is busy trying to get a church cooled for the evening service, and she thinks it might be a charge problem. You find a set of R-22 gauges and a thermometer. The supervisor asks you to call her back with indoor and outdoor temperatures and suction line and liquid line temperatures, and to hook up the gauges to determine if the unit is charged by recording the high and low side pressures. Once the information is collected, you call your supervisor with the temperatures and pressures, and she diagnoses the system as being severely undercharged. She asks you to turn off the equipment until she can arrive to check the system for leaks and take care of the problem.

Tech Tip

This chapter will discuss a series of customer issues that you will be handling in the field. The outcome of these events can make you a hero or make you lose a customer. The professional practices stated in this chapter will help smooth out or hopefully avoid negative experiences to preserve your customer and your image. The study of heat transfer will help you relate air conditioning operations in layman's terms. For example, you can tell the customer that heat transfers from hot to cold without getting into how heat transfers from the warm return air to the cold refrigerant in the evaporator.

British Thermal Units, Sensible Heat, and Latent Heat

To be able to fully understand and apply these principles, we must be able to measure temperature changes and the amount of heat transferred. Heat is energy, just as electricity is energy. Energy does work or causes things to happen. Energy is not a solid, liquid, or gas; it cannot be measured in inches, gallons, or cubic feet. It must be measured by what it does or the effect it produces.

British Thermal Units

Heat content is measured in the British thermal unit or BTU, not degrees Fahrenheit or Celsius. *The BTU is defined as the amount of heat required to raise 1 pound of pure water 1 degree Fahrenheit at atmospheric pressure.* Figure 4–1 illustrates that raising the temperature of 1 pound of water 1 degree, from 68 to 69 degrees Fahrenheit, would require 1 BTU of heat energy.

BTUs may be added or removed from water or another substance, as shown in Figure 4–2. Added heat increases the temperature, while removing heat drops the temperature. Adding 1 BTU will increase or decrease the temperature of 1 pound of water 1 degree Fahrenheit. Adding 2 BTUs will change the temperature 2 degrees Fahrenheit, etc.

British scientists long ago decided on that standard unit of measuring heat, so it has come to be called a British thermal unit; remember that "thermal" refers to heat. Because much of the American system of standard weights and measures comes from the British, we use the BTUs as our standard heat unit.

So the BTU is a measure of the amount of heat, and the degree of temperature is a measure of the effect of that heat on 1 pound of water. Therefore, the amount of water must be known because it would take twice as much heat (2 BTUs) to

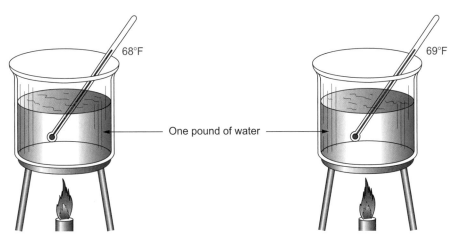

Figure 4–1
One BTU added to 1 pound of water causes its temperature to rise 1 degree Fahrenheit.

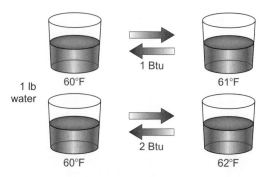

Figure 4-2
Heat can be added or removed from a substance by increasing or decreasing its temperature. The number of BTUs added or removed will change the temperature of the water. (Courtesy of Trane)

warm 2 pounds of water 1 degree as it would to warm 1 pound of water 1 degree. Moreover, it would take twice as much heat (2 BTUs) to warm 1 pound of water 2 degrees as it would to warm 1 pound of water 1 degree.

We can see how warming 5 pounds of water 5 degrees Fahrenheit would require 25 BTUs to be added to the 5 pounds of water. So the amount of heat, in BTUs, to warm any amount of water by any number of degrees of temperature is found by multiplying the number of pounds of water by the number of degrees Fahrenheit that the temperature is to be raised. The answer is the number of BTUs of heat that must be added to the water.

Let us not forget that the heat (energy) added to the water increases the movement of the water molecules to make them more active, to give them more rapidity of motion, and this increased rapidity of motion produces the effect that we call a rise in temperature.

But materials vary in the amount of heat required to warm them 1 degree. Compared with other materials (solid, liquid, or gaseous), water requires a great deal of heat energy to warm it. Oil requires only about one-half as much as water, so it takes only $\frac{1}{2}$ BTU to warm 1 pound of oil 1 degree. Gasoline, kerosene, and crude oil also require about $\frac{1}{2}$ BTU per pound per degree to warm them. Mercury requires only about $\frac{1}{30}$ BTU; alcohol about $\frac{3}{5}$ BTU; and chloroform about $\frac{1}{4}$ BTU. Different materials require different levels of heat to increase their temperature.

Gases vary a great deal in the amount of heat required to warm them, depending upon their temperature and pressure. At atmospheric pressure and room temperature, air, which is a mixture of oxygen, nitrogen, and carbon dioxide, requires 0.24 BTU per pound to raise its temperature 1 degree Fahrenheit.

How are BTUs used in our profession? The unit BTUs per hour is abbreviated "BTUH." Figure 4-3 represents a ton of ice. Melted over a 24-hour period, a ton of ice equals 12,000 BTUs of heat absorbed per hour. One ton (2,000 pounds) of ice melting in 24 hours is equal to 1 ton of refrigeration. Therefore, a 3-ton system will absorb three times that much heat, or 36,000 BTUs per hour: 12,000 × 3 tons = 36,000 BTUH.

Heating systems are also rated in BTUH, but this is actually in reference to the amount of heat added to the air. The nameplate on a gas furnace will indicate the BTUH input required. Because the byproducts of combustion are vented outside, the gas furnace heats the space at less than the rated input. A 90% efficient gas furnace with a 100,000 BTUH input rating will produce 90,000 BTUH to heat the space.

Sensible Heat

Sensible heat is heat that is added or removed from a substance and that causes a temperature change without a change of state, as shown in the previous Figure 4-2. For example, heating or cooling the air is a sensible heat process. Sensible heat is heat that can be measured by a thermometer or detected by touch. Another example

Figure 4-3
One ton of refrigeration is equal to melting 2,000 pounds of ice over a 24-hour period. Adding 12,000 BTUs per hour will melt all the ice in one day. This is equal to removing a total of 288,000 BTUs. (Courtesy of Trane)

of sensible heat is illustrated in Figure 4-4. To bring the temperature of 1 pound of 60°F water up to 212°F, 152 BTUs are added to the water. At 212°F, the water is hot but not boiling. At this temperature, water can be one of three states: 100% liquid, 100% vapor, or a combination of vapor and liquid.

Latent Heat

Latent heat is associated with BTU transfer that causes a change of state without a change of temperature. It takes a lot of energy to change states from a liquid to a vapor or back from vapor to liquid. Enough heat (BTUs) can be added to 1 pound of 212°F water to cause it to explode into tiny particles of vapor. As shown in Figure 4-5, it takes about 970 BTUs to convert 1 pound of 212°F water to 1 pound of 212°F steam or vapor. The temperature does not change if exactly 970 BTUs are added. Heat added above this point will cause the temperature to rise above 212°F. Heat added above the boiling point is sensible heat.

The latent heat process is also seen when water is cooled to make ice or ice is warmed to make water. It takes a removal of 144 BTUs to change one pound of 32°F water to 32°F ice or the addition of 144 BTUs to change one pound of 32°F ice to 32°F water. The latent heat process involves more heat transfer than the sensible heat process.

Figure 4-4
Sensible heat is heat added or removed from a substance that causes a change in temperature without a change of state. In this case, 152 BTUs are added to 1 pound of water, increasing its temperature from 60°F to 212°F. (Courtesy of Trane)

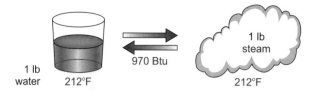

Figure 4-5
The latent heat process is associated with a change of state. In this example, the hot water adds 970 BTUs to convert to steam, and 970 BTUs can be removed from 212°F steam to convert back to water. (Courtesy of Trane)

HEAT QUANTITY AND INTENSITY

Heat intensity is measured by its temperature, commonly in either degrees Fahrenheit (°F) or degrees Celsius (°C). If all heat were removed from an object, the temperature of the object would decrease to −460°F, or −273°C. This temperature is referred to as absolute zero and is the temperature at which all molecular activity stops.

The quantity of heat contained in an object or substance is not the same as its intensity of heat. For example, the hot sands of the desert in Figure 4-6 contain a large quantity of heat, but a single burning candle has a higher intensity of heat. Figure 4-7 is another example of quantity and intensity on a small scale.

TEMPERATURE SCALES

The common temperature scale used in the United States is Fahrenheit, with a boiling point of 212 degrees and a freezing point of 32 degrees. *Celsius, sometimes known as centigrade, is the temperature scale used by most other countries in the world.*

Figure 4-6
Heat can be stated in quantity and intensity. The desert (left) represents a large quantity of heat, and the picture to the right represents the candle intensity of heat. (Courtesy of iStockphoto)

Figure 4-7
These two different masses of water contain the same quantity of heat, yet the temperature of the water on the left is higher. Why? The water on the left contains more heat per unit of mass than the water on the right. In other words, the heat energy within the water on the left is more concentrated, or intense, resulting in the higher temperature. Note that the temperature of a substance does not reveal the quantity of heat that it contains. (Courtesy of Trane)

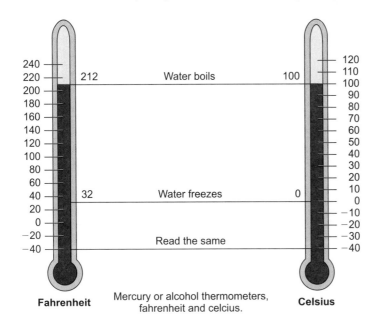

Figure 4-8
Compare the Fahrenheit and Celsius temperature scales. Water boils at 212°F or 100°C. Water freezes at 32°F or 0°C. At −40°, the scales are the same. (Courtesy of RSES)

The Celsius boiling point is 100 degrees and the freezing point is zero degrees. Figure 4-8 compares the Fahrenheit and Celsius scales.

Rankine (R) and Kelvin (K) scales are also used in scientific circles. Absolute zero for both Rankine and Kelvin is zero degrees. A comparison of Fahrenheit and Rankine is shown in Figure 4-9. Absolute temperature is the measurement of a temperature from absolute zero, −460°F. When converting absolute temperature to Fahrenheit, 460° are added to the Fahrenheit temperature. For example, what is the absolute temperature of 100°F? Add 100°F to 460° = 560° absolute temperature. Absolute temperature will be used in the discussion of gas laws.

HEAT FLOW

How does heat flow? Does cold air blowing across water in your freezer create ice cubes? Or is the heat in the water released into the freezing air until it gets cold enough to freeze the water?

We can define heat as the energy of the molecules' motion. This motion is transmitted to other molecules that have less motion. Some of the molecules give up some of their energy to other molecules that have less energy. The greater the molecular motion is, the greater the heat content of a substance.

Another way of saying this is that heat flows from the material with a higher temperature to a material with a lower temperature. Heat, therefore, always flows

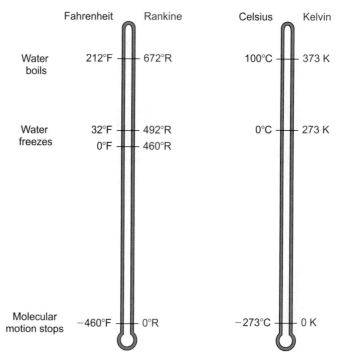

Figure 4-9
This chart compares Fahrenheit, Rankine, Celsius, and Kelvin scales. Scientific and engineering publications use Rankine or Kelvin scales when citing temperature examples. Absolute temperatures are measured in Rankine.

from hot to warm, warm to cool, and cool to cold—never from a low temperature to a high temperature.

In fact, that is how we get heat to flow. We place a hot object near or touching a colder one and let the heat flow from the hot object to the colder one. We do not actually "add heat" or "remove heat." Heat is simply transferred.

Heat flows in three different methods: conduction, convection, and radiation. What type of heat transfer is involved in the following scenarios:

- sun rays into a car?
- heat moving upstairs in a two-story building?
- heat moving from a gas furnace metal heat exchanger into the air passing through the furnace?

We are going to answer these questions in the next section.

Conduction

If we heat one side of building material, as shown in Figure 4-10, the heat travels through it from the hot side to the cooler side. This heat, in turn, may be conducted to another cooler object touching it or radiated to a cooler object some distance away.

This transfer of heat through a material is conduction. The hot side gives motion to the molecules, which give motion to nearby molecules and so on through the material. In doing so, some of the heat energy is given up to the molecules and stays there as heat energy, so all of the heat does not get through.

A material that transmits heat easily, with little loss, is called a conductor of heat. Some of the best conductors are copper, silver, aluminum, and steel.

A material that does not conduct heat through itself easily is called an insulator. Some of the better insulators (poor conductors) are cork, cotton, air, and many other materials that are composed of thousands of tiny air cells. In many of the insulating materials, the number of trapped air pockets is the key to reducing heat transfer.

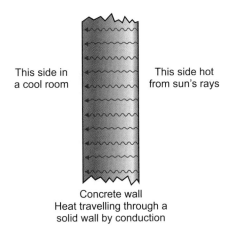

Figure 4–10
Conduction is created by molecules exchanging heat. Heat travels from warmer to cooler surfaces. (Courtesy of RSES)

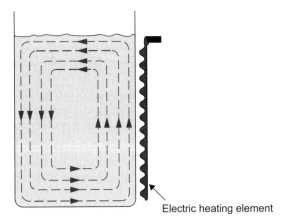

Figure 4–11
Heat being transferred through the wall by conduction creating convection currents inside the room. As the air heats, it becomes lighter and rises in the space. As the air cools, it becomes more dense and falls. (Courtesy of RSES)

Convection

Convection occurs only in fluids, although that could include both liquids and gases. We know that when a material is warmed, it expands in volume and becomes lighter. In the case of fluids, the cooled fluid becomes heavier and, as a result, crowds out the lighter, warmer fluid, thus pushing it upward. As illustrated in Figure 4–11, this sets up a cycle of fluid circulation, which carries heat upward on one side of the room and, as the fluid cools, downward on the other side of the room. This method of transferring heat is called convection. It is very important in refrigeration and heating, where a large part of the process is cooling or heating fluids like water, oil, water vapor, air, etc., although in many cases the equipment is intended only to cool the fluids as a means of carrying heat away from or to foods, human beings, or other objects.

The example depicted in Figure 4–11 is called natural convection. *Natural convention is a slow movement of fluid as it is heated*; therefore, heat transfer occurs very slowly. An example of forced convection is when a fan is installed to blow air across a heat exchanger such as an evaporator or condenser coil. A great quantity of heat can be exchanged using the forced convection design.

Radiation

If we place a hot object near a cooler one, heat jumps across the space between them to warm the cooler object. There does not have to be any gas or other material in the space. An excellent example of radiation, in Figure 4-12, is the heat from the sun, which radiates heat through 93 million miles of vacuum to reach the Earth. Another example is the heat radiated from a fire or radiant heat from hot pipes buried in the floor of a building. Another good example of radiation is sitting at a campfire. The side of you facing the campfire will feel warm because it is feeling radiant heat from the fire.

CHARACTERISTICS OF SOLIDS, LIQUIDS, AND GASES
Motion of the Molecules

Mechanical refrigeration is a physical rather than a chemical process, so we deal with molecules and their movements and rarely need to go into chemical processes, which involve breaking down the molecule into its constituent atoms or the union of atoms to form molecules. Nevertheless, it is necessary to have an elementary understanding of the composition of matter in order to more easily understand how gases, liquids, and solids behave under various conditions. All matter, whether gaseous, liquid, or solid, consists of billions of molecules.

If the material is in solid form, such as iron, wood, stone, or ice, the molecules are held together by their attraction for one another. This mutual attraction of like molecules is called cohesion. The molecules are not tightly jammed together, nor are they motionless. There are spaces between them and they move somewhat, but their motion is quite limited, being more of a shifting than free motion.

The colder a solid is, the less motion the molecules have. If the material had absolutely no heat in it—that is, if it were at a temperature of absolute zero, which is about –460°F—there would be no motion of the molecules. If the very cold material were slightly warmed by adding a little heat, the molecules would begin to move somewhat and the motion of the molecules would become greater as the material got warmer. It takes energy to cause movement or to do work. The molecules must be given heat, which is one form of energy, to give them motion. So the more heat energy the molecules get, the greater their motion and the faster they move.

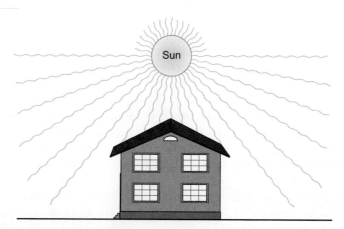

Figure 4-12
Radiation or sun rays transfer heat by electromagnetic waves through space. Radiation is converted to heat when it strikes a solid surface on Earth. Any difference in surface temperature greater than 1 degree causes heat to be lost or gained by the human body. (Courtesy of RSES)

The heat that we add to raise the temperature of the solid material and give the molecules more movement is called "sensible heat," for we can tell that heat has been added by our sense of feeling, which tells us that the solid is warmer than it was before. Sensible heat is heat associated with a change in temperature, and sensible heat change can be measured with a thermometer.

Melting and Fusion

If we continue to add heat energy to the solid, it gets warmer and the molecules move faster but still within a very limited space, for they are still held to one another by their mutual attraction.

Finally, however, they get enough heat energy to partly overcome their attraction for one another. Then the molecules can move freely about and the material becomes liquid. This process of the molecules' breaking away from one another and changing the material from a solid to a liquid is called melting or fusion.

The attraction of the molecules in a solid for one another is strong. The molecules must receive a great deal of heat energy to allow the solid to become a liquid. As you would expect, the heat energy required to melt a solid is more than the amount of heat required to warm it a few degrees. We learned earlier that it takes only 1 BTU to increase or decrease the temperature of 1 pound of water, but it takes 144 BTUs to change cold water to ice or ice to water without changing the temperature of 32°F.

We call the heat required to break up the mutual attraction of the molecules and to change the material from a solid to a liquid the latent heat of melting. Latent heat is the heat consumed in the change of state and does not affect temperature. The temperature of the solid immediately before it melts and immediately after, when it has become a liquid, are exactly the same. The latent heat of fusion has caused a change in the state of the material from the form or condition of a solid to the state of a liquid. Whether melting 1 pound of 32°F ice or freezing 1 pound of water, 144 BTUs are required to accomplish this latent heat process.

The water molecules are now moving more freely and are not held together. The material can no longer stand rigidly by itself and must have some container such as a cup, tank, or other vessel to support it. The speed of the molecules' movement is much greater but not yet enough to overcome the force of gravity. They are held downward in the vessel, but the liquid can be poured from a higher container to a lower one or pumped from a lower to a higher container.

A material in the liquid state must have more heat in it than when it was in the solid or frozen state and, except at the exact melting temperature, must always be warmer.

Because the molecules are freer as a liquid than as a solid, they become more separated. The molecules in a liquid have considerably more heat energy, so they move about in a lively manner and at a more rapid speed. They bump into one another and into the sides of the container.

As heat energy is added to the liquid, it becomes warmer and the speed of its molecules increases. As in the case of the solid, energy is required to increase the motion of the molecules, and the heat energy that is added to do this is again called sensible heat. But this time, it is sensible heat of the liquid. It is the heat added to a liquid that causes the liquid to become warmer.

Sublimation

We have said that the molecules in a solid are held rather closely together by their cohesion. However, some of the molecules near the surface break away from the rest of the molecules and jump clear away from the solid. That is, they become vapor.

This process is called sublimation and is somewhat similar to evaporation of a liquid to a vapor. In sublimation, the solid turns directly into a vapor without going through the liquid state, that is, without melting.

- Wet clothes on an outside line in below-freezing weather are said to "freeze dry." Actually, the water freezes into ice, which gradually sublimes into the air as water vapor, leaving the clothes dry.
- Ice on sidewalks in zero-degree weather gradually disappears because it sublimes into water vapor and diffuses into the air.
- An ice cube in a home freezer will shrink in size if left in this cold environment for weeks.
- One of the outstanding examples of sublimation is that of solid carbon dioxide, as shown in Figure 4–13. Under normal conditions, it does not melt into a liquid but turns to a gas directly from a solid. Because it does not melt and become wet, it is known as dry ice.

Evaporation

All molecules do not move at the same speed; in fact, some molecules in the top part of the liquid may get enough speed to fly entirely out of liquid and escape into the space or air above the liquid. Some of them get out of the liquid temporarily but do not have enough speed or energy to entirely escape, so they fall back into the liquid.

Figure 4–13
Dry ice, or frozen carbon dioxide, sublimes from a solid to a gas. Dry ice evaporates at a temperature higher than −109.4°F. (Courtesy of Photos.com)

Some of these molecules do get away and get mixed into the air or other gas above the liquid. Thus, some of the molecules are constantly escaping. They form a gas or vapor blanket above the liquid, and this vapor tends to diffuse into and mix with the air above the liquid.

We call this process of some of the molecules' escaping from the surface of the liquid evaporation, for it is the process of forming a vapor.

An example of evaporation of a liquid is water in an open vessel. Some of it evaporates into a gas "water vapor," and eventually all of the water disappears. The warmer the water is, the faster it evaporates. When more heat energy is added, more molecules get enough velocity to escape from the liquid.

Boiling Point

If we keep adding heat to the liquid beyond its melting temperature, the liquid becomes warmer and warmer and the molecules move faster and faster. When enough heat energy (BTUs) has been added, the molecules move so fast that they lose all restraint and fly out of the liquid, much the same as in evaporation, but in far greater numbers.

At this temperature, the liquid "disintegrates" and breaks loose from even the force of gravity. The molecules fly in all directions. We now call this condition a gas or a vapor. As a vapor, the material takes up a great deal more room than it did as a liquid or a solid, for the molecules are flying about and are therefore widely separated. Thus, the volume of the vapor is much greater than when the material was a liquid.

We call this process of rapid changing from a liquid to a vapor boiling or vaporization. The temperature at which it occurs is called the boiling temperature. A great deal of heat energy is required for the molecules to get enough speed or velocity to break away from one another, to overcome the power of mutual attraction and of gravity, and to fly out and away. In fact, the amount of this energy must be rather tremendous as compared with the amount normally required to warm the solid or the liquid a degree or so or even to change the solid to a liquid. How much energy is required to break the liquid bond? One pound of water at 212°F takes 970 BTUs to change it to 212°F vapor or steam.

This very large amount of heat energy required to get the liquid to boil and to give the molecules enough energy to escape from the liquid and to form a vapor is called the latent heat of boiling or, more correctly, the latent heat of vaporization. As in the case of melting, this heat is latent or hidden, for it goes toward overcoming the forces that hold the molecules together as a liquid and to giving them enough energy to become free. The latent heat process involves changing water to vapor or vapor back to water without a change in temperature.

If the liquid is water in an open vessel, the molecules that escape from the liquid into the air form what is known as water vapor, which is also called moisture in the air. Any liquid can and does have its own vapor that forms just above the surface of the liquid but that also diffuses or spreads through the space above the surface of the liquid.

Sensible Heat of Vapor

A vapor or gas can be warmed, just as a solid or liquid can. If heat energy is added to the vapor molecules, their speed, or velocity, increases and we say that the vapor is warmer. The heat that is added to a vapor and that causes it to become warmer is called the sensible heat of the vapor. It is also called superheat. *Superheat is heat added to a substance once it boils off to a vapor.* For example, under standard atmospheric pressure, water will boil at 212°F. Adding BTUs to the hot water vapor will increase its superheated temperature. This is a sensible heat process because the vapor temperature increase can be measured with a thermometer.

More on Latent Heat

As we stated before, *latent heat is the heat added or removed from a substance that causes a change of state without a change in temperature.*

The terms "vapor" and "gas" mean about the same, for any gas is really the vapor from some material that can exist as a liquid or a solid. Oxygen, hydrogen, nitrogen, and the other gases that normally exist as gases can also exist, under the proper conditions, as liquids or solids.

If we cool pure oxygen enough—that is, take enough heat away from it—the molecules will not have enough energy to remain free and will have to go back into liquid form. If the liquid oxygen is cooled further, it cannot remain as a liquid and will have to return to a solid.

The temperatures required to do this are extremely low: −297°F to cause oxygen to liquefy and −361°F to cause it to solidify.

A more familiar example is water vapor or steam, which is the usual name for very hot water vapor. If it is cooled, it turns to water, a liquid. If the liquid is further cooled, it becomes ice, the solid form of water.

Applying Sensible and Latent Heat

Here are two examples of calculating sensible and latent heat using 1 pound of water.

Example One: The heat that we must remove from water vapor to return it to a liquid (water) will be as follows:

1. The sensible heat of the hot vapor to cool it down to the boiling temperature, 212°F. This is approximately 0.5 BTUs per pound of steam.
2. The latent heat of vaporization or condensation to change the vapor at 212°F to water at 212°F. This is approximately 970 BTUs per pound of water.

Then, we can further remove sensible heat from the liquid and thus cool it down to as low as 32°F, but it will remain a liquid.

The heat that we must remove from water to return it to a solid (ice) will be as follows:

1. The sensible heat of the water to cool it down to the melting temperature of 32°F. This is 1 BTU per pound of water.
2. The latent heat of fusion or solidification to change the water at 32°F to ice at 32°F. This is 144 BTUs per pound of water.

If we wish, we can further cool the ice below 32°F, down to about any temperature we want, by removing the sensible heat of the ice, but of course it will still remain as ice. This requires about 0.5 BTU per pound of ice.

Example Two: "Total heat" is a term that refers to the amount of sensible and latent heat required to cool air. Figure 4-14 is an example of adding heat to 1 pound of ice at a starting temperature of 0°F up to the boiling point of 212°F.

So the total heat from 0°F to 212°F would be the total of these steps:

1. The sensible heat of ice from 0°F to 32°F is 16 BTUs.
2. The latent heat of melting ice at 32°F to 32°F water requires 144 BTUs.
3. The sensible heat of water from 32°F to 212°F is 180 BTUs.
4. The latent heat of vaporization from 212°F water to 212°F steam is 970 BTUs.

The total sensible and latent heat requirement for this to occur is 1,310 BTUs. The process can be reversed with the same results of dropping the temperatures.

Specific Heat

You are asked to calculate the refrigeration capacity of a walk-in freezer for a restaurant. You must consider many things, but two important pieces of information are what items will be stored in the freezer and the temperature at which the items

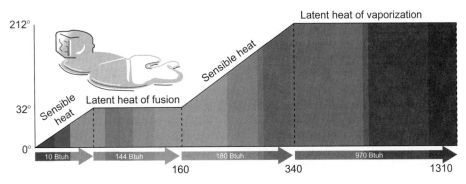

Figure 4-14
The total BTUs required to heat 1 pound of ice from 0°F to 212°F is 1,310 BTUs. It will take the removal of 1,310 BTUs to cool the 212°F steam to 0°F. (Courtesy of Dick Wirz, Refrigeration Training Services)

will be entering the box. Most items stored in a freezer require different rates of heat removed to cool or freeze the product. For example, it takes about 0.5 BTU to drop the temperature of ice 1 degree. As we said earlier, it takes 144 BTUs to convert 32°F cold water to the 32°F ice. It takes about 0.41 BTU to drop the temperature of frozen beef 1 degree. Look at this example and you will notice that it will take more heat removal to store water or ice than beef. This leads us to a discussion of specific heat because specific heat is the amount of BTUs removed to cool 1 pound of a substance 1 degree Fahrenheit.

The amount of heat required to raise 1 pound of material 1 degree Fahrenheit is called specific heat. For water, the specific heat is one because it takes 1 BTU to raise 1 pound of water 1 degree Fahrenheit. Tables of specific heats are usually given in decimals, as ice at 20° is 0.48, water vapor 0.46, apples 0.87, and green peas 0.79.

To calculate the amount of heat required to raise a material from one temperature to a higher temperature, we first figure the number of BTUs just as if the material were water: we multiply the number of pounds of the material by the number of degrees it is to be cooled or warmed. Then we multiply this amount by the specific heat of that particular material.

Example: How many BTUs are required to raise the temperature of 1 pound of ice from 20° to 32°? The specific heat of ice is 0.48.

$$\text{BTUH} = \text{Weight} \times \text{Temperature Difference} \times \text{Specific Heat}$$
$$1 \text{ pound} \times 12°F \text{ temperature difference} \times 0.48 = 5.76 \text{ BTUs}$$

It will take 5.76 BTUs to warm 1 pound of ice from 20°F to 32°F. The ice would not have changed states to water. It would still be 32°F ice.

How many BTUs are required to chill 1,000 pounds of apples from 70°F to 40F°? The specific heat of a pound of apples is 0.87. Use the same formula to determine the BTUs required to chill water.

$$\text{BTUH} = \text{Weight} \times \text{Temperature Difference} \times \text{Specific Heat}$$
$$26,100 \text{ BTUH} = 1,000 \text{ pounds} \times 30°F \text{ temperature difference} \times 0.87 \text{ specific heat}$$

It would take a refrigeration system 26,100 BTUH to cool 1,000 pounds of apples from 70°F to 40°F in one hour. This calculation does not include cooling the air around the apples or the heat gain through the refrigerated box.

Latent Heat of Fusion

To change 1 pound of ice at 32°F to water at 32°F requires that we add the latent heat of fusion of ice, which is 144 BTUs.

It will take 144 BTUs × 20 pounds = 2,880 BTUs to melt 20 pounds of ice.

If we now want to warm the 20 pounds of 32°F water to 50°F (an additional 18°), we will have to supply another 20 pounds × 18°F × 1 BTU = 360 BTUs.

If we want to heat the water from 32°F to the boiling point of 212°F (a difference of 180°), we will have to multiply 20 pounds × 180°F × 1 BTU = 3,600 BTUs of heat required to increase the temperature.

The latent heat of fusion also varies with the material.

Latent Heat of Vaporization

To boil the water at 212°F and turn it into steam also at 212°F requires latent heat of vaporization. For water in an open pan, it is 970 BTUs per pound, so 20 pounds of water at 212°F requires 20 pounds × 970 BTUs/pound = 19,400 BTUs to turn it into steam or water vapor, also at 212°F. This is a latent heat process because there is a change in state without a change in temperature.

Then, if we want to superheat this steam to 300°F, that is, raise its temperature above 212°F, we must multiply: 20 lbs × 88°F temperature difference × 0.46 specific heat of steam = 809.6 BTUs.

To warm 20 pounds of ice from 20°F to 32°F, change it to water, heat the water to 212°F, change the water to water vapor (steam), and heat the steam to 300°F, we will require the following:

To warm 20 pounds of ice from 20°F to 32°F:	(20 × 12 × 0.48) =	115.2 BTUs
To change the 32°F ice to water at 32°F:	(20 × 144) =	2,880.0 BTUs
To warm 20 pounds of water from 32°F to 212°F:	(20 × 180 × 1) =	3,600.0 BTUs
To change the 212°F water to steam at 212°F:	(20 × 970) =	19,400.0 BTUs
To warm 20 pounds of steam from 212°F to 300°F:	(20 × 88 × 0.46) =	809.6 BTUs
Total to change 20 pounds of ice at 20°F to steam at 300°F =		26,804.8 BTUs

From these figures, it can be seen that the latent heats of fusion and of vaporization are very large compared with the sensible heats (to raise or lower their temperatures). Moreover, the latent heat of fusion (solid to liquid) and vaporization of water (liquid to vapor) are quite large compared with other materials. Most materials have far less heat capacity than ice, water, and steam.

We can accurately calculate how much heat will be required for these processes under specified conditions by obtaining specific heat values from tables and charts.

Transferring Heat

We have emphasized that heat is energy "heat energy." There are other kinds of energy, the most common of which are electrical energy, mechanical energy, and chemical energy. Energy can be changed from one kind to another. Chemical energy in coal can be changed into heat energy in the steam, and heat energy in the steam can be changed into mechanical energy in the turbine and then into electrical energy in the generator. Electrical energy can then be changed back into heat energy in the toaster, to mechanical energy in the motor, or to chemical energy in the storage battery. In each of those steps, all of the electrical energy in the motor does not go into mechanical energy or power because some of it was "lost" as heat. We say "lost" because we did not get any use out of the heat of the motor, but actually it was not lost; the heat was energy transformed from electrical energy.

So efficiency is the useful output energy from a machine divided by the input energy to the machine, and expressed as a percentage. The difference is lost as far as doing useful work is concerned, but it actually is not lost because energy can be neither destroyed nor made. There is as much energy now, but no more, as there was thousands of years ago. We have changed some of it, but it still exists, although in a different form.

> **Tech Tip**
>
> So energy cannot be destroyed, nor can it be created. It can merely be transformed to or from some other kind of energy. This principle is known as the law of conservation of energy. It is wise to remember it because it explains many things.
>
> It explains efficiency, for example; in changing from electrical energy to the motor, some went into mechanical energy (or power) and some into heat. The efficiency is that part of the electricity that becomes power expressed in a percentage. That is, the mechanical energy (the output energy) determines the efficiency. If three-fourths of the electrical energy became mechanical energy, then the efficiency was 75% (3 ÷ 4 = 0.75 or 75%).

WHAT IS PRESSURE?

Pressure is force over a given area. In the case of air conditioning, we are looking for pounds of pressure per square inch, designated as "psi." For example, pressure can be used to measure the force of a column of water being pumped from a water-cooled condenser to a cooling tower. It is important to know that the pump capabilities are high enough to force the water from the condenser to the cooling tower. Next are other instances where pressure is involved.

Gauge Pressure

A type of gauge pressure is read when you check tire pressure. Many car and truck tires have a recommended pressure of around 35 psi. This means that the air pressure inside the tire is 35 pounds of force per square inch of the inside tire surface. It would be the same as cutting open a tire and placing the flat surface of one end of a 1-square inch rod that weighs 35 pounds on the tire rubber. The weight of the 1-square-inch rod would create a pressure of 35 psi on the rubber surface.

Pressure is important to measure when checking the refrigeration cycle operation. The high and low pressures are checked to help determine the charge of a system. Almost every refrigerant-related job that a service technician works on requires hooking up the manifold gauge set and measuring pressure.

Gauge pressure is also stated as psig, which means "pounds per square inch gauge." If the abbreviation "psi" is used, it is assumed to mean pounds per square inch gauge, or psig.

Absolute Pressure

We live at the bottom of a sea of air mixed with water vapor, some 600 miles deep. Air and the water vapor in it have weight. As illustrated in Figure 4–15, a 1-square-inch column of this air and water vapor extending upward from sea level weighs nearly 14.7 (exactly 14.696) pounds, so every square inch of the sea and of land level with the sea, as well as every ship, person, and article, is carrying this weight of almost 14.7 pounds on each square inch of its surface.

If this 1-inch-square column of air was based in Denver, which is 1 mile above sea level, the column would not be as long and would therefore weigh less. We would have to subtract the weight of the air and water vapor 1 mile long from sea level up to Denver. Thus, the normal atmospheric pressure at Denver is about 12 pounds per square inch absolute.

The farther up we go above sea level, the less the atmospheric pressure. Moreover, the air and water vapor become rarified or thinner, for they do not have as much weight of air and water vapor on them from above, which tends to compress them.

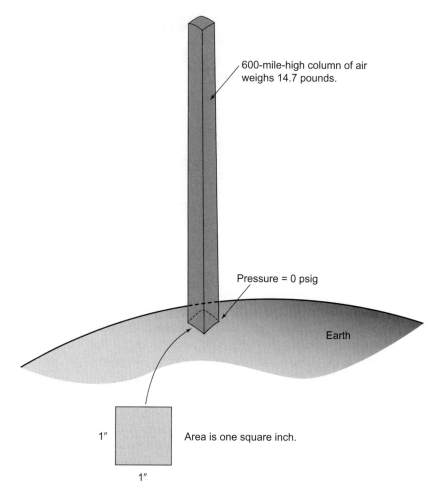

Figure 4–15
A column of air pushes down on the Earth's surface at a pressure of 14.7 psi at sea level.

At sea level, not only is the pressure of the air and water greater, but the density of the air and water vapor is also greater. This is shown by the fact that people have more trouble breathing enough air at high altitudes, where there is less air per cubic foot. Its density is less than that at sea level.

If there were no air or water vapor surrounding the Earth (as is true on the moon), there would be no atmospheric pressure; there would be a perfect vacuum, and the pressure would be absolute zero.

We are so accustomed to being and working in this atmospheric pressure that our pressure gauges start at atmospheric pressure, so the atmospheric pressure of 14.7 pounds per square inch absolute (abbreviated psia) is zero pounds per square inch gauge, as shown in Figure 4–16.

In some types of work, atmospheric pressure is referred to in inches of mercury, abbreviated "in. Hg." The letters "in" mean "inches," and "Hg" stands for the element "mercury." One cubic inch of mercury weighs 0.491 pounds, so it takes 2.037 cubic inches of mercury to reach a weight of 1 pound. A pressure of 14.696 pounds is therefore equivalent to 29.921 (2.036 × 14.696) inches of mercury.

This is shown by making a U-tube, each leg of which is 30 inches or more. Mercury is poured into the tube to about halfway, as shown in Figure 4–17. Then all of the air is pumped out to create a vacuum in one end of the tube. The air pressure of 14.696 psia will therefore push the mercury upward into the tube having the vacuum so that the difference in the two mercury levels is 29.921 inches.

This instrument is called a barometer and is used to measure atmospheric pressure. Atmospheric pressure is different according to the altitude of the place

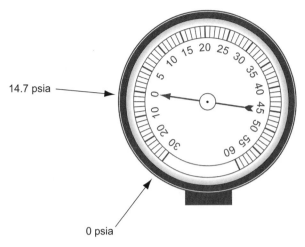

Figure 4–16
Compound pressure gauge, calibrated in psig and in. Hg. vacuum. (Courtesy of RSES)

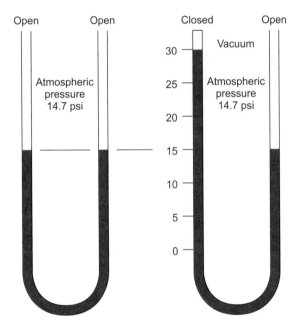

Mecury U tube and barometer to measure atmospheric pressure

Figure 4–17
This is shown by making a U-tube, each leg of which is 30 inches or more. Mercury is poured into the tube to about halfway, as shown in the figure. Then all of the air is pumped out to create a vacuum in one end of the tube. The air pressure of 14.696 psia will therefore push the mercury upward into the tube having the vacuum so that the difference in the two mercury levels is 29.921 inches. (Courtesy of RSES)

and varies from day to day according to the density of the air and water vapor at that time.

In refrigeration work, we often refer to pressures below atmospheric pressure as "inches of vacuum," meaning inches of mercury pressure less than atmospheric pressure, and our compound gauges are calibrated in inches of mercury vacuum below zero on the gauge as shown in Figure 4–18. We need to create a deep vacuum in a refrigeration or air conditioning system to remove all the air and other impurities prior to charging it with refrigerant.

Figure 4–18
Compound gauge used to measure suction pressure and vacuum. Readings above 0 psig are pressure readings. Readings below 0 psig are in a vacuum and stated as inches of mercury, or "in. Hg."

WHAT IS MATTER?

We are surrounded by matter. We walk on matter. We eat, drink, and breathe matter. When we use the term "matter," what are we talking about? A quick review of matter finds that the smallest component is called an atom. Atoms collect to create molecules.

Matter can exist in any one of three physical states: solid, liquid, or gas. *A solid is a substance that has definite volume and mechanical strength to stay together.* A solid exerts forces in a downward direction either toward the surface on which it is resting or toward the Earth, as shown in Figure 4–19. Liquids have definite volumes but do not have definite shapes. The shape a liquid takes depends on the shape of the container that holds it.

Gases have neither definite volume nor definite shape because gas molecules have little attraction for each other. The gas completely fills any vessel that contains it and exerts pressure in all directions against the walls of the container, as illustrated in Figure 4–20. The gas laws that determine the behavior of contained gases are known by their inventors:

- Boyle's law
- Charles's law
- Dalton's law

Before we review these important gas laws that relate to HVACR, let's review what is meant by the scientific term "law."

WHAT IS A SCIENTIFIC LAW?

A scientific law is an absolute or truth that can be tested over and over again with the same outcome—no exceptions! The same results will be achieved if you apply the

A block of ice and water   A piece of wood

Figure 4–19
A solid exerts force in a downward direction toward either the surface it is resting on or toward the Earth, as in the left figure. Liquids have definite volumes but do not have definite shapes. The shape a liquid takes depends on the shape of the container that holds it. Water can be neatly stored in cylindrical glass. When it spills, the water seeks the shape of the surface it falls on, which sometimes seems quite large.

Gases

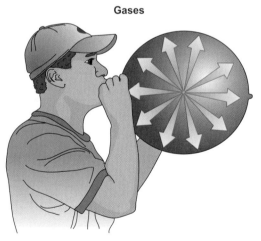

Figure 4–20
Gas molecules exert pressure on the walls of a balloon.

same methods to achieve those results. For example, a common law that affects us every day is the law of gravity. If you repeat dropping the same object from the same height to the ground, it will take the same amount of time to hit the ground every time. Gravity forces the object to the Earth's surface. Gravity does not vary at the same place on the Earth, but it can vary if you move to another part of the Earth.

Understanding the five basic laws of nature will aid in your understanding of air conditioning, heating, and refrigeration systems:

Law 1: Heat exists in the air down to absolute zero, which is −460°F.
Law 2: Heat flows from a higher temperature to a lower temperature regardless of how small the temperature difference might be.
Law 3: Due to friction between molecules, all gases become warmer when compressed.
Law 4: Matter can be in a solid (ice), liquid (water), or gas (vapor) state.
Law 5: The temperature at which a material changes from a liquid to a gas or from a gas to a liquid depends on the pressure at which it is contained.

The Laws of Thermodynamics

There are two laws of thermodynamics:

- Energy cannot be created or destroyed, but it can be converted from one form of energy to another. When one form of energy is generated, it is generated at the expense of another form of energy.
- Energy degrades into low-level heat energy. Heat energy is a byproduct of energy conversion and flows from a warmer substance to a cooler substance. In other words, heat goes to cold. The greater the temperature differences are between two materials, the faster the heat transfer rate is.

All substances have molecules in motion: slow movement for solids, faster movement for liquids, and the fastest movement of molecules for vapor.

Question: When a tank of refrigerant is placed in warm water, why does its pressure rise?

Answer: Heat makes molecules move faster. The faster they move, the higher the pressure.

Boyle's Law of Volume and Pressure

Boyle's law states that the pressure and volume of a gas are inversely related. What does this mean? It means that if the volume decreases, the pressure will increase, and if the volume increases, the pressure will decrease. They are opposite provided the temperature is held constant. It is important to note that this relationship is proportional only with absolute pressure, expressed as psia.

For example, Figure 4–21 shows a cylinder with a piston. With the position of the piston shown on the left side, the volume of the cylinder above the piston is 1 cubic

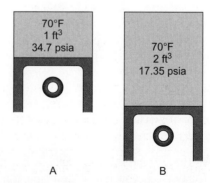

Figure 4–21
Boyle's law expansion example: The volume of the cylinder above the piston on the left is 1 cubic foot. The pressure is 20 pounds per square inch gauge, or 34.7 psia (34.7 − 14.7 = 20 psig). The piston on the right has been slowly lowered twice as far, so now the volume is 2 cubic feet and half the pressure, or 17.35 psia. (Courtesy of RSES)

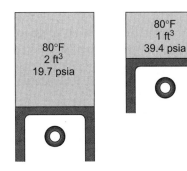

Figure 4-22
Boyle's law compression example: The piston is pushed upward slowly to one-half its height and the volume becomes one-half. Keeping temperature constant, the absolute pressure is doubled. The cylinder above the piston is originally 2 cubic feet, and its pressure is 5 psig or 19.7 psia. The piston is slowly pushed up halfway.
To reduce 39.4 psia to gauge pressure, we subtract 14.7, leaving 24.7 psig. The absolute pressure was doubled by halving the volume, the gauge pressure was increased from 5 psig to 27.7 psig, over five times its starting point. (Courtesy of RSES)

foot. The pressure is 20 pounds per square inch gauge or 34.7 psia (34.7 − 14.7 = 20 psig). The piston on the right has been slowly lowered twice as far, so now the volume is 2 cubic feet. What happens to the pressure? It goes down in the same proportion as the volume went up.

So the pressure drops to 17.35 psia or 2.65 psig (17.35 − 14.7). Therefore, with the volume doubled, the absolute pressure is one-half, but the gauge pressure drops from 20 psig to a little over 2.5 psig.

Obviously, if the piston is pushed upward slowly to one-half its height, the volume becomes one-half. Keeping temperature constant, the absolute pressure is doubled. The cylinder above the piston is originally 2 cubic feet, and its pressure is 5 psig or 19.7 psia. The piston is slowly pushed up halfway as shown in Figure 4-22.

To reduce 39.4 psia to gauge pressure, we subtract 14.7, leaving 24.7 psig. So although the absolute pressure was doubled by halving the volume, the gauge pressure was increased from 5 psig to 27.7 psig, over five times its starting point.

Charles's Law of Volume and Pressure

During the afternoon, a leak on a refrigeration system is located and repaired. The technician then pressurizes the system with dry nitrogen to a pressure of 150 psig. The technician notes that the ambient temperature is 95°F. The technician allows the system to set overnight, and upon returning in the morning, the standing nitrogen pressure is 141 psig. At first glance, the technician might believe that the system has a leak, and then Charles's law pops into his head. Charles's law states that "for a constant volume, the pressure changes as the absolute temperature changes." The morning temperature is 65°F, and that explains the pressure drop as a result of the temperature dropping from 95°F. The constant volume of the system takes into consideration the refrigerant piping, condenser, evaporator coil, compressor, and any other accessories.

Charles's law states that at a constant pressure, the absolute volume changes directly with the absolute temperature, and at a constant volume, the absolute pressure changes directly with the absolute temperature. Keeping the pressure of a gas constant means that when the temperature increases, so does the volume. The opposite is also true: as the temperature decreases, the volume decreases. Charles's law also recognizes the relationship between pressure and temperature of a gas if the volume is kept constant. If the temperature increases, the pressure increases, and if the temperature decreases, the pressure decreases.

In other words, a gas expands as it heats, and it contracts as it cools. Due to the fact that a gas adapts itself to its container, regardless of size or the amount of gas in the container, another factor is introduced—pressure. If the container is already filled, then a rise in temperature from the addition of heat cannot cause an increase in volume, but it can result in an increase in the pressure of the gas against the inner walls of the cylinder.

We know that this is true, and we know that the change in pressure with a change in temperature can be easily calculated as long as the volume stays the same, as of course it will in a gas cylinder.

It is very simple: the gas pressure goes up at the same rate as the temperature. If the temperature rises 25%, the pressure goes up 25%. If the gas cools down to half of its temperature, the pressure goes down to 50% of what it was.

There are two things to remember when using this law. The first is that temperatures and pressures must be expressed in absolute temperatures and pressures. These are easy to determine by adding 14.7 to the pressure reading and converting the Fahrenheit temperature to Rankine by adding 460 degrees. The second thing to remember is that these laws apply only to pure gases.

Dalton's Law of Volume and Pressure

Air or another gas not removed from an air conditioning system causes higher than normal operating pressures. Air or nitrogen in an air conditioning system are known as noncondensibles. The foreign gas present in the refrigeration system has pressure that, along with the refrigerant pressure, produces an even higher pressure. These higher pressures reduce the capacity of the refrigeration system. The additive effect of pressures is the principle of Dalton's law.

Dalton's law states that the total pressure of a confined mixture of different gases is equal to the sum of the pressures of each gas in the mixture. Figure 4-23 is an example of Dalton's law. If there is a container with oxygen at 30 psia and a separate container with nitrogen at 40 psia, then when we put the two in one container, the pressure will become 70 psia.

What Happens When Temperature, Pressure, and Volume All Change?

When a compressor squeezes or compresses a refrigerant gas, the temperature and pressure of the gas increases and the volume decreases, as indicated in Figure 4-24.

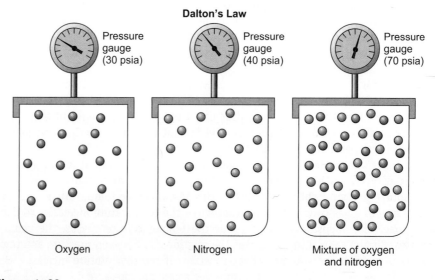

Figure 4-23
Dalton's law states that if the volume is constant, pressures of different gases will add to create an overall higher pressure. The pressures are not averaged as some might expect.

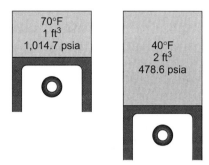

Figure 4-24
Combining the gas laws and putting them into practice. Temperature, volume, and pressure all vary as the conditions of a compressor. (Courtesy of RSES)

1. According to Boyle's law, the temperature remains constant and the pressure changes with a change in volume.
2. According to Charles's law, the volume remains constant and the pressure changes with a change of the absolute temperature. The second part of Charles's law states that, at a constant pressure, the volume of a confined gas varies directly as the absolute temperature.

Both of these help us in understanding how pressures, volumes, and temperatures change in containers of gas, and in fact we can calculate each of them, as shown in the preceding examples.

Notice, however, that in each of these three laws, something remained constant: the pressure, the volume, or the temperature. Gases are not always so considerate. All three (pressure, volume, and temperature) may change at the same time; that is, none of them remain constant.

SUMMARY

This chapter is full of new terms and information that may not seem important to your job at hand, but the concepts are important to understand. For example, the gas laws showed us that as a gas is added to a system that has a set volume, the pressure increases. As we add refrigerant to an air conditioning or refrigeration system, the pressure increases. During the compression cycle, the pressure and temperature increase while the volume decreases.

We learned that heat travels from hot to cold and that heat cannot be created or destroyed but is transferred from place to place. This chapter discussed the three methods of heat transfer stated below:

- *Conduction:* Transfer of heat through a solid material by molecule-to-molecule contact. Example: Heat transfer in a metal rod.
- *Convection:* Transfer of heat by means of movement of air or water. The warm fluid will rise because it is lighter and less dense, and the cooler fluid will fall because it is heavy. This creates convection currents or a flow pattern. Example: The warm air rising from a gas space heater.
- *Radiation:* Transfer of heat through space without heating the air. The radiant transfer is converted to heat when it hits a solid object. Example: Radiation traveling through space, passing through a window, and heating the furniture in a room.

Finally, you learned about the difference between gauge pressure and absolute pressure and how heat transfer is measured in BTUs and not degrees Fahrenheit.

FIELD EXERCISES

1. Boil a pot of water at a low heat setting while measuring the temperature of the water with a high-temperature probe that can measure at least 220°F. Keep the probe in the upper half of water to prevent radiation effects from the bottom of the pot. At sea level, you will notice that the boiling temperature is 212°F. If water is heated at a higher elevation, it will boil at a temperature lower than 212°F. Now increase the amount of heat from the burner. You will notice that the water will boil faster but the temperature will not increase. The point of the exercise is to show that adding heat at the same pressure does not increase the temperature of the water, just the rate of boiling. The temperature of the water will increase if the area above the water is pressurized, such as placing a lid on the pot.
2. Locate an example of each of the three methods of heat transfer. Describe how each method transfers heat.

REVIEW QUESTIONS

1. How many BTUs will it take to heat a pound of water from 50°F to 100°F?
2. How many BTUs will it take to heat 10 pounds of water from 50°F to 100°F?
3. How many BTUs will it take to cool 10 pounds of water from 100°F to 50°F?
4. Give an example using Boyle's law.
5. Give an example using Charles's law.
6. Give an example using Dalton's law.
7. What is the difference between sensible heat and latent heat?
8. Give an example of sensible heat and latent heat.
9. What is specific heat? Where is specific heat applied?

10. Describe three methods of heat transfer. Give an example of each method.
11. What is pressure?
12. What is the difference between gauge pressure and absolute pressure?
13. Convert a 300 psig pressure reading to psia.
14. What is thermodynamics?
15. What is a ton of air conditioning? How many BTUH will a 10-ton air conditioning system remove?
16. What is the absolute temperature of 50°F?
17. What is the difference between boiling, evaporation, and sublimation?
18. What is matter? What are the three states of matter?
19. What are the five basic laws of nature?
20. What are the two laws of thermodynamics?

CHAPTER 5

Basic Refrigeration Cycle

LEARNING OBJECTIVES

The student will:
- Describe the basic refrigeration cycle.
- List the basic components that make up a refrigeration system.
- Describe the function of a compressor.
- List the various types of compressors.
- Describe the function of a condenser.
- Describe the function of a metering device.
- List three types of metering devices.
- Describe the function of an evaporator.
- Use a pressure-temperature chart to find refrigerant pressures and temperatures.
- Read manifold gauge pressures and temperatures.
- Define "superheat."
- Define "subcooling."
- Define "saturation."
- Describe the operation of two types of service valves.

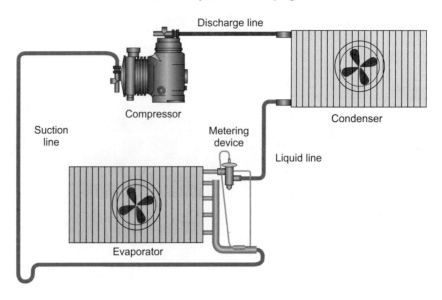

Basic Components and Piping

The refrigeration cycle with four basic components. (Courtesy of Dick Wirz, Refrigeration Training Services)

INTRODUCTION

As was introduced in Chapter 2, this chapter will further discuss the refrigeration cycle. The chapter will introduce pressures and temperatures found in the refrigeration cycle. You will learn how the manifold gauge can be used as a troubleshooting tool. New terms such as "saturation" and "subcooling" will be introduced. Also, the use of gauges and the measurement of temperatures are important to ensure optimal system performance and comfort. These measurements will help you charge a system accurately. Knowing how to read system pressures and temperatures will also assist in troubleshooting refrigerant-side problems. These are the tools of the trade.

Field Problem

During your busy day as a technician helper, you helped to recover refrigerant from systems with three different refrigerants, namely R-22, R-134a, and R-410A. The refrigerants were placed into three new and unmarked recovery cylinders. At the end of the day, the lead technician asked which bottle had which refrigerant. He had a label for each of the refrigerants. Unfortunately, you did not mark the cylinders and they all seem to be the same, with a yellow top and a light gray body. The recovery cylinders were only partially filled, not enough to be accepted at the supply house for exchange of an empty cylinder. The cylinders needed to be filled with additional refrigerant; therefore, you needed a way to determine which cylinder had which refrigerant. So how should you determine which recovery cylinder has which refrigerant?

Place the three cylinders in an environment with a constant temperature for 12 hours or more. When set in a room with a constant temperature, the three refrigerants have enough pressure difference for you to determine which one is R-22, R-134a, or R-410A. You will learn that R-134a will have the lowest pressure and that R-410A will have the highest pressure in the group. R-22 will have a pressure between these two refrigerants. Problem solved and refrigerant identified. Using this method will help you to remember to label recovery cylinders prior to using them.

BASIC REFRIGERATION CYCLE

The basic refrigeration cycle includes four basic components connected by copper tubing:

- Compressor
- Condenser
- Metering device
- Evaporator

Let's trace the refrigerant cycle in Figure 5-1. The refrigerant flows from the compressor to the condenser top, through the metering device, through the evaporator, and back to the compressor.

- Letter A represents the outlet or discharge of the compressor.
- Letter B is the inlet to the metering or expansion device.
- Letter C is the inlet to the evaporator.
- Letter D represents the return piping or suction line to the compressor.

It is important to understand the names of the pieces of interconnecting copper tubing. You will be able to attach pressure gauges to the refrigerant lines recommended in the equipment instructions or information given to you by your supervisor. Reading pressures or temperatures from the wrong refrigerant lines results in inaccurate results and confusion. Identify the refrigerant lines in Figure 5-1 by their location in the circuit:

- The refrigerant line that connects the compressor and condenser is labeled as the discharge line. The discharge line handles the hot refrigerant vapor from the compressor.
- The refrigerant line that connects the condenser and metering device is called the liquid line. The liquid line handles the warm liquid created by the condenser.
- The refrigerant line that connects the expansion device and evaporator is called the distributor. In some designs, the distributor is very short or not present.
- The refrigerant line that connects the evaporator and compressor is the suction line. The suction line carries cool, low-pressure gas to the compressor.

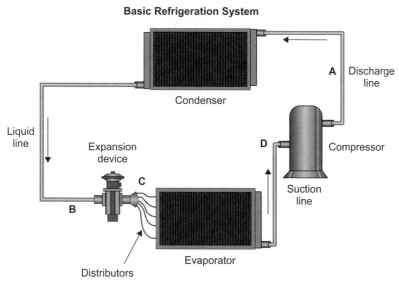

Figure 5-1
Follow the basic cycle by tracing the refrigerant flow from the compressor to the condenser, to the expansion device, and through the evaporator as it returns to the compressor. (Courtesy of Trane)

SATURATION, SUPERHEAT, AND SUBCOOLING

Before we dive into the individual components that make up the refrigeration cycle, we will introduce new terms that relate to the condition of the refrigerant. Heat transfer and pressure terms introduced in the chapter on thermodynamics will be applied in this chapter. All of these new terms relate to the condition of the refrigerant. We will start with the term "saturation" first.

Saturation

What is saturation? When used with regard to HVACR, *"saturation" means that both liquid and vapor refrigerant exist together at the same pressure and temperature.* Where does this normally occur in the refrigeration system?

- In the condenser
- In the evaporator

How do you know when the refrigerant is saturated? For a specific pressure, the refrigerant has a specific saturation temperature. A pressure gauge will need to be hooked to the low and high sides of the system in order to obtain the pressure readings used with a pressure-temperature (P-T) chart, which is used to convert pressure to temperature or temperature to pressure. We will then locate saturation conditions by matching the temperature to the pressure on a refrigeration gauge face.

Using the Pressure-Temperature Chart

The refrigerant we use has a pressure and temperature relationship that can be determined by the use of a P-T chart as shown in Figure 5-2. The pressure-temperature relationship is valid only when liquid and vapor exist in the same container or same area of the refrigeration system. From the head pressure, you can determine the condensing temperature. From the suction pressure, you can determine the evaporator

RED (in of Hg) = Vacuum   BLACK (psig) = Vapor   BOLD (psig) = Liquid

°F	R-12	R-22	R-134a	R-401A	R-402A	R-404A	R-407C	R-407C Liquid	R-408A	R-409A	R-410A	R-502	R-507A AZ-50
-50	15.5	6.1	18.7	17.9	1.1	0.1	11.0	2.7	2.0	18.7	4.9	0.8	0.9
-48	14.6	4.8	18.0	17.2	1.9	0.7	9.8	1.2	0.2	18.0	5.9	0.3	1.7
-46	13.8	3.4	17.3	16.4	2.8	1.6	8.6	**0.2**	0.6	17.2	7.0	1.1	2.6
-44	12.9	1.9	16.5	15.6	3.7	2.4	7.4	**1.0**	1.4	16.5	8.2	2.0	3.5
-42	12.0	0.4	15.7	14.7	4.7	3.4	6.0	**1.9**	2.2	15.6	9.4	2.8	4.4
-40	11.0	0.6	14.8	13.9	5.7	4.3	4.6	**2.7**	3.1	14.8	10.7	3.7	5.4
-38	10.0	1.4	13.9	12.9	6.8	5.3	3.2	**3.7**	4.0	13.9	12.0	4.6	6.4
-36	8.9	2.2	13.0	11.9	7.8	6.3	1.7	**4.6**	5.0	13.0	13.3	5.6	7.5
-34	7.9	3.1	12.0	10.9	9.0	7.4	0.1	**5.6**	5.9	12.0	14.7	6.6	8.6
-32	6.7	4.0	10.9	9.8	10.1	8.5	0.8	**6.6**	7.0	11.0	16.2	7.7	9.8
-30	5.5	4.9	9.8	8.7	11.4	9.6	1.6	**7.7**	8.0	0.9	17.7	8.7	10.9
-28	4.3	5.9	8.7	7.5	12.6	10.8	2.5	**8.8**	9.1	9.8	19.3	9.8	12.2
-26	3.0	6.9	7.5	6.3	13.9	12.0	3.4	**10.0**	10.3	7.6	20.9	11.0	13.5
-24	1.7	8.0	6.3	5.0	15.3	13.3	4.4	**11.2**	11.5	6.4	22.6	12.2	14.8
-22	0.3	9.1	5.0	3.6	16.7	14.6	5.4	**12.4**	12.7	5.1	24.4	13.5	16.2
-20	0.5	10.2	3.7	2.2	18.2	16.0	6.5	**13.7**	14.0	3.8	26.23	14.7	17.6
-18	1.3	11.4	2.3	0.8	19.7	17.4	7.5	**15.1**	15.3	2.4	28.1	16.1	19.1
-16	2.0	12.6	0.8	0.3	21.2	18.9	8.7	**16.5**	16.7	1.0	30.0	17.4	20.6
-14	2.8	13.9	0.0	1.1	22.9	20.4	9.9	**17.9**	18.1	0.3	32.0	18.9	22.2
-12	3.6	15.2	1.1	1.9	24.5	22.0	11.1	**19.4**	19.6	1.0	34.0	20.3	23.8
-10	4.5	16.5	1.9	2.8	26.3	33.6	12.3	**20.9**	21.1	1.8	36.3	21.9	25.5
-8	5.3	17.9	2.8	3.6	28.0	28.0	13.7	**22.5**	22.7	2.6	38.5	23.4	27.2
-6	6.2	19.4	3.6	4.5	29.9	27.0	15.0	**24.2**	24.3	3.5	40.8	25.1	29.0
-4	7.2	20.9	4.6	5.4	31.8	28.8	16.4	**25.9**	26.0	4.4	43.4	26.7	30.9
-2	8.1	22.4	5.5	6.4	33.8	30.7	17.9	**27.7**	27.7	5.3	45.7	28.4	32.8
0	9.1	24.0	6.5	7.4	35.8	32.6	19.4	**29.5**	29.5	6.3	48.2	30.2	34.8
2	10.1	25.7	7.5	8.5	37.9	34.6	21.0	**31.4**	31.3	7.3	50.8	32.1	36.8

Figure 5-2
P-T chart. (Courtesy of Emerson Climate Technologies)

°F	R-12	R-22	R-134a	R-401A	R-402A	R-404A	R-407C	R-407C	R-408A	R-409A	R-410A	R-502	R-507A
4	11.2	27.4	8.5	9.5	40.0	36.6	22.6	33.3	33.2	8.3	53.5	34.0	38.9
6	12.3	29.1	9.6	10.7	42.3	38.7	24.3	35.3	35.2	9.4	56.3	35.9	41.1
8	13.4	31.0	10.8	11.8	44.6	40.9	26.1	37.4	37.2	10.5	59.2	37.9	43.3
10	14.6	32.8	11.9	13.0	46.9	43.1	27.9	39.5	39.3	11.6	62.2	40.0	45.7
12	15.8	34.8	13.1	14.2	49.4	45.4	29.7	41.7	41.4	12.8	65.2	42.1	48.0
14	17.0	36.8	14.4	15.5	51.9	47.8	31.7	44.0	43.6	14.0	68.4	44.3	50.5
16	18.3	38.8	15.7	16.9	54.4	50.2	33.7	46.3	45.9	15.3	71.6	46.5	53.0
18	19.6	40.9	17.0	18.2	57.1	52.7	35.7	48.7	48.2	16.6	74.9	48.9	55.6
20	21.0	43.1	18.4	19.6	59.8	55.3	37.9	51.2	50.6	18.0	78.4	51.2	58.2
22	22.4	45.3	19.9	21.1	62.6	58.0	40.1	53.8	53.1	19.4	81.9	53.7	61.0
24	23.9	47.6	21.3	22.6	65.5	60.7	42.3	56.4	55.7	20.8	85.5	56.2	63.8
26	25.3	50.0	22.9	24.2	68.5	63.5	44.7	59.1	58.3	22.3	89.2	58.8	66.7
28	26.8	52.4	24.5	25.8	71.5	66.4	47.1	61.9	61.0	23.9	93.1	61.4	69.6
30	28.4	55.0	26.1	27.4	74.7	69.3	49.6	64.7	63.7	25.5	97.0	64.2	72.7
32	30.0	57.5	27.8	29.1	77.9	72.4	52.1	67.7	66.6	27.4	101.1	67.0	75.8
34	31.6	60.2	29.5	30.9	81.2	75.5	54.8	70.7	69.5	28.8	105.2	69.8	79.0
36	33.3	62.9	31.3	32.7	84.6	78.7	57.5	73.8	72.5	30.5	109.5	72.8	82.3
38	35.1	65.7	33.1	34.6	88.0	82.0	60.3	77.0	75.6	32.3	113.9	75.8	85.7
40	36.9	68.6	35.0	36.5	91.6	85.4	63.2	80.2	78.7	34.2	118.4	78.9	89.2
42	38.7	71.5	37.0	38.5	95.3	88.8	66.1	83.6	81.9	36.1	123.0	82.1	92.7
44	40.6	74.5	39.0	40.5	99.0	92.4	69.2	87.0	85.3	38.0	127.7	85.4	96.4
46	42.6	77.6	41.1	42.6	102.9	96.0	72.3	90.6	88.7	40.1	132.6	88.7	100.1
48	44.6	80.8	43.2	44.8	106.8	99.8	75.5	94.2	92.2	42.1	137.5	92.1	104.0
50	46.6	84.1	45.4	47.0	110.8	103.6	78.8	97.9	95.7	44.3	142.6	95.6	107.9
52	48.7	87.4	47.7	60.4	115.0	107.5	82.2	101.7	99.4	46.5	147.9	99.2	111.9
54	50.8	90.8	50.0	63.0	119.2	111.6	85.7	105.6	103.1	48.7	153.2	102.9	116.1
56	53.1	94.4	52.4	65.7	123.6	115.7	89.3	109.6	107.0	51.1	158.7	106.6	120.3
58	55.3	98.0	54.9	68.4	128.0	119.9	93.0	113.7	110.9	53.4	164.4	110.5	124.6
60	57.6	101.6	57.4	71.2	132.6	124.2	96.8	117.9	115.0	55.9	170.1	114.4	129.1
62	60.0	105.4	60.0	74.1	137.2	128.7	100.7	122.3	119.1	58.4	176.0	118.5	133.6
64	62.4	109.3	62.7	77.0	142.0	133.2	104.7	126.7	123.3	61.0	182.1	122.6	138.3
66	64.9	113.2	65.4	80.0	146.9	137.8	108.8	131.2	127.6	63.6	188.3	126.8	143.0
68	67.5	117.3	68.2	83.1	151.9	142.6	113.0	135.8	132.0	66.4	194.6	131.1	147.9
70	70.1	121.4	71.1	88.3	157.0	147.4	117.3	140.5	136.6	69.2	201.1	135.5	152.9
72	72.7	125.7	74.1	89.5	162.2	152.4	121.7	145.4	141.2	72.0	207.7	140.0	157.9
74	75.4	130.0	77.1	92.8	187.5	157.5	126.2	150.3	145.9	75.0	214.5	144.7	163.1
76	78.2	134.5	80.2	98.2	173.0	162.7	130.9	155.4	150.8	78.0	221.4	221.4	268.5
78	81.1	139.0	83.4	99.7	178.5	168.0	135.6	160.5	155.7	81.1	228.5	154.2	173.9
80	84.0	143.6	86.7	103.2	184.2	173.4	140.5	165.8	160.8	84.2	235.8	159.1	179.5
82	87.0	148.4	90.0	106.8	190.1	179.0	145.5	171.2	165.9	87.5	243.2	164.1	185.1
84	90.0	153.2	93.5	110.6	196.0	184.6	150.6	176.8	171.2	90.8	250.7	169.2	190.9
86	93.2	158.2	97.0	114.4	202.1	190.4	155.9	182.4	176.6	94.2	258.5	174.5	196.9
88	96.3	163.2	100.6	118.2	208.3	196.4	161.2	188.2	182.2	97.7	266.4	179.8	202.9
90	99.6	168.4	106.3	122.2	214.6	204.4	166.7	194.1	187.7	101.3	274.5	185.3	209.1
92	102.9	173.7	108.1	126.2	221.1	208.6	172.3	200.1	193.5	104.9	282.7	190.8	215.4
94	106.3	179.1	112.0	130.4	227.7	214.9	178.1	206.3	199.3	108.7	291.2	196.5	221.9
96	109.8	184.6	115.9	134.6	234.4	221.3	184.0	212.5	205.3	112.5	299.8	202.3	228.5
98	113.3	190.2	120.0	134.9	241.3	227.9	190.0	219.0	211.4	116.4	308.6	208.2	235.2
100	116.9	195.9	124.2	143.3	248.3	234.6	196.1	225.5	217.6	120.4	317.6	214.3	242.1
102	120.6	201.8	128.4	147.8	255.5	241.5	202.4	232.2	224.0	124.5	326.7	220.4	249.1
104	124.4	207.7	132.7	152.4	262.8	248.5	248.5	239.0	230.5	128.7	336.1	226.7	256.2
106	128.2	213.8	137.2	157.1	270.2	255.6	215.4	245.9	237.1	133.0	345.7	233.1	263.5
108	132.1	220.0	141.7	161.9	277.8	262.9	222.2	253.0	243.9	137.3	355.4	239.6	271.0
110	136.1	226.4	146.4	166.8	285.6	270.4	229.0	260.3	250.7	141.8	365.4	246.3	278.6
112	140.2	232.8	151.1	171.8	293.5	278.0	236.1	267.6	257.8	146.4	375.5	253.0	286.3
114	144.3	239.4	156.0	176.8	301.6	285.7	243.3	275.1	264.9	151.1	385.9	260.0	294.2
116	148.6	246.1	160.9	182.0	309.8	293.8	250.6	282.8	272.2	155.8	396.5	267.0	302.3
118	152.9	253.0	166.0	187.3	318.2	301.7	258.1	290.6	279.7	160.7	407.3	274.2	310.5
120	157.3	260.1	171.2	192.7	326.7	309.9	265.8	298.6	287.2	165.7	418.3	281.5	318.9
122	161.8	274.3	176.5	198.2	335.5	318.3	273.6	306.7	295.0	170.8	429.5	288.9	327.5
124	166.3	281.7	181.8	203.8	344.3	326.3	281.6	315.0	302.9	176.0	441.0	296.5	336.2
126	171.0	289.2	187.4	209.5	353.4	335.5	289.8	323.4	310.9	181.3	452.7	304.3	345.1
128	175.7	296.9	193.0	215.3	362.6	344.4	298.1	332.0	319.1	186.7	464.6	312.1	354.2
130	180.5	304.7	198.7	221.2	372.0	353.5	306.6	340.7	327.4	192.2	476.8	320.2	363.5
132	185.5	312.6	204.6	227.2	381.6	362.8	315.4	349.7	335.9	197.8	489.2	328.4	372.9
134	190.5	320.7	210.6	233.4	391.4	372.2	324.2	358.7	344.6	203.6	501.9	336.7	382.6
136	195.6	329.0	216.7	239.7	401.4	381.9	333.3	368.0	353.4	209.4	514.8	435.2	392.4
138	200.8	337.4	222.9	246.0	411.6	391.7	342.6	377.4	362.4	215.4	528.0	353.8	402.4
140	206.0	345.9	229.2	252.5	421.9	401.7	352.1	387.0	371.5	221.5	541.4	352.6	412.7
142	211.4	354.6	235.7	259.1	432.5	412.0	361.7	396.7	380.9	227.7	555.2	371.1	423.1
144	216.9	363.5	242.3	265.9	443.3	422.4	422.4	406.6	390.4	234.1	569.2	380.7	433.8
146	222.5	222.5	249.0	272.7	454.2	433.1	371.7	416.7	400.0	240.6	583.5	390.0	444.7
148	228.1	372.5	255.9	279.7	465.4	443.9	392.0	427.0	409.9	247.2	598.1	399.4	455.8
150	233.9	281.7	262.9	288.8	476.9	455.1	402.5	437.5	419.9	253.9	613.0	409.0	467.2

Figure 5–2
Continued

temperature. These temperatures will be important when learning how to check superheat and subcooling. We will also learn how to use the P-T chart to determine refrigerant type.

Look at Figure 5-3 to see how to use the P-T chart and work though a few examples:

1. If the suction pressure of R-22 is 69 psig, what is the evaporator temperature? Locate the R-22 refrigerant column. Find the desired pressure, which in this example is 69 psig. The corresponding temperature to 69 psig is 40°F. What we can learn from this exercise is that the evaporator is 40°F if there is liquid present in the coil. It is important to be able to calculate the evaporator temperature because you will learn how to find suction superheat using this low-side saturation temperature.
2. If the suction pressure of R-410A is 123 psig, what is the evaporator temperature? On the P-T chart, you will find 42°F.
3. If the evaporator is supposed to be 36°F, what is the suction pressure for R-22? The answer is 63 psig.

These examples converted low-side pressures and temperatures. Review Figure 5-4 to calculate examples of high-side pressures and temperatures. These calculations will help you determine the condensing pressure and temperature, which we will later use to find subcooling.

1. If the head pressure of R-22 is 260 psig, what is the condensing temperature? The P-T chart shows that the corresponding temperature is 120°F. If the system is operating, this means the refrigerant in the condenser is condensing at 120°F.
2. If the head pressure of R-410A is 365 psig, what is the condensing temperature? The answer found on the P-T chart is 110°F.
3. If the condensing temperature is 124°F, what is the head pressure for R-22? For R-410A? R-22 is 281 psig, and R-410A is 441 psig. As shown by these examples, the R-410A refrigerant has a much higher pressure than R-22.

Using the Manifold Gauge Set to Measure Pressures and Temperatures

The manifold gauge set is designed for accessing the system and to measure the pressures in the refrigeration system. As shown in Figure 5-5, the gauge on the right is connected to the high-pressure side of the system. The low-pressure gauge on the left is connected to the suction side of the system. The center hose goes to a refrigerant cylinder, vacuum pump, or recovery unit.

Temp. °F	Pressure R22	Pressure R410A
24	48	86
26	50	89
28	52	93
30	55	97
32	58	101
34	60	105
36	63	110
38	66	114
40	69	118
42	72	123
44	75	128
46	78	133
48	81	138

If the suction pressure of R22 is 52 psig, what is the evaporator temperature?
    Answer: 28°F

If the suction pressure of R410A is 123 psig, what is the evaporator temperature?
    Answer: 42°F

If the evaporator is supposed to be 36°, what is the suction pressure for: R22?
    Answer: 63 psig

R410A? Answer: 110 psig

Figure 5-3
Low-pressure examples of using the P-T chart. (Courtesy of Dick Wirz, Refrigeration Training Services)

Temp.	Pressure	
°F	R22	R410A
100	196	318
102	202	327
104	208	336
106	214	346
108	220	355
110	226	365
112	233	376
114	239	386
116	246	397
118	253	407
120	260	418
122	267	430
124	274	441
126	282	453
128	289	465
130	297	477
132	305	489
134	313	502
136	321	515
138	329	528
140	337	541

If the head pressure of R22 is 260 psig, what is the condensing temperature?
 Answer: 120°F

If the head pressure of R410A is 365 psig, what is the condensing temperature?
 Answer: 110°F

If the condensing temperature is 124°, what is the head pressure for: R22?
 Answer: 274 psig

R410A? Answer: 441 psig

Figure 5-4
High-pressure examples of using the P-T chart. (Courtesy of Dick Wirz, Refrigeration Training Services)

Tech Tip

The pressure-temperature relationship holds true only for saturated refrigerants. A saturated refrigerant has liquid and vapor in the same container or in the same space. The refrigerant can be undergoing change from liquid to vapor or vapor to liquid. The refrigerant can be at a steady state if it is at a constant temperature for a long period of time. Superheated and subcooled refrigerants do not have the pressure-temperature relationship.

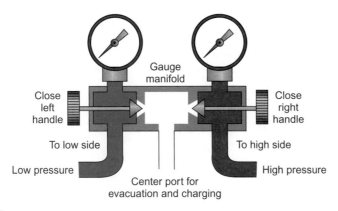

Figure 5-5
This illustration depicts the workings of a manifold gauge set. (Courtesy of Dick Wirz, Refrigeration Training Services)

Figure 5–6
Schrader valve tee. This tee is shown with the Schrader core removed. (Courtesy of Carrier Corporation)

An access valve is used to measure the high- and low-side pressures. There are several varieties of access valves. Figure 5–6 and Figure 5–7 show the common Schrader valve, which is similar to a tire valve. The fitting on the refrigeration gauge hose has a depressor that pushes down on the Schrader valve stem and allows pressure to be measured at the gauge.

Other types of access valves are shown in Figure 5–8 and Figure 5–9. Figure 5–8 is a common service valve found on residential systems, and Figure 5–9 represents an access valve that is more commonly found on commercial systems. These valves have three basic positions. Refrigerant flows through the valve when the valve stem is extended all the way out, or "back seated." The gauge access port has no refrigerant pressure in this condition. Slightly opening or "cracking" the valve does not interfere with the refrigerant flow but allows pressure to the gauge hose. The closed, or "front seated," position blocks the flow of refrigerant through the valve while allowing pressure to the hose. Even though this blocks the refrigerant flow,

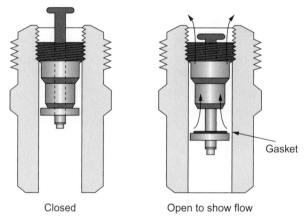

Figure 5–7
Schrader valve cross-section. The left side is sealed shut. The right side is pressure open to relieve or accept pressure. The center valve can be unscrewed and replaced if it leaks. The core should be removed when the Schrader valve is exposed to the heat of brazing or the gasket will burn open. (Courtesy of Carrier Corporation)

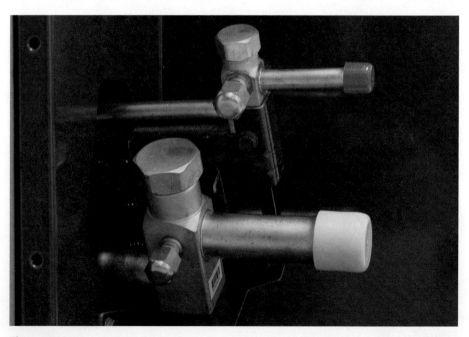

Figure 5–8
Common liquid-line service valve found on residential condensing units. (Photo by Susan Brubaker)

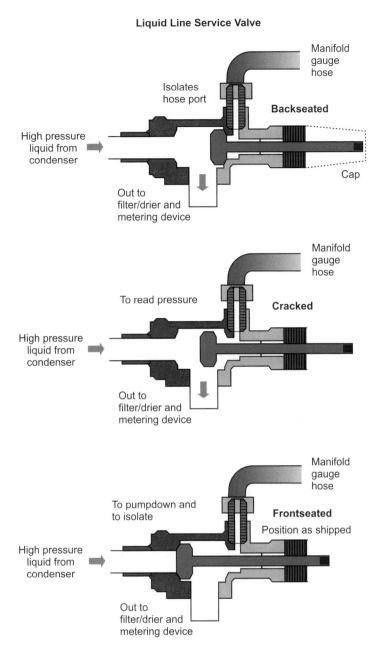

Figure 5-9
Cutaway views of a liquid-line service valve. There are discharge and suction valves with the same construction. (Courtesy of Carrier Corporation)

there is pressure on both sides of the valve and at the hose access. To "front seat" a discharge service valve with compressor operation can cause serious damage to the compressor and severe injury to the technician. Figure 5-10 details the hook-up of a gauge set to a commercial system with service valves.

The manifold gauge set has two gauges, as illustrated in Figure 5-11. The gauge on the right is the high-pressure gauge. The outer ring will measure discharge or liquid-line pressure. The inner temperature rings measure the corresponding condensing temperature for the system refrigerant when the unit is operating. Line up the gauge needle with the pressure and system temperature, as shown in Figure 5-12. The gauge is reading 278 psig; when measuring a system with R-22, this is equal to a condensing temperature of 125°F. Another example in this figure is 210 psig at 105°F.

The gauge on the left in Figure 5-12 is a compound gauge that measures pressures and vacuum. Pressures are measured above zero psig, and vacuum is measured

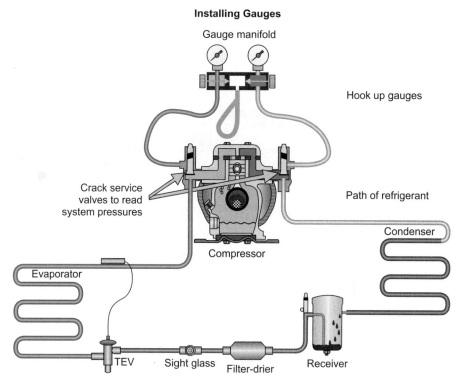

Figure 5–10
The high-side gauge is connected to a discharge line, and the compound gauge is connected to the suction valve. Both valves need to be cracked in order to obtain pressure readings. (Courtesy of Dick Wirz, Refrigeration Training Services)

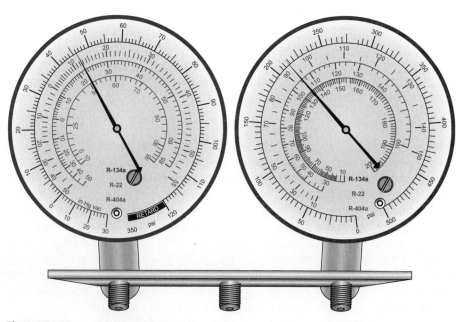

Figure 5–11
Pressure Gauges (Courtesy of Dick Wirz, Refrigeration Training Services)

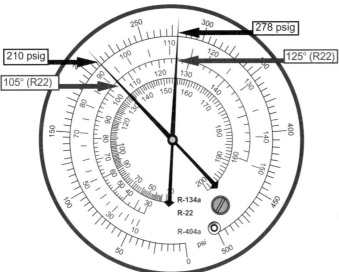

Figure 5-12
Two examples of the pressure-temperature relationship of R-22. At 210 psig, the corresponding saturation condensing temperature is 105°F. The 278 psig, corresponding condensing temperature is 125°F. (Courtesy of Dick Wirz, Refrigeration Training Services)

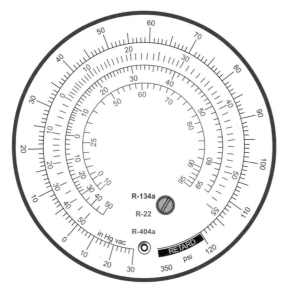

Figure 5-13
This is the faceplate of a compound gauge that is hooked to the low-pressure side of the system. The outer ring measures the evaporating pressure. The inner temperature rings are the corresponding evaporating temperatures. If the temperature is something other can what is on the gauge faceplate, a P-T chart can be used. (Courtesy of Dick Wirz, Refrigeration Training Services)

below zero psig. The pressure is in the outer ring and temperatures are in the inner rings when removing the needle from the compound gauge, as shown in Figure 5-13.

When the system shuts down, the compound gauge pressure rises. The compound gauge has a "retard" region in which the gauge pressure measures between 120 and 350 psig. This gauge safety region allows the off-cycle pressure to rise without damaging the gauge with excess pressure.

An example of the use of the low-side gauge is illustrated in Figure 5-14, using R-22 at 69 psig at 40°F or 49 psig at 25°F.

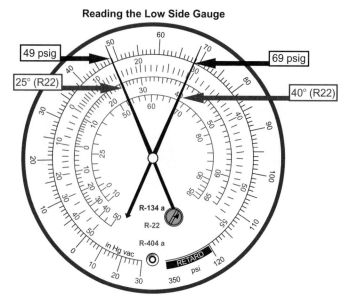

Figure 5-14
Measuring an operating system with R-22, a pressure reading of 69 psig will correspond to an evaporating pressure of 40°F. R-22 at 49 psig will correspond to an evaporating temperature of 25°F. (Courtesy of Dick Wirz, Refrigeration Training Services)

Superheat

Understanding and measuring superheat will help you determine the correct charge and operation of the refrigeration system. In later training, you will learn to use superheat as a troubleshooting tool.

Superheat is heat added to the refrigerant after it has changed from a liquid to a vapor. Sometimes it is defined as heat added to the refrigerant above its saturation temperature. Superheat is measured by subtracting the evaporating temperature from the suction line temperature. Figure 5-15 is an example of how to calculate superheat. The following example shows a calculated superheat of 10 degrees:

1. Take the pressure reading and convert it to the evaporating temperature by using a P-T chart or referring to the gauge temperature of the refrigerant in use.
2. Subtract the evaporating temperature from the suction line temperature. In this example using R-22, 75 psig is converted to 44°F and the suction line temperature is measured at 54°F.
3. This is calculated as follows:

   Superheat = Suction line temperature − Evaporating temperature
   10°F superheat = 54°F suction line − 44°F evaporating temperature

Figure 5-16 is another example of measuring superheat using R-410A. The calculated superheat is 20°F.

Subcooling

Subcooling is defined as heat removed from the liquid refrigerant below its condensing (saturation) temperature. Subcooling takes place toward the end of the condenser.
  Subcooling = Condensing temperature − Liquid line temperature
  The advantages of subcooling are as follows:

- Increased system efficiency (0.5% per 1°F of subcooling)
- Ensures a full column of liquid at the metering device

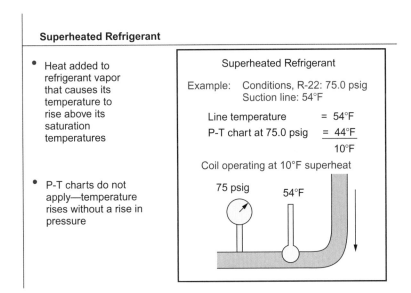

Figure 5-15
Superheat refrigerant is heat added to the refrigerant after it has boiled off in the evaporator. Any additional absorption of heat causes the superheat to rise. Notice how superheat is calculated in the example.

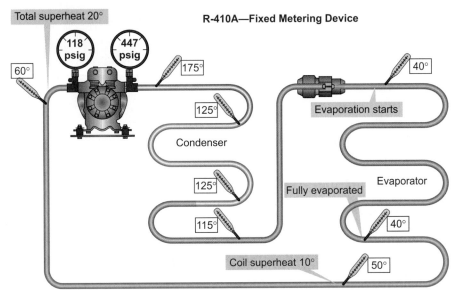

Figure 5-16
In order to calculate superheat, convert the suction pressure of 118 psig to the corresponding temperature of 40°F. Subtract 40°F from the suction temperature of 60°F. The superheat is 20°F. (Courtesy of Dick Wirz, Refrigeration Training Services)

Figure 5-18 is a field example of measuring subcooling. The condensing pressure for R-22 is 178 psig, which is converted to a condensing temperature of 94°F. The liquid line temperature is 80°F. To calculate subcooling, use the formula given previously:

14°F SC = 94°F condensing temperature − 80°F liquid line temperature

Example from Figure 5-17:

        125°F condensing temperature
      −115°F liquid line temperature at condenser outlet
         10°F subcooling

Sub-cooling	
• Heat removed from the liquid refrigerant that causes its temperature to drop below its saturation temperature • P-T charts do not apply—temperature drops without a drop in pressure	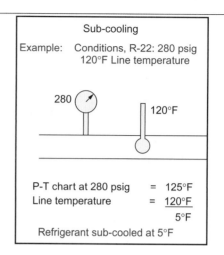 Sub-cooling Example: Conditions, R-22: 280 psig 120°F Line temperature P-T chart at 280 psig = 125°F Line temperature = 120°F 5°F Refrigerant sub-cooled at 5°F

Figure 5–17
Subcooling is removing heat from the liquid refrigerant in the condenser. Subcooling occurs near the end of the condenser coil.

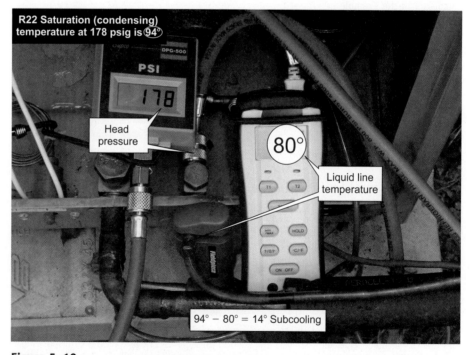

Figure 5–18
Calculating subcooling by subtracting the liquid line temperature from the condensing temperature. (Courtesy of Dick Wirz, Refrigeration Training Services)

COMPRESSORS

The purpose of the compressor is to create a circulation motion and increase pressure so heat can be given up. *The compressor is a vapor pump and is considered the heart of an air conditioning system.* The compressor receives the cool, low-pressure, superheated vapor from the suction line. The suction line is the larger, cool refrigerant line that may be insulated. The compressor increases the refrigerant pressure, which causes a refrigerant temperature rise as the refrigerant molecules are compressed or squeezed. The internal motor heat and compressor friction also cause a rise in refrigerant temperature.

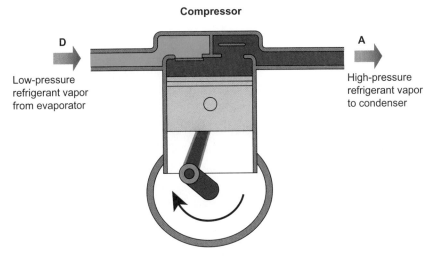

Figure 5-19
The reciprocating compressor uses a piston to pressurize and force the direction of the refrigerant flow to the condenser. (Courtesy of Trane)

The common types of compressors are the reciprocating, scroll, and rotary compressors, which will be covered in this chapter. Larger industrial compressors, covered later, are the rotary screw compressor and centrifugal compressor.

Reciprocating Compressors

Reciprocating compressors use one or more pistons to compress the vapor refrigerant and send it to the condenser through the discharge line. Reciprocating compressors are positive displacement compressors. Figure 5-19 shows the compressor receiving low-pressure vapor from the suction side. Leaving the compressor is hot, high-pressure vapor.

The two types of compressors are the sealed hermetic compressor, illustrated in Figure 5-20, and the semi-hermetic compressor. The hermetic compressor is sealed except for the discharge, suction, and process tube fittings and is not repairable in the field. These are disposable compressors and are replaced when there is a compressor problem.

Operation of the Reciprocating Compressor

Figure 5-21 illustrates the operation of the reciprocating compressor. There are four stages, or strokes, to the reciprocating cycle:

- *Suction stroke:* Starting from a standstill, the intake valve (suction valve) and exhaust valve (discharge valve) are both closed. On the downstroke represented on the left side of Figure 5-21, the piston expands the area available in the cylinder. When the cylinder pressure is lower than the suction pressure, the intake valve opens, pulling in refrigerant. For example, if the suction pressure is 70 psig, the pressure in the cylinder will need to be lower than 70 psig in order to open the suction valve. After the piston reaches its full stroke, or bottom dead center, the suction valve shuts when the pressure begins to increase. Think of the compressor valves as check valves allowing refrigerant to pass one way and blocking flow in the opposite direction.
- *Compression stroke:* The piston rises and compresses the refrigerant gas until it reaches the pressure-opening level of the exhaust or discharge valve.
- *Discharge stroke:* The discharge valve opens up at a very high pressure, as shown in the right side of Figure 5-21. In this example, the pressure can

> **Safety Tip**
>
> High temperatures are involved when working with the discharge line of the compressor. Additionally, the switch to new, alternative refrigerants has resulted in an increase in operating pressures, some in excess of 500 psi when using R-410A. Remember to wear your safety glasses and butyl-lined protective gloves when working around refrigerant-bearing equipment.

Figure 5-20
This is a sealed hermetic compressor. The suction connection is the larger fitting. One of the two smaller fittings is the discharge fitting. (Courtesy of Bristol Compressors International, Inc.)

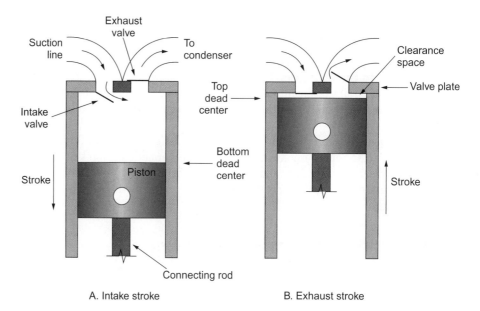

Figure 5-21
Suction and discharge strokes. The intake stroke pulls in the suction vapor; the valve closes; and when the pressure is high enough, the exhaust valve opens, pushing out the hot, high-pressure, superheated vapor. (Courtesy of RSES)

range from 150 psig to 400 psig. The piston pushes out most of the refrigerant from the cylinder, leaving a small amount trapped in the clearance space between the piston head and valve plate. This is also called top dead center. The suction and discharge pressure depend on the refrigerant type, system application, load, and operating conditions.

- *Reexpansion:* There is high-pressure vapor trapped above the top of the piston when the discharge valve shuts due to a lack of high pressure to keep it open. This high-pressure vapor must reexpand or drop in pressure before the suction valve will open again. If the pressure trapped above the piston is

300 psig, it must drop below the operating suction pressure before the lower-pressure activated suction valve will open. This reexpansion of high-pressure refrigerant is an operating efficiency loss but a necessary part of the reciprocating cycle.

Finally, the suction intake is from the shell of the compressor. The shell of the reciprocating compressor contains low-pressure refrigerant. Because the compressor shell contains refrigerant vapor, the body is either cool or near ambient temperature. The discharge vapor goes through piping to a fitting outside the compressor. In summary, the reciprocating compressors use one or more pistons to pull and push refrigerant through the refrigeration piping.

Semi-hermetic Compressors

The semi-hermetic compressor is similar to the sealed hermetic except that it is bolted together and repairable. The field service consists of minor site repairs such as replacing the valve plate, changing the oil pump, or changing oil. These compressors can be unbolted from the discharge and suction line, removed, and re-manufactured. In many cases, the replacement compressor is readily available off the shelf. Another advantage of a semi-hermetic compressor is that it can be opened up and examined to find clues as to what caused its failure. Most compressor failures are caused by problems elsewhere in the system, and unless corrected, will cause additional compressor failures.

Scroll Compressor

The scroll compressor uses stationary and driven scroll elements to compress low-pressure refrigerant located in the shell of the compressor. The scroll compressor uses two tightly machined scrolls to squeeze and compress the refrigerant, as seen in the construction of the scroll compressor and scroll elements in Figure 5-22, Figure 5-23, and Figure 5-24.

1. The shell of the scroll compressor is cool, suction vapor pressure. Vapor enters the outer opening as one scroll orbits the other, as shown in Figure 5-24.
2. The open passage is sealed as vapor is drawn into the compression chamber.
3. One scroll continues orbiting as the vapor is compressed into increasingly smaller pockets.
4. Vapor is continually compressed to the center of the scrolls, where it is discharged through precisely machined ports and sent to the discharge line.
5. During actual operation, all passages are in various stages of compression at all times, resulting in near continuous intake and discharge.

Tech Tip

It is important to check the accuracy of your manifold gauges. One way to do this is by placing cylinders of R-134a or R-22 in a stable temperature for at least 12 hours. After measuring the temperature of the room, hook up the low-side gauge to the R-134a refrigerant and determine if the pressure and temperature correspond to the pressure on the gauge by using a P-T chart. For example, if the stable room temperature is 75°F, the gauge pressure will read about 79 psig. Adjust the pressure setting on the gauge if necessary. Do the same for the high-side gauge, hooking it to R-22 cylinder pressure and calibrating the gauge's corresponding pressure and temperature. R-22 pressure at 75°F is equal to about 132 psig.

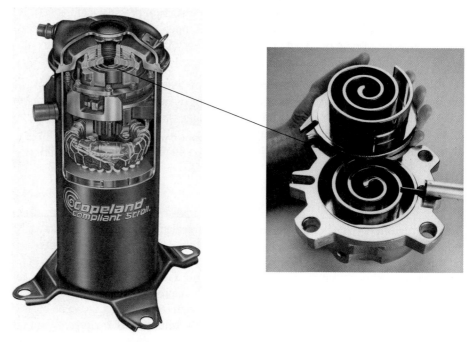

Figure 5-22
A scroll compressor has the scroll compression elements on the right side. The scrolls are located in the upper part of the compressor housing. (Courtesy of Emerson Climate Technologies)

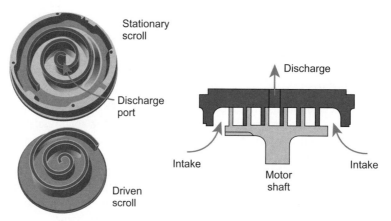

Figure 5-23
Stationary and driven scroll. The driven scroll is connected to the motor shaft. The stationary scroll is located over and loosely meshes with the driven scroll and moves back and forth. Vapor discharges in the center section. (Courtesy of Trane)

Rotary Compressor

Figure 5-25 shows a simple one-vane type of rotary compressor. The rotary compressor, like the reciprocating compressor, is a positive-displacement compressor. Instead of pistons that move up and down, it uses an off-center rotor that rotates inside a compression chamber fitted with suction and discharge ports. The rotary compressor uses a rotating roller to compress vapor and push it out of the chamber. A spring-loaded blade, similar to a vane, is held snugly against the outer ring of the compression chamber by spring tension. The vane protrudes from the outer ring and seals against an eccentric rotor. Rotation of the center lobe squeezes the refrigerant from the low side to the high side of the compressor. The multiple-vane rotary compressor, as shown in Figure 5-26, has four compartments to gradually increase the

1. Vapor enters an outer opening as one scroll orbits the other.
2. The open passage is sealed as vapor is drawn into the compression chamber.
3. As one scroll continues orbiting, the vapor is compressed into an increasingly smaller "pocket."
4. Vapor is continually compressed to the center of the scrolls, where it is discharged through precisely machined ports and returned to the system.
5. During actual operation, all passages are in various stages of compression at all times, resulting in near-continuous intake and discharge.

Figure 5–24
Shows progressive cycle of the scroll compressor. (Courtesy of Copeland Corporation)

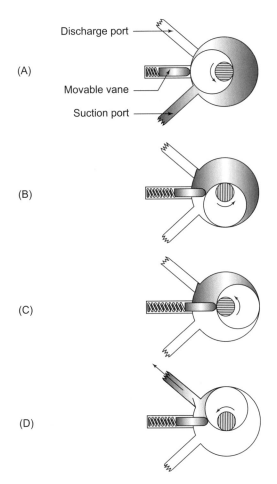

Figure 5–25
A simple one-vane rotary compressor. Most rotary compressors have more than one vane for compression.

pressure and discharge it into the shell of the compressor. Multivane compressors are more common than single-vane units.

Rotary compressors do not have suction valves, but some have a check valve in the suction or discharge line. This keeps discharge vapor from driving the compressor in reverse during the off cycle, which would allow refrigerant back into

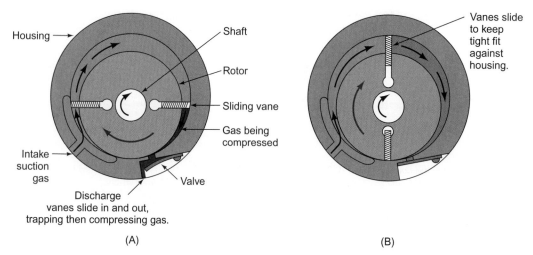

Figure 5-26
The rotary compressor uses an elliptical orbit to gradually squeeze the refrigerant vapor and build up its pressure prior to being discharged into the shell of the compressor.

Figure 5-27
The rotary compressor is protected against liquid flooding by an accumulator attached to the right side of the compressor shell. The accumulator collects liquid refrigerant and allows it to boil off to a vapor prior to entering the compressor. The replacement compressor will have an attached accumulator. (Photo by Susan Brubaker)

the low side of the system. The suction enters the accumulator that is attached to the compressor, as illustrated in Figure 5-27. From the accumulator, the vaporized refrigerant is piped directly into the rotary compression chambers. In comparison, the suction vapor in a reciprocating piston-type compressor is dumped directly into the compressor shell. The discharge line in a reciprocating compressor is piped directly to the outer shell fitting. The hot discharge of the rotary compressor goes directly into the compressor shell, causing the rotary compressor to run hot. The motor windings are designed to handle these high-discharge temperatures.

CONDENSER

The function of the condenser is to remove heat from the refrigerant. In order for heat to transfer, the temperature of the refrigerant needs to be hotter than the air passing over it. The refrigerant absorbs heat from air passing over the evaporator, while some additional heat is absorbed from the heat of compression. The refrigerant enters the condenser as a hot, high-pressure, superheated vapor and leaves as a warm, high-pressure, subcooled liquid. The pressure drop in the condenser coil is minimal.

Three phases of the condenser's heat removal are as follows:

1. De-superheat
   - Responsible for about 15% of condenser's total heat rejection at the beginning of the condenser
   - Removes sensible heat from discharge vapor
2. Condense
   - Responsible for about 80% of condenser's total heat rejection
   - Removes latent heat absorbed in the evaporator
   - Is a saturated mixture where liquid and vapor are present together in the middle 80 percent of the condenser, mostly vapor at first and becoming more liquid as heat is removed from the mixture
3. Subcool
   - Responsible for about 5% of condenser's total heat rejection at the end of the condenser
   - Removes sensible heat from condensed liquid

Types of Condensers

There are three basic types of condensers:

- Air cooled
- Water cooled (Figure 5-28)
- Evaporative cooled (Figure 5-29)

Of the three listed condensers, the air-cooled condenser is the most common, followed by the water-cooled condenser. No matter what type of condenser it is, the principle of heat removal is the same.

Finally, the condensing unit, as shown in Figure 5-30, is a combination of the compressor, the condenser coil, and several electrical controls. This is the most standard residential type of condensing unit.

METERING DEVICE

The metering device has many generic names, such as the flow control, feeder device, expansion device, refrigeration control, and restrictor device. The metering device receives the warm, high-pressure, subcooled liquid from the condenser. The metering device is a restriction that is sized to drop the refrigerant pressure, as seen in Figure 5-31, thereby dropping its temperature.

Heat transfers from a hot to a cold surface. In order for a heat exchange to occur in an air conditioning system, the refrigerant must be cooler than the air

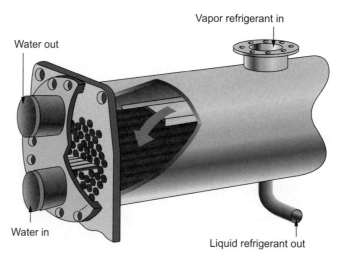

Figure 5–28
Horizontal shell and tube condenser. The water passes through the tubing, and the refrigerant is located in the condenser shell. (Courtesy of Trane)

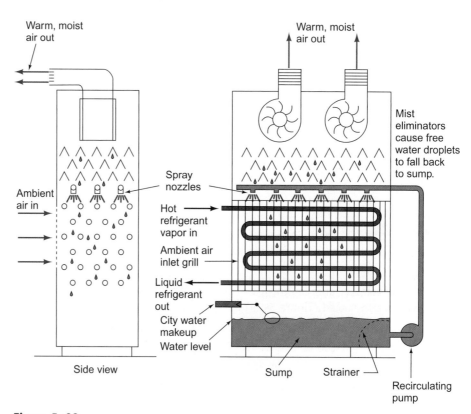

Figure 5–29
Water circulates in the evaporative condenser.

passing over it. The metering device receives the warm, high-pressure, subcooled refrigerant and, by reducing its pressure, creates a cold, low-pressure, saturated mixture. *"Saturated mixture" means that liquid and vapor are present in the same section of refrigerant piping.* At this point, the saturated mixture is a ratio of about 75% liquid and 25% vapor at the outlet of the metering device. The make-up of this saturated mixture may vary, depending on the design and application of the system.

Figure 5-30
Condensing unit. (Courtesy of Carrier Corporation)

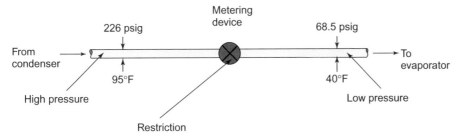

Figure 5-31
Warm, high-pressure, subcooled liquid feeds the metering device. The metering device creates a restriction, reducing the pressure and temperature of the refrigerant.

For an air conditioning system, the temperature of the saturated refrigerant mixture is about 40–45°F. This cold saturated refrigerant mixture is directed to the evaporator through the distributor piping. For refrigeration systems, the operating pressures and temperatures are much lower.

Types of Metering Devices

There are three types of common metering devices:

- Fixed metering capillary tube or piston/orifice
- Thermostatic expansion valve (TEV or TXV)
- Automatic expansion valve (AEV)

Fixed Metering Devices

The capillary tube or piston devices are shown in Figure 5-32 and Figure 5-33. These are fixed-bore metering devices that restrict the flow of refrigerant. The restriction drops the pressure and temperature of the refrigerant before it enters the evaporator.

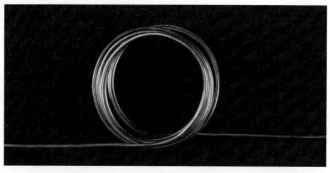

Figure 5-32
The capillary tube is so named because of its small capillary size. The capacity of the cap tube is created by the size of the inside diameter, the length of the tube, and the condenser pressure.

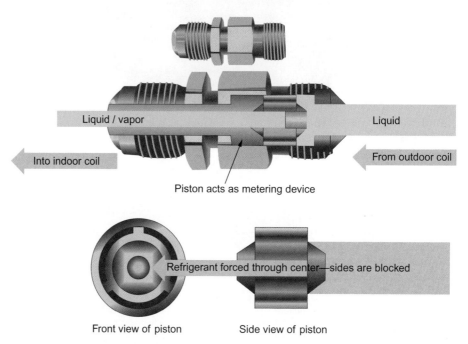

Figure 5-33
The fixed metering piston has an opening that reduces the pressure and temperature of the refrigerant. The orifice opening is drilled to allow a specific flow and capacity matched to the system. (Courtesy Carrier and Dick Wirz, Refrigeration Training Services)

The capacity of these restrictive devices is controlled by the following:

- Size of the opening
- Length of restriction
- Condenser pressure

The BTU capacity of the capillary tube is created by the size of tube opening, the length of the tubing, the coils of capillary tube, and the condenser pressure. The "cap tube" is sized by inside diameter, expressed as "ID." Standard copper tubing used for suction, liquid, and discharge lines are measured by outside diameter, or OD. The smaller the ID is, the greater the restriction and the less refrigerant available to the evaporator. The longer the cap tube is, the greater the restriction, which also means less refrigerant to the evaporator. Lower levels of refrigerant reduce the capacity of the evaporator.

The condenser pressure also plays an important role in the cap tube capacity. As the condensing pressure increases, more refrigerant is forced through the cap tube. Greater flow equals more BTU capacity.

The piston device has a drilled orifice opening for the refrigerant flow. It is similar to a cap tube in that it controls refrigerant capacity by the size of the opening and the amount of condenser pressure forcing refrigerant through the system.

Thermostatic Expansion Valve

The thermostatic expansion valve (TXV or TEV), as shown in Figure 5-34, is designed to maintain a constant refrigerant superheat at the outlet of the suction line. The function of the TXV is shown in Figure 5-35.

The TXV operates on three independent pressures:

- Valve spring pressure
- Suction pressure
- Sensing bulb pressure

Review Figure 5-35 for the following discussion on TXV operation. The spring pressure is factory-adjusted to maintain the required superheat of the system of 10°F. To calculate superheat, this example subtracts the evaporating temperature from the evaporator outlet temperature:

Superheat = Evaporator outlet temperature − Evaporating temperature
10°F Superheat = 50°F evaporator temperature − 40°F evaporating temperature

In this example, the spring pressure is 15 psig. The spring pressure and the evaporator pressure (at the outlet of the TXV) push against the bottom of the valve

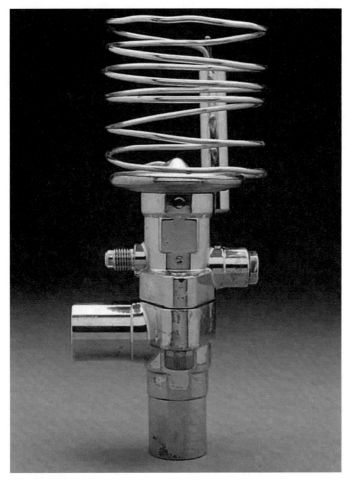

Figure 5-34
Thermostatic expansion valve controls the superheat at the outlet of the evaporator. (Courtesy of Emerson Climate Technologies)

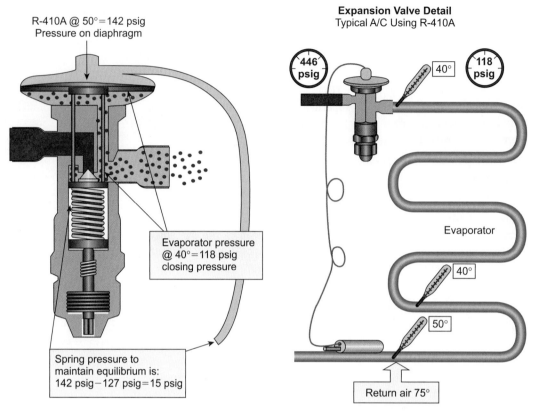

Figure 5-35
TXV operation. (Courtesy of Dick Wirz, Refrigeration Training Services)

diaphragm. The sensing bulb is attached to the TXV through a capillary line that connects to the cap of the valve. The sensing bulb is also charged with refrigerant and attached to the outlet of the evaporator; the response to the outlet temperature is part of the superheat calculation. If the superheat increases, the pressure in the bulb increases and pushes the diaphragm down, opening the valve to lower the superheat.

Low temperature output or low superheat at the outlet of the evaporator will reduce the pressure in the sensing bulb. The reduction of pressure is transferred to the top of the diaphragm, and the closing valve restricts the flow of refrigerant into the evaporator, causing the superheat at the evaporator outlet to rise.

Automatic Expansion Valve

The automatic expansion valve (AEV), as shown in Figure 5-36, tries to maintain a constant evaporator temperature around an adjusted pressure set point. The evaporator pressure and spring pressure control the valve operation, which will be discussed in more detail in your second-year training.

EVAPORATOR

The evaporator is a heat exchanger similar to a condenser. The function of the evaporator is to absorb heat from the air or water passing over it. As heat is removed from the air or water passing over the coil, the temperature will drop. Moisture is condensed on the surface of the coil, drying the air. In order for heat to transfer, the refrigerant temperature must be lower than the medium passing through it. For example, in order to create 50°F air, the refrigerant temperature needs to be about 40°F. The 50°F air will transfer heat to the colder, 40°F refrigerant.

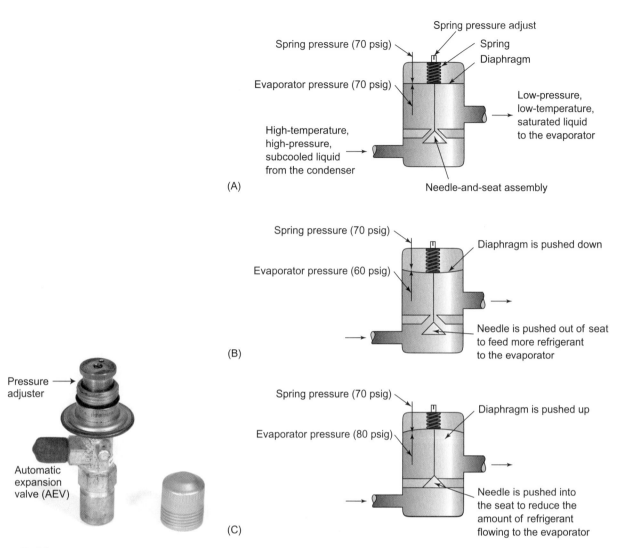

Figure 5-36
The left image is an actual automatic expansion valve. (Courtesy of Bill Johnson) The sequence on the right shows the operation of the AEV under varying load conditions.

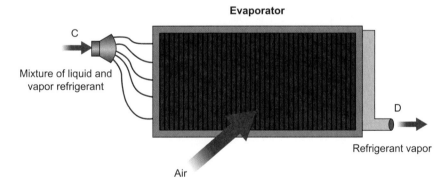

Figure 5-37
The evaporator receives a cold mixture of liquid and vapor refrigerant and boils off as it removes heat from the air passing through it. The refrigerant leaves the coil as a cool, low pressure, superheated vapor. (Courtesy of Trane)

The evaporator receives a cold, low-pressure mixture of liquid and vapor refrigerant from the metering device, as seen in Figure 5-37. As the refrigerant passes through the evaporator piping, it absorbs heat from the air passing through it. The refrigerant boils off or evaporates as it absorbs heat, dropping the temperature and moisture content of the air.

Near the leaving end of the evaporator, the liquid refrigerant will be converted to 100% vapor. The vapor will continue to absorb heat as it moves from the last passes of the evaporator piping and into the suction line. Most of the heat is absorbed by the cold liquid refrigerant boiling or evaporating to a vapor. Heat that is added to the vapor refrigerant beyond its boiling temperature is known as superheat. It is important that the evaporator return superheated vapor because liquid returning to the compressor will damage it.

Most evaporator coils are constructed with copper or aluminum tubing and aluminum fins as shown in Figure 5–38. The smaller tube feeding the coil is the liquid line. The larger line is the suction line, which is insulated because it is cold and will create condensation that will drip onto the structure, causing water damage.

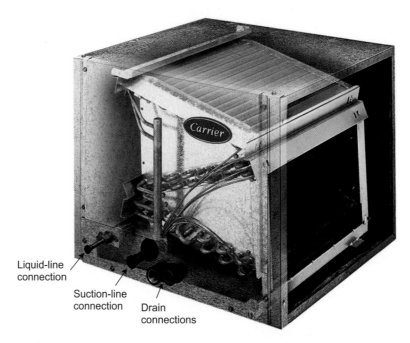

Figure 5–38
Side view of an evaporator with a copper coil and aluminum fins to improve heat exchange from the air passing through it. The large tube is the suction line, and the smaller tube is the liquid-line connection. (Courtesy of Carrier Corporation)

SUMMARY

The vapor-compression refrigeration cycle is a repeating cycle consisting of a compressor, condenser, metering device, and evaporator. The compressor is a pump and is considered to be the heart of the system. It changes the refrigerant from a low-temperature, low-pressure, superheated vapor to a high-temperature, high-pressure, superheated vapor. The act of compression and internal motor heating causes the temperature of the vapor to rise. The hot, high-pressure refrigerant makes its way to the condenser, where heat will be extracted. Three common types of compressors are reciprocating, scroll, and rotor.

The condenser rejects heat from the system and changes the refrigerant from a high-temperature, high-pressure vapor into a lower temperature, high-pressure subcooled liquid.

Condensers used on residential systems are typically air cooled. When the temperature of a substance changes, a sensible heat transfer takes place, which can be measured by a thermometer. This occurs in the first part of the condenser as the hot vapor superheat is removed. In the middle of the condenser, the refrigerant is condensed from a vapor to a liquid, which is a latent heat process. Latent heat transfer cannot be measured with a thermometer because the temperature of the refrigerant does not change. In this section of the condenser, this is where a majority of the heat is removed, although there is no change of temperature in the refrigerant. The last part of the condenser subcools the refrigerant.

The metering device controls the flow of refrigerant to the evaporator. It changes the high-temperature, high-pressure liquid into a low-temperature, low-pressure liquid/vapor mixture. Two common metering devices are the fixed metering device, like the capillary tube or piston, and the thermal expansion valve. The cold liquid refrigerant mixture absorbs heat in the evaporator, as seen in Figure 5-39.

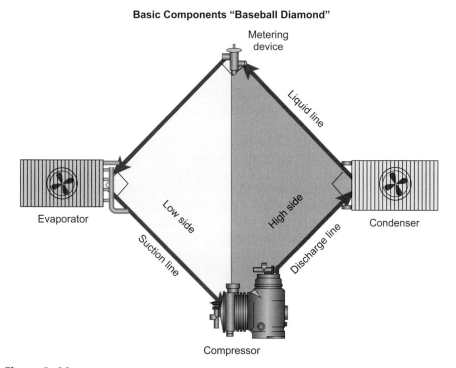

Figure 5-39
Shows the breakdown between the high- and low-pressure sides. (Courtesy of Dick Wirz, Refrigeration Training Services)

Knowing the names of the refrigerant lines is important when communicating problems to your supervisor or when selecting replacement parts. There is a difference between the suction service valve and the discharge service valve.

It is important to be able to use the pressure-temperature chart along with the manifold gauge pressures and/or suction and liquid tube temperatures in order to calculate superheat and/or subcooling to help evaluate system performance.

FIELD EXERCISES

1. Follow the refrigeration cycle on a split system and a package unit. Identify the major components and label the refrigerant lines.
2. Locate the distributor in a split system and on a package unit.
3. Place a cylinder of refrigerant in a room with a constant temperature and compare the temperature and pressure relationship of the refrigerant to a P-T chart and the gauge temperature.

REVIEW QUESTIONS

1. Refer to Figure 5-39. On a separate piece of paper, redraw, connect, and label the refrigerant lines.
2. Draw the refrigeration cycle. Label the major components and refrigerant lines.
3. Which components create a pressure change in the refrigeration cycle?
4. Which component rejects heat?
5. Which component is known as the heart of the system?
6. Which component absorbs heat from the air?
7. Which component uses cool refrigerant in order to condense moisture from the air?
8. Which component condenses the refrigerant in order to reject heat?
9. Which components add heat to the refrigerant?
10. Which component has hot, superheated vapor at its outlet?
11. Which component has warm, subcooled liquid at its outlet?
12. For R-22, what is the corresponding pressure for the boiling point at 35°F?
13. For R-410A, what is the corresponding boiling point temperature for 130 psig?
14. For R-22, what is the corresponding boiling point temperature for 300 psig?
15. For R-410A, what is the corresponding boiling point temperature for 400 psig?
16. What is superheat?
17. What is the superheat of an R-22 air conditioning system if the suction pressure is 70 psig and the suction line measures 50°F?
18. What is subcooling?
19. What is the subcooling of an R-410A air conditioning system if the liquid line pressure is 477 psig and the liquid line temperature is 110°F?
20. Draw the refrigeration cycle. Label all the components and the direction of refrigerant flow. Label the conditions of the refrigerant as it enters and leaves each component in the cycle.

CHAPTER 6
Scheduled Maintenance

LEARNING OBJECTIVES

The student will:
- Describe the difference between scheduled maintenance and service.
- Describe twelve different things to check on a scheduled maintenance check.
- Compare clean and dirty filter operation.
- Discuss the costs of not doing maintenance.
- Describe the effects of a lack of maintenance on the equipment and comfort.

INTRODUCTION

Scheduled maintenance has many names, such as maintenance and inspection, preventive maintenance, maintenance contracts, planned maintenance, clean and check, and other descriptions too numerous to mention. A scheduled maintenance program can be something simple like changing out filters and generally looking over the equipment for existing or potential problems, or it can also be planned to be so extensive that it may take several hours to complete. We will discuss the differences between scheduled maintenance programs, service contracts, the reasons for scheduled maintenance, and numerous examples of maintenance possibilities.

SCHEDULED MAINTENANCE IS NOT SERVICE

Scheduled maintenance work is not a service call. The equipment must be working properly before maintenance can be attempted. Maintenance is scheduled and the work to be accomplished is specified by an agreement or contract with the homeowner or equipment owner. Minor services or repairs may be completed at the time of the appointment, but major repairs may need to be scheduled with the dispatcher or supervisor. If your company is not too busy, a major repair could be completed on the same day as the scheduled maintenance.

Some companies handle scheduled maintenance as an activity that can be rescheduled because the priority of maintenance is lower than service and can wait a day or more. Maintenance can be handled by an individual or a small group trained in a checklist procedure. When servicing equipment that has not been maintained, it is likely that scheduled maintenance will need to be conducted after making a repair. Maybe no repair is actually required, but simply a good cleaning of filters and coils will get the system performing to specifications.

WHY DO SCHEDULED MAINTENANCE?

Maintenance is required for several reasons. First, the system will operate more efficiently if the coils and filters are clean, the motor bearings lubricated, and the system has proper airflow and is charged properly. More efficient operation translates into a lower electric bill.

Field Problem

The new technician has been doing scheduled maintenance work for several weeks. The technician was instructed on what to do and shown several times what was expected. The workday's appointments had three scheduled maintenance jobs to be completed. Usually it takes about an hour and a half to do the required inspection, filter change, oil the motors, and clean the coils. This day is a different story. The technician arrives at the customer's house and completes the routine scheduled maintenance activities. Before the technician leaves the job, he is required to turn on the system to make sure it operates because, when doing scheduled maintenance, something could get knocked loose and the system may not function. When the technician turns on the power and sets the thermostat to cool, he notices that the system is not cooling. The indoor and outdoor blowers are operating, but the refrigerant lines are both the same temperature. He touches the compressor and notices that it is not vibrating. The compressor seems to be off.

The technician calls the office, and another technician is immediately dispatched to verify the problem. It turns out that the compressor is locked up. The senior technician informs the customer of the problem, and the customer is upset because she thinks that her system was working fine prior to the technician's arrival.

The lesson learned from this is that the equipment should be operated and checked for normal operation prior to starting the maintenance work. The compressor was probably locked up before the technician arrived, but the customer did not know it. Do not get caught in a situation like this. It is important to check the full operation of the cooling or heating system prior to conducting general maintenance on equipment. It is equally important to make sure the system is working properly before leaving the job.

Second, the comfort of the customer will be improved. The appropriate temperature will be reached and the humidity level maintained without excessive run time, and/or appropriately heat and humidify during the winter season.

Third, the system will last longer and have fewer breakdowns. In some cases, a defective or potentially defective component is spotted and changed before failure. For example, a small refrigerant leak, an oil leak on a capacitor, or worn-out bearings on a motor can be scheduled for service before the equipment is totally shut down during the heat of the summer or the coldest day of the winter.

Fourth, scheduled maintenance assures the safety of the equipment, occupants, and structure. Improperly maintained equipment can create equipment fire hazards and, in the case of fossil-fuel heating equipment, carbon monoxide poisoning. Thermal and high- and low-pressure safeties should be checked to ensure that they are functioning properly.

Fifth, the manufacturer warranty requires maintenance.

Maintenance is good for you and your company because work can be scheduled during slow periods of the year and service opportunities can be generated if required. These visits allow the technician to get familiar with the system operation so that upon returning, the same technician knows the location and equipment. She will spend less time getting oriented to where everything is located and how to access it. The visit is also an opportunity for the technician to discuss upgrades and improvements with the customer. The customer will remember your company when major problems develop.

Figure 6-1 shows the results of one of several studies conducted on system problems. The top diagonal line represents a correctly installed system with the correct charge, proper airflow, properly sized ducts, and no duct leaks. The diagonal line below it represents a 10% undercharge. The SEER rating, which stands for Seasonal Energy Efficiency Ratio and is an efficiency rating for air conditioners similar to miles per gallon for cars, drops about 1.5, which is about a 15% drop. This means the system will cost 15% more to operate and that the system will operate 15% longer trying to cool the space. Following the diagonal lines down reflects reduced efficiency and increased operating cost and run time. This chart illustrates the reason it is important to do scheduled maintenance.

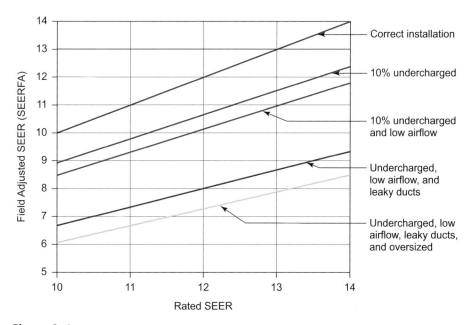

Figure 6-1
Impact of installation and maintenance factors on a fixed-orifice, split-system central air conditioner. The best efficiency and output is shown by the top diagonal line. The lower lines represent problems such as undercharged, low airflow, and leaky ducts. (Courtesy of Advanced Energy)

SCHEDULING MAINTENANCE AND INSPECTION WORK

Maintenance is planned work. Service work is troubleshooting and repair work and is not usually planned because equipment breakdowns are random. Your company's dispatcher or service supervisor schedules maintenance, normally when service work is slow in coming. For example, a good time to schedule air conditioning service work is in the spring or fall prior to the start of the peak heating or cooling season. In colder climates, this might be May, and in warmer climates, late March or April. Maintenance for commercial refrigeration should be scheduled at about the same time because HVACR companies do not want to miss service opportunities when maintenance and inspection can be set up during a slow period. Commercial refrigeration service work is year-round, while maintenance is scheduled in the spring or fall for residential and many commercial air conditioning customers.

Improvement Opportunities

Many technicians may not think of themselves as salespeople, but they really are. The technician is selling a service, and that service is either to repair the equipment or, in the case of scheduled maintenance, to make it more safe and efficient to operate. Doing your job correctly and efficiently sells the customer on your skills and on your company; in the eyes of the customer, you, the technician, are the company. It is a big responsibility to represent the company, but it is a reality.

A maintenance call is the best opportunity to make aftermarket sales. It is difficult to sell customers anything additional after handing them a repair service invoice. The maintenance contract is a soft sell. It can be an opportunity to suggest a filter upgrade or the installation of a programmable thermostat. The system may be old and inefficient, and the best time for a scheduled replacement should be mentioned to the customer. Too many options should not be offered because they will overwhelm the customer and divert her attention from the purpose of the visit. In addition, a hard sell will turn a customer away from your company and into the hands of the competition. Stating your recommendations on the invoice is a good way of reminding the customer of them. It plants a seed on what needs to be done in the future.

WHAT TO CHECK IN A SCHEDULED MAINTENANCE AND INSPECTION PROGRAM

As stated before, there are many maintenance options. Maintenance agreements are spelled out in clear and concise terms. Your company will have the maintenance requirements for the job. Residential and commercial agreements will differ as commercial agreements will most likely have more items to check, clean, or change and conditions of replacement or repair of systems.

The maintenance agreement will identify the name and address of the customer and important contact information. Many states want the companies' license numbers on the agreement as well. Some contractors offer customers an agreement, priority service, and invoice discounts when they call for service and repair.

Like a checklist, the agreement should have boxes to check once each part of the job has been completed. For example, if the contract requires changing filters, a box will be checked to indicate that task was completed and pressure and temperature readings are recorded in the appropriate spaces. The checklist helps ensure that the technician does not forget anything required in the agreement and assures the customer that the job was done correctly.

There is no right or wrong agreement as long as the customer agrees to the amount of work to be completed, the conditions of the agreement, and the technician completes the work stated on the contract. Some information on handling customer complaints follows, as do some examples of activities that could be included in a residential or commercial maintenance agreement.

Customer Complaints

Prior to starting the job, it is a good idea to review what you are going to do with the customer. Sometimes customers have expectations that are greater than the actual work you are going to complete. Maintenance agreements vary widely, and the customer may be comparing your work with what is done for a neighbor by a different company with a different maintenance plan.

The technician should ask the customer if she has had any problems with her system. The technician can try to identify the problem or refer it for scheduled service. This is not a time to conduct extensive troubleshooting. Asking a customer about problems is a good way to break the ice and show her that you are concerned about her equipment, safety, and comfort.

Outdoor and Indoor Conditions

During a maintenance job, the dry bulb air temperature entering the condenser coil is measured. *A dry bulb thermometer is a common measuring device used to measure temperature.* The temperature should be measured near the entrance of the condenser coil, as shown in Figure 6-2. The temperature measured at a distance away from the condenser may be different from that of the air entering at the condenser. For example, the condensing unit may be in the shade, so the temperature in the sun may be 10°F higher than at the condenser.

The two important indoor conditions are the dry bulb and wet bulb readings recorded by a psychrometer, as shown in Figure 6-3. The dry bulb temperature is important when calculating the temperature drop in the system. The temperature drop, or delta-T, is the difference between the return air temperature and the supply air temperature, as shown being measured in Figure 6-4 and Figure 6-5.

The wet bulb temperature will be useful if checking the charge of a system with a fixed metering device or determining the relative humidity of the conditioned space.

Air Filters

It is a good idea to check the air filter even if it is not a replacement item on the scheduled maintenance agreement. The primary purpose of the air filter, shown in

Figure 6-2
Measuring the temperature of air entering the condensing unit. Measure the air temperature without touching the coil or metal. (Photo by Joe Moravek)

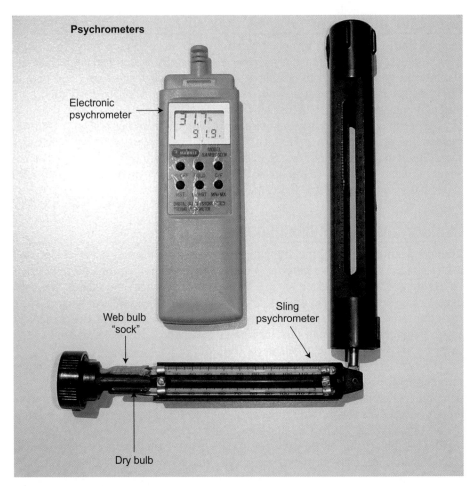

Figure 6-3
Digital and standard psychrometers used to measure dry bulb and wet bulb temperature. (Courtesy of Dick Wirz, Refrigeration Training Services)

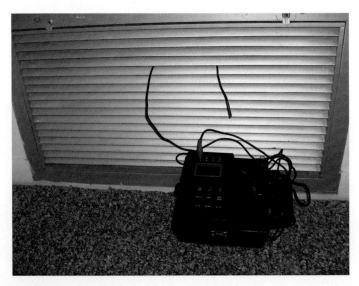

Figure 6-4
Measuring the dry bulb temperature at the return air grille. The best spot to measure this is near the air entering the coil. (Photo by Joe Moravek)

Figure 6–5
Measuring the supply air temperature. To find the differential temperature, subtract the return air temperature from the supply air temperature. (Photo by Joe Moravek)

Figure 6–6
Air filter placed in a filter grille. (Photo by Joe Moravek)

Figure 6–6, is to keep the evaporator coil clean. The secondary purpose is to reduce dust in the air. A dirty air filter will interfere with the airflow and cooling and heating operation. Watch for the installation of two air filters in series. For example, you might find one in a filter grille and one inside the air handler or furnace. Two filters in series are not recommended unless they are part of a designed filtration system such as a prefilter, found in some high-efficiency filtration packages. For filters in a grille, remove the filter and look into the ductwork returning to the air mover. Use a light to inspect for leaks or cracks. Duct leaks and dirt in the duct system should be reported to the building owner. Leaky ducts or ducts with a large dirt accumulation may need to be rescheduled as a service job. A certain amount of vacuuming dirt is expected on most jobs, so a portable, rechargeable vacuum cleaner will come in handy.

All air filters create a pressure drop when they are new. The high-efficiency filters create more of a pressure drop than the least efficient fiberglass filters. Filter efficiency is measured as a MERV rating, which stands for Minimum Efficiency Report Value. The

Figure 6-7
The arrow indicates the direction of the airflow through the filter. Installing the filter in the correct direction ensures the best dirt removal. (Photo by Joe Moravek)

higher the MERV rating a filter has, the greater its filtering ability. This rating reports a filter's ability to capture larger particles, between 0.3 and 10 microns. This value is helpful in comparing the performance of different filters. When replacing filters, it is important that the replacement filter to be of the same size and specifications.

Air filters are the least restrictive when they are new. As the filter collects dirt, the pressure drop increases across the filter, reducing airflow in the duct system. As filters collect dirt, they become more efficient because the surface dirt acts like an additional filter, but the reduced airflow is a detriment to system operations. The value of the increased filtering capacity of a dirty filter does not outweigh the lack of system airflow.

When recommending a filter upgrade, it is important to consider the reduction in airflow that may create system problems. This should be handled by a senior technician with experience in duct design issues. In some instances, changes to the return air system may be required.

As shown in Figure 6-7, most filters are directional and should be installed with the airflow arrow pointing into the return air duct system. In a commercial filter rack, the V-shape design is less restrictive to airflow and allows more filter surface area to trap more dirt before the filter needs replacing.

Cleaning Coils

Both indoor and outdoor coils should be inspected. In most cases, the condenser coil should be cleaned unless it has been flushed in the past few months. Figure 6-8 shows the foaming action of one type of condenser coil cleaner. The outdoor coil should be washed in the opposite direction of the airflow, thereby preventing dirt from being pushed and impacted into the middle of the coil.

Condenser coil cleaner may be different from evaporator coil cleaners. Condenser coil cleaner is to be used in the outdoor coil only and evaporator coil cleaner is to be used in the indoor coil. Do not use condenser coil cleaner on the indoor coil. Vapor from the wrong cleaner can be dispersed into the air or duct system, causing breathing problems for a technician in a tight attic, closet, or basement space, as well as affecting the occupants.

If you use condenser coil cleaning chemicals or soap, it is important that they be thoroughly washed away. Leaving these residues can create corrosion or chemical reactions that will deteriorate the coil. This is especially true where the coil touches the condenser base pan because it tends to collect rinse water. Be sure to familarize yourself with the Material Safety Data Sheet (MSDS) for the product you are using.

What to Check in a Scheduled Maintenance and Inspection Program   **147**

Figure 6-8
Foaming action of condenser coil cleaner forcing dirt and debris out of the fins. The foam must be thoroughly washed off with water to prevent coil damage. (Courtesy of Refrigeration Technologies)

After cleaning the condenser coil, it is important to allow the system to run long enough to dry the water off the coil before checking the charge; otherwise, the reading will be misleading. In most cases, the condenser will need to run 15–20 minutes in order to evaporate all the water trapped between the fin surfaces.

Figure 6-9 shows an evaporator coil cleaner. If the evaporator is excessively dirty, the main debris should be removed with a stiff brush prior to applying coil cleaner. Do not let the debris fall into the drain pan because it might block the drain line. Figure 6-10 illustrates what the evaporator should like after it is cleaned. Part of the evaporator coil check is ensuring that the coil drain pan is clean and free flowing. A wet-dry vacuum can be used to suck up rust, slime, and anything collecting in the drain pan and can be used to blow out the drain lines. After cleaning the coil, water should be run in the drain pan to check for good drainage.

Safety Tip

When using any type of chemical, it is important to follow the use directions. Coil cleaners are generally available for outdoor use or for indoor use. In some cases, the coil cleaner used to clean the outdoor coil creates vapors that are hazardous to humans or pets if used indoors. If you have any doubt, simply use a mild soap and wash out the residue prior to placing the equipment in service. Always familarize yourself with the Material Data Safety Sheet (MSDS) when using any chemicals on the job.

Figure 6-9
Evaporator coil cleaner is not as strong as condenser cleaner. Some evaporator cleaners do not need to be rinsed because the operation of the coil will automatically rinse the coil. (Courtesy of Refrigeration Technologies)

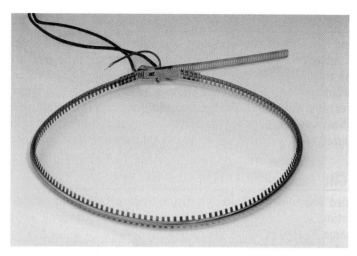

Figure 6-13a
This crankcase heater is wrapped around the bottom of the compresser shell. The heater keeps the oil warm and keeps refrigerant from condensing in the compressor. (Photo by Susan Brubaker)

Figure 6-13b
Check the wiring for burnt or charred connections. Replace the whole wire if damaged. Look for overheated coil and bulging or leaking capacitors. (Photo by Joe Moravek)

failures. It would be unfortunate for the customer and the company if the customer needed to call again for service a few days after the scheduled maintenance.

Emergency Drain Pan and Float Switch

The emergency drain pan, sometimes called the auxiliary drain pan, is installed under the evaporator coil housing to collect condensation overflow should the main drain line become plugged. It is required when an evaporator is installed in a location where an overflow condition would create structural water damage. The auxiliary drain pan must be piped separately and usually drips someplace where it will be visible to the building occupant. The drain pan should be dry and connected to drain piping. Some codes suspend the requirement for drain piping if the overflow pan has a float switch that shuts off the system if water collects in the pan. The float switch should be checked to ensure that the system will turn off if the water level rises in the drain pan.

Tech Tip

Emergency drain pans should be leak checked before they are installed. In order to do this, plug the drain pan fitting and fill the pan with water. The pan should hold the water without leaks or drips. If the pan leaks, repair or replace the drain pan before installing. If a drain pan test is done on an installed unit, be prepared to vacuum the water out with a wet vacuum should it leak.

Duct Air Leaks

Duct air leaks are difficult to detect. Air leaks should be checked when the system is operating. Feel for air leaks at the equipment, duct joints, and duct transitions. The most difficult duct leaks to detect are those in the return air trunk line. It is more difficult to detect air going into a duct system than air coming out as positive pressure leaks. Suspected return air duct leaks can be verified by placing a piece of paper over the suspected leak. If a leak exists, it will create a suction that may hold the paper to the duct. Also, "smoke puffers" are available that develop a synthetic and harmless cloud that allows the technician to check for visible positive or negative air leaks.

Check for rips in the duct insulation. Torn insulation will cause moisture to contact the cold duct surface and condense water vapor that can drip onto the building structure. Most air leaks are minor and can be repaired with approved duct tape or mastic, as shown in Figure 6-14. Extensive duct damage may need to be rescheduled for full replacement at a separate service call.

Figure 6-14
This illustrates all three common types of duct material. Starting at the top, you see sheetmetal duct with flexduct connecting the duct board. Approved mastic is used to seal duct joints after they are mechanically connected.

Thermostat Operation

Checking the thermostat operation can be the simplest part of the scheduled maintenance process. Measure the temperature near the thermostat with your temperature tester. Most thermostats have a thermometer on their display. Compare the reading of the thermometer temperature with your thermometer. Some mechanical thermostats have an adjustment screw for the thermometer readout. The temperature display on a digital thermostat may be adjustable as well. If the digital temperature display is out of calibration and not adjustable, little can be done except to notify the owner that the reading is either too high or too low by the amount of degrees difference and to offer to replace or upgrade the thermostat.

Lubricate and Clean Blowers and Fans

Some condenser and evaporator fan motors need periodic lubrication to keep the bearings from running dry and seizing. Lubrication is more common with commercial systems and less common on residential motors. The type of oil or grease used is important. Follow manufacturers' lubrication recommendations. Most motors have sealed bearings that do not require lubrication. Larger horsepower motors require grease for bearing lubrication.

Part of scheduled maintenance is checking and cleaning the blower or fan blade if required. A clean blower, like the one shown in Figure 6-15, moves more air than blades that are packed with dirt. Blowers with V-belts should have the proper tension so that the airflow will be correct. Overtightening the belts will cause shorter belt, pulley, and motor bearing life.

Although not a standard maintenance item, applying a special, weather-resistant lubricant on the motor shaft of a replacement motor, as shown in Figure 6-16, will make it easy to remove the blower or fan blade next time the motor needs replacement.

Measure Amperage and Voltage

The amperage of the compressor motor and blower motors should be measured and compared to recommended amp readings. For the single-phase compressor, the amperage on the wire connected to the common terminal is checked as shown in

Figure 6-15
A dirty blower wheel will need to be cleaned by removing the blower housing and scraping or brushing off the blades. Care should be taken not to dislodge balancing weights. A clean blower will move more air and attract less dirt to the surface. At this time, the blower wheel should be carefully inspected to check for loose blades. (Photo by Joe Moravek)

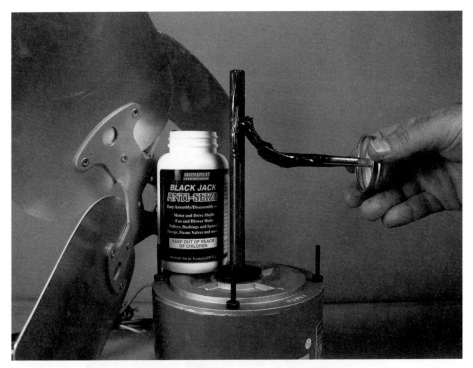

Figure 6-16
Applying a permanent shaft lubricant when replacing a blower wheel or fan blade will make the next removal easier and prevent rusting of the shaft. Consult the MSDS sheet when using lubricants. (Courtesy of Refrigeration Technologies)

the previous Figure 6-12. It some instances, it may be difficult to obtain this information. Some manufacturers print this information on the nameplate, and sometimes is it printed on the motor itself. Even without the recommended amp reading, the information can be used as a baseline for comparison from year to year or when doing future service.

The voltage should also be checked. Most condensing units like the one in Figure 6-17 should be within ±10% of the rated voltage. For example, if the condensing unit is rated for 230 volts, a range of 207 to 253 volts is acceptable. This is calculated as follows:

$$230 \text{ volts} \times 0.10 = \pm 23 \text{ volts}$$

$$230 \text{ volts} + 23 = 253 \text{ volts}$$

$$230 \text{ volts} - 23 = 207 \text{ volts}$$

A 205-volt reading in Figure 6-17 would be considered a reportable problem for a unit designed to operate on 230 volts.

Additional Comments

Many maintenance reports have a section for additional comments. This is an opportunity to make recommendations to the customer. It is also a spot to record problems and future service requirements. If these reports are reviewed prior to each of the scheduled maintenance activities, information can be transferred between technicians over a period of years. For example, if the access to the attic is located in a hard-to-find location, recording this information in the comments section could be helpful to other technicians.

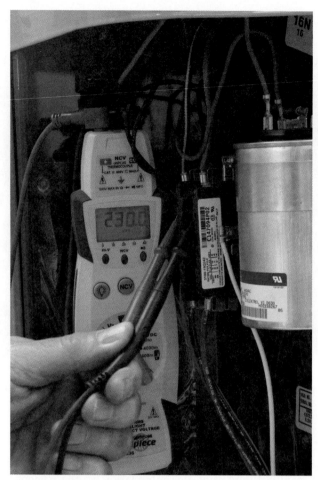

Figure 6-17
Measure the voltage with the system operating. The voltage reading should be within ±10% of the rated voltage requirements. (Photo by Susan Brubaker)

Go Beyond Expectations

Finally, try to do something that goes beyond the expectations of the customer. Accomplishing this should be strived for at each inspection and service activity. Going beyond expectations would include waxing the condensing unit. Remove trash around the condensing unit or around the furnace in the attic or basement.

The easiest way to go beyond expectations is to tell the customer what you are going to do before you start and review what you did after completing the work. Communicate with the customer, and your friendly attitude will be a way to stand out.

Review the Report with the Customer

Good customer service means reviewing the scheduled maintenance report with the customer. A short review will go a long way in continuing the customer's interest in your company's service.

MAINTENANCE CONTRACTS

Maintenance contracts vary widely from company to company. Maintenance contracts have different price schedules and different amounts of work requirements. You should not judge the contract but rather fulfill it completely.

TOOLS, EQUIPMENT, AND SUPPLIES REQUIREMENTS

The tools, equipment, and supplies will vary with the type of maintenance program the technician is working to complete. The technician will require the standard HVACR tools such as hand tools, manifold gauges, and an assortment of meters.

It is good policy to have more than one set of gauges in case one breaks or in case a job involves maintaining more than one system. Common meter stock should include a multimeter, clamp-on ammeter, multi-probe temperature tester, and psychrometer. Having a backup of each meter is advisable but not the norm until a technician has been in the field a few years. A wet-dry vacuum cleaner is a requirement when doing maintenance jobs, and a cordless vacuum is best for residential jobs. A 100-foot garden hose and sprayer will be required to clean coils, and a cordless drill speeds removal and reassembly of equipment panels.

General supplies required for maintenance include the following:

- Lubricants
- Coil cleaners
- Rags
- Trash bags
- Air filters

Many of these common stock items are found on standard service vehicles because one of the reasons a service call may be required is that the system had not been maintained and cleaned on a regular basis.

SUMMARY

Scheduled maintenance is not service work but a way of cleaning and checking the equipment to maintain peak performance, low operating cost, and long equipment life. In some cases, the technician will spot a problem and prevent unscheduled downtime by fixing it before it worsens.

Maintenance is a way to gain experience with no pressure to get a system back on line. Normally, the system will be working and the building cool. The equipment will be off while it is being maintained but usually not long enough for the temperature to get too high in the space. It is still important to have a plan to quickly do the job and move to the next one. In most jobs, it takes 10 or 15 minutes to unload and load tools, instruments, and supplies for the job. This can be a useful opportunity to see if the equipment is functioning properly. Greet the customer, brief her on what you are going to do, and adjust the thermostat to a cool setting to see if the system is operating properly. It will not hurt to drop the temperature in the building a degree or two lower than normal because the downtime will cause a rise in temperature. While the building is cool, gather what is needed and place supplies by the outdoor section and indoor sections.

The scheduled maintenance report should be legible and understandable for the customer and for the possible use of another technician who may visit the site. Print if your cursive writing is not legible. Many companies are switching to laptops to record and forward the information to the office, but it will be a few more years before this is the norm for everyone.

Maintenance is a good opportunity for you to learn about the equipment. It will also prove that you are a good technician worthy of added responsibility and other challenging jobs.

FIELD EXERCISES

1. Develop a maintenance list of the many items that can be used on a residential system.
2. Develop a maintenance list of the many items that can be used on a light commercial system.
3. This chapter covered many of the checks that can be covered in a scheduled maintenance program. What items not covered would you include on a maintenance program?
4. Make a detailed inventory list of the tools, equipment, and supplies you will need to complete a thorough maintenance check.

REVIEW QUESTIONS

1. Why is it important to check the air filter?
2. Why is it important to check the system pressures?
3. Why is it important to check for air leaks?
4. What are two reasons for scheduled maintenance?
5. What is meant by "exceeding the expectations of the customer"?
6. What device is used to measure the dry bulb and wet bulb temperatures?
7. What are important considerations when using coil-cleaning chemicals?
8. How are air filters rated? What is the size of the filtered particles in this rating?
9. What is the acceptable voltage range (percentage) measured at a condensing unit?
10. List ten important things to check in a scheduled maintenance program.

CHAPTER 7
Systematic Problem Solving

LEARNING OBJECTIVES

The student will:

- Describe the importance of communication between the customer, dispatcher, and technician.
- List ways the customer can give information on the problem.
- Discuss three quick checks to narrow down the troubleshooting process.
- Describe the electromechanical sequence of an HVAC system.
- List what instruments will be used for problem solving.
- Describe the diagnostic process of troubleshooting.
- Discuss the importance of productivity when it comes to keeping your job.
- Discuss how to communicate the problem with the customer.
- Describe how information in this chapter can be used in the field.

INTRODUCTION

This chapter will discuss the process of solving problems, otherwise called troubleshooting. In many cases, there is more than one way to solve an HVACR problem. The key of problem solving is to select the path that will create the quickest, most accurate results and not require a second service call.

Troubleshooting is learned through a good foundation of training in HVACR, including developing an understanding of electrical and mechanical sequences, airflow, and systems. A systematic approach to troubleshooting will be emphasized.

COMMUNICATION IS IMPORTANT

A strong communication link between the customer, the dispatcher, and the service technician is critical. The customer contacts your company and discusses a problem. The dispatcher contacts the technician, discusses the customer's problem, and gives an appointment time and location for a service call. The technician should write down, or print out if e-mail is used, the dispatcher's comments along with the address and directions.

Let's look at the process from the customer's viewpoint. First, the customer may not understand the problem. The customer believes there is a problem but may not be expressing it correctly. The customer will not always state the problem in technical terms. The dispatcher passes the customer's problem as stated, right or wrong, to the assigned technician.

It is important that the dispatcher discuss the information about the problem with the technician prior to assigning the job. From this discussion, the technician might be able to think about or list the symptoms and possible causes. Additionally, does the technician have the equipment needed for the job? A job may require an extension ladder for equipment access, but the assigned technician may not have the appropriate ladder on the vehicle. Discussing the job in advance gives the technician an opportunity to list the symptoms, to develop possible causes, and to collect tools and equipment needed for successful completion, or the technician may request that the job be assigned to someone else who has the necessary knowledge and equipment.

Once the technician arrives, it is important for him to once again ask the customer to discuss the problem and prior problems or service calls. The explanation of the problem will be useful, and having the customer express the problem is a good icebreaking technique. The important things to remember are that the customer perceives that there is a problem and that it is the technician's job to find out what the problem is and repair it. In rare instances, the problem is a result of operator error, and it is up to the technician to educate the customer on the system's operation. Listening to the customer allows venting of the concern and clues to the discovery of the problem. The customer may be angry or simply uncomfortable. Allowing the customer to vent relieves some tension for him, and the listening technician is seen as caring and empathetic with the ability to solve the problem. The technician is the HVACR doctor who needs to be sensitive to patient (customer) needs.

Obtain some history on the problem. Ask the customer if the problem has occurred before. If so, ask who repaired it and what they did, if known. If available, examine the previous invoice or bill that relates to this problem. The problem you are called to resolve may be unrelated to what is on the previous bill, or it may be a recurring problem not truly solved by the former service call.

To Obtain the Symptoms, Listen to the Customer

The customer can usually give the technician a little history on the equipment and express problems that need to be addressed. If there is an existing problem, the customer can offer symptoms such as poor cooling in the afternoon or that the unit runs too long or short-cycles.

In most instances, listening to the customer can give the technician information needed to help narrow down the list of possible problems (causes). For example, the customer may state that the "heating system blows warm air sometimes and cold air at other times." If it blows both warm air and cold air, it seems that the heating system is turning on and off. Judging from the discussion with the customer, there are several aspects to the problem:

- Power
- Blower moves air
- Thermostat works, calling for blower operation and heating
- System heats sometimes

Another example is the customer's complaint of no cooling. It is a split system. The customer states that the indoor section is not blowing air and notes that the outdoor unit comes on and goes off on a regular basis, about every 4 or 5 minutes. The dispatcher asked the customer to turn the system off at the thermostat until the technician arrived. This recommendation reduces possible system damage due to short-cycling of the condensing unit. When the technician arrives, the technician turns the thermostat to the cooling mode and notices no indoor air movement but the condensing unit starts. The technician suspects an indoor blower problem and therefore turns the cooling system off and switches the thermostat to "fan on" to continue troubleshooting the airflow problem.

By verifying the customer's observations, the technician has narrowed the problem to the following possible causes:

- Blower control section of the thermostat
- Blower motor
- Blower motor control components

POSSIBLE CAUSES

After determining the symptoms, list all the possible causes of the problem. Listing the symptoms provides the technician with a means of eliminating many possible causes before beginning the troubleshooting process. Possible causes include mechanical, electrical, or airflow. It could also be a combination of these problems. The next section gives the technician logical steps to troubleshoot a system.

Tech Tip

Narrowing down the problem is part of systematic troubleshooting. Prior to breaking out tools, mentally or on paper determine the symptoms and then list all the possible causes. For example, on a "no cooling call," the technician makes certain that the blower is operating continuously and finds that the condensing unit is short-cycling on and off. The following is a possible problem list:

1. The problem is not the indoor blower section.
2. The problem could be electrical. The condenser fan is not operating and compressor is cycling on the high-pressure control. Compressor is short-cycling on overload protection. A mechanical thermostat could be rocking between cooling and off.
3. Check for causes of cycling on an overload such as a bad run capacitor, bad connections, internal overload, etc.
4. The problem could be mechanical. If someone worked on it recently, it could be the charge. Is it over- or undercharged? The compressor can be cycling on high- or low-pressure switch.
5. Not likely, but the compressor could be defective by drawing excessive amperage and cutting out on internal overload.

Finding Opens in the Control Sequence

Figure 7-1 represents a simple series line voltage control circuit. This is a series circuit with voltage applied to a light through control components: switch, fuse, and thermostat. If one of the control components opens, then the circuit is broken and the light will go out.

To locate the open component, use the voltmeter in parallel with the load method of troubleshooting. This method systematically steps through the control circuit, searching for a loss of voltage that would indicate an open control component. Figure 7-1 demonstrates this process. Start by attaching one voltmeter lead to the right side of the light. Next, measure the 120 volts above the switch. Measure the input voltage at the bottom of the switch and on both sides of the fuse and thermostat. At the left side, 120 volts should be present. After this sequence of checks, if the light is not operating with 120 volts applied, check the light bulb or light bulb socket.

A more complex circuit is shown in Figure 7-2, but the sequence is the same. For this example, assume the low-pressure switch (LPS) is open due to a lack of refrigerant charge. Attach one probe of the voltmeter to the right side of the line voltage circuit L2. This is most likely a 240-volt circuit. With the other meter probe, 240 volts will be measured at the top of the L1 and on both sides of the control relay and the high-pressure switch, and 240 volts will be measured on the left side of the LPS and zero on the right side of this pressure switch. The open LPS will break the power to the contactor coil (C), keeping the two-pole contactor open and not allowing the compressor to operate. The final check in the process is to measure across the suspected open component. The supplied voltage will be measured across an open switch such as a set of contacts, thermostat, or pressure switch. In this case, we stated that the LPS was open due to a low charge; therefore, 240 volts will be measured across the open LPS.

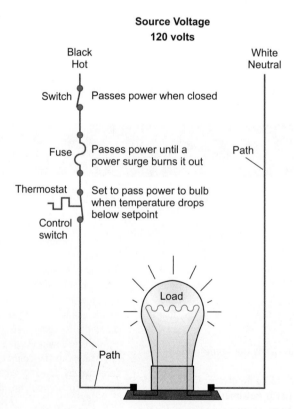

Figure 7-1
Simple series electrical circuit that is complete when the switch, fuse, and thermostat are closed to bring voltage to the light.

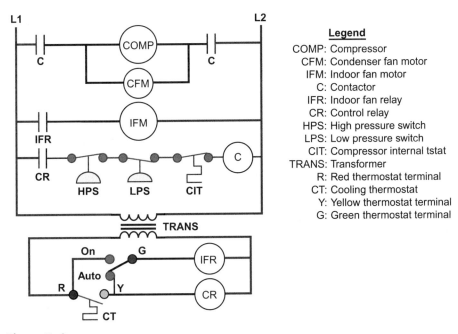

Figure 7-2
This is a functional diagram—a package unit schematic with legend. (Courtesy of Dick Wirz, Refrigeration Training Services)

Is It a Mechanical, an Electrical, or an Airflow Problem?

After taking steps to isolate the symptoms, the technician needs to determine common solutions that relate to mechanical, electrical, or airflow problems.

Diagnosing mechanical problems includes checking some of the following areas:

- Undercharged or overcharged condition
- Locked up compressor or fan motors
- Indoor or outdoor blower hitting or bound against motor housing or brackets

A visual check of the electrical wiring and components is the recommended first step. An open fuse, burned wire or electrical component, or bulging capacitor may be found in a visual inspection. This may be the problem or the symptom of a problem.

After the visual check, mechanical problems can quickly be checked by hooking up the gauges and reviewing the high- and low-side pressures. A pressure difference between the high side and low side will show that the compressor is working, creating a refrigerant flow. The final charge will be later determined when the superheat and subcooling have been checked. Next, see if the fans are turning. If they are not, are they hitting or bound against the housing bracket? In a multifan condensing unit, one fan can be turning backward, appearing that it is operating. Is there a seized or dragging motor bearing? If the fan is not working, turn off the power and turn the blower by hand to see if it moves freely. If the problem does not fall into one of these categories, it may be an electrical problem.

Electrical problems include the following areas to check:

- Improper voltage to the indoor and outdoor sections
- Control voltage, usually 24 volts
- Thermostat operation
- Components that operate the motors such as a relay, contactor, or capacitor

The first electrical check is to measure the input voltage to the indoor and outdoor sections. Next, check power to the primary and secondary side of the transformer. Loss of any of these voltages will prevent the cooling system from operating. The electric or gas heating systems need voltage only to the indoor section and control transformer to operate.

Airflow problems include the following things to check:

- Air movement from all supply registers and into return grilles
- Clean filter
- Indoor and outdoor blower moving air at rated airflow
- Cool or warm air coming from the registers

This is a preliminary inspection—checking the airflow is a simple part of the airflow check. Using airflow instruments to measure the exact airflow is the best way to determine the airflow, but these instruments are not used unless an airflow problem is suspected.

Problems in new cooling installations or system replacements are generally charging or airflow related. If a new installation has both of these problems, the airflow should be corrected first.

Electromechanical Sequence

It is important to understand the basic electromechanical sequence of an HVAC system. Without a fundamental understanding of the sequence of operation, it will be difficult to list the symptoms and possible causes and determine whether there is a problem or what the problem is. Next is a review of this sequence.

Simple Sequence of Operation

In a call for cooling, the start sequence has the thermostat energizing the indoor blower and condensing unit at the same time. A system with an indoor and outdoor section is identified as a split system, as illustrated in Figure 7-3. A package unit contains both the indoor and outdoor sections in one housing, as shown in Figure 7-4. All components operate simultaneously:

- Indoor blower
- Condenser fan
- Compressor

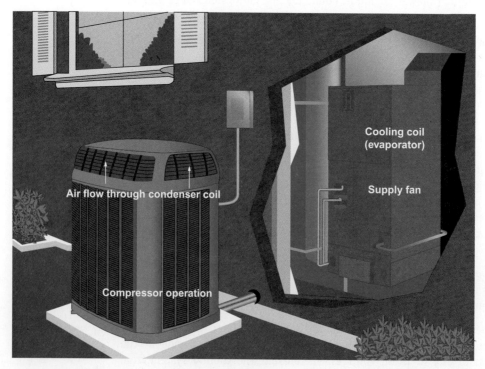

Figure 7-3
Checking operation of outdoor section by feeling for warm air and compressor operation. (Courtesy of Trane)

Figure 7-4
A package unit contains a complete air conditioning system minus the ductwork and thermostat. (Photo by Susan Brubaker)

Basic power diagrams are shown in Figure 7-5 and Figure 7-6. On a split system, there are two separate required voltage sources. This is important to remember because both power supplies are required to operate a split system. Turning off power to the outdoor section, such as pulling out the disconnect in Figure 7-7, will not deenergize the indoor section. The indoor section will be electrically "hot," and a technician should take precautions not to get shocked.

Most package units have one power supply, as shown previously in Figure 7-4, and it is split into different voltages inside the unit if necessary.

Figure 7-8 represents a simple heating-cooling diagram. The supply voltage is 240 volts, and the control voltage is 24 volts. The line voltage section includes the compressor, heating source, and indoor fan motor. The control voltage section includes the thermostat, heating relay coil, contactor coil, indoor fan relay coil, and three pressure-temperature safeties.

In order for the cooling circuit to operate, the thermostat completes a circuit from R to Y and R to G. The R to Y circuit energizes the contact coil C, which closes the two C contacts in the line voltage section starting the compressor.

When the R to G circuit is complete, it energizes the indoor fan relay coil and closes the indoor fan relay, starting the indoor fan motor.

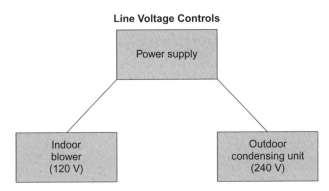

Figure 7-5
In a split system, two separate power supplies and disconnects are required. In some installations the two sections will be the same voltage or a different voltage. Turning off the power to the indoor blower will not kill the power to the condenser; therefore, care must be taken when working around the outdoor section that seems to be deenergized but is not.

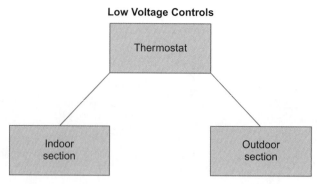

Figure 7-6
The indoor section normally provides low voltage, 24 volts, through the thermostat to the outdoor condensing unit. The thermostat is the switch that controls the indoor blower and condensing unit.

Figure 7-7
The condensing unit's fused disconnect is removed. Power may still be present at the indoor section. It is a good practice to check for voltage even if the power has been removed. (Courtesy of Dick Wirz, Refrigeration Training Services)

The heating circuit operates when the thermostat completes the circuit from R to W, energizing the heating relay coil, closing HR contacts, and starting the heating source. As the heat exchanger warms up, the thermal fan switch closes, starting the indoor fan motor. The circuit is used as an example and does not include the condenser fan circuit, disconnect, circuit protection, and capacitors usually associated with motors.

Required Tools, Instrumentation, and Supplies

The technician will be required to carry standard hand tools including wrenches, screwdrivers, nut drivers, pliers, wire cutter/stripper, hammer, and hex wrenches. Other hand tools will be required depending on the job.

Instruments include the refrigeration manifold gauge set, multimeter, and clamp-on ammeter. Many technicians carry equipment that can be used for repairs

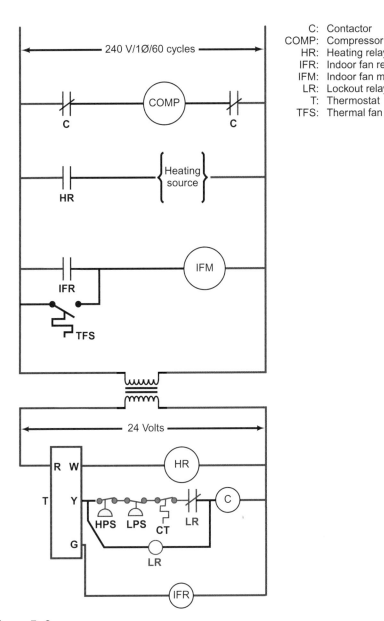

Figure 7–8
Simple cooling and heating diagram. The upper section is the 240-volt input. The lower section is the 24-volt control section.

such as a brazing rig, vacuum pump, micron gauge, digital scale, recovery unit, and recovery cylinders.

Supplies include coil cleansers, sprayer, duct tape, electrical tape, wire, and electrical accessories. Some technicians carry a small stock of common replacement items like capacitors, relays, and contactors.

Tech Tip

Tools are used following the collection of symptoms and listing of possible causes. The voltmeter is the most valuable for troubleshooting, and the ohmmeter is used for isolating the problem to determine what is defective. Installation of gauges is only done after all electrical, mechanical, and airflow problems have been addressed.

Safety Tip

With the variety of refrigerants in today's equipment, the technician needs two or three different sets of manifold gauges. The technician should be careful about which set of gauges are installed on what equipment. Accidentally installing an R-22 gauge set on an R-410A system can create a dangerous situation. R-410A operates at approximately 50% higher pressure than R-22. In extreme cases, when hooked to an R-410A system, the R-22 gauges could go beyond their pressure limits, damaging the gauge and causing it to blow out. The R-22 hoses are also rated for a lower burst pressure than those for R-410A. Finally, it is not a good practice to mix gauge sets between refrigerants because trapped oil can drain from the hose and into the system. R-22 and R-410A use different refrigerant oils and should be kept separate.

Field Problem

After completing the first job of the day, the technician called the dispatcher for the next service call. The technician was told that a customer's air conditioner was not cooling. When the technician arrived on the job, the customer said that the system stopped cooling in the middle of the night and the temperature inside was rising. The customer stated that the indoor section was blowing air and the outdoor section seemed to be working also. The technician observed that it was a split system and checked the thermostat setting. There was air flow from several checked registers, and the return air grille was returning air. Removing the grille revealed that the air filter was clean.

The technician went outside and listened for the condenser fan operation. The technician was not sure that the compressor was operating. The suction and liquid lines were the same temperature, and that was close to the ambient temperature. This indicated that the compressor was not operating.

A quick inspection of the electrical components revealed that a wire was burned off the run capacitor, preventing the compressor operation. This may have been caused by a loose connection. The complete wire from the contactor to the capacitor was replaced, and the compressor turned on. The technician verified that the amperage on the compressor was within range and the refrigerant charge was correct. The technician was not sure what caused the problem, but the systematic sequence of troubleshooting helped to find and correct the problem.

In summary, the technician received instructions from the dispatcher, listened to the customer, and gained more information on the problem. Next, the technician took general steps to see what was working. The thermostat, indoor air flow, and air filter were first checked. The outdoor blower was operating, and it seemed that the compressor was off with both refrigerant lines being the same temperature. The technician then observed that a wire was burned off the run capacitor and was able to repair the problem.

DIAGNOSIS

When you go to a doctor with a problem, you receive a diagnosis and possible treatment.

A diagnosis determines what is causing the problem, as in troubleshooting. The solution follows the diagnosis. When determining the solution, the technician should explain to the customer what is going to be done to correct the problem. The customer should approve the repair prior to its being started. On costly repairs, the customer may wish to get a second option from another technician at your company or another company. The technician should not be offended if a second option is requested. Remember, you may be asked to give a second opinion on another date.

PRODUCTIVITY

In order for companies to survive in today's market, they must be able to produce quality service or product at a reasonable profit. For that, they need employees who take pride in their work and promote the company in a positive manner. In many cases, the technician will represent the face of the company, and your actions will be the basis of the customer's opinion of the company. If you show up when expected and present yourself as a well-organized professional, the customer will be more trusting and view your professionalism as an added value. If you cannot be on time, call ahead and inform the customer of the delay. This shows respect for the customer's time.

A company with a good reputation does not have to have the lowest prices. People are willing to pay a little extra for a reputable company with a history of fair dealings. In fact, lowering bid prices to obtain work or the promise of future work is probably the worst course of action. Satisfied customers will tell a few of their friends and family members about the good service they received, but unhappy customers will tell everyone about a bad experience.

As a valued employee, you must perform your work as quickly and professionally as possible. The best way to do this is to preplan your work when possible. After you have listed the symptoms, listed possible causes, and used systematic troubleshooting to identify the problem, think about the job and list the steps necessary to complete it. Then select the tools needed to get the job done correctly and as quickly as possible. Systematic troubleshooting and preplanning are steps to increasing your productivity.

EVALUATION

"Evaluation" in this instance means communicating what was wrong to the customer. The sequence of troubleshooting is as follows:

1. Listening to the dispatcher and customer
2. Identifying symptoms and possible causes and discovering what the problem is
3. Notifying the customer of the problem and obtaining authorization for the repair
4. Reviewing with the customer the work conducted and a summary of the invoice bill, and as a courtesy, giving the customer an option to keep the defective parts

This evaluation is the final time the technician has to make an impression on the customer. The evaluation period is short but will carry a lasting impact on your service.

CASE STUDIES

This section discusses various problems and solutions the technician may experience in the field. At this point in your career, you are not expected to be able to solve all of these problems, but you should be able to develop a sequence to narrow down what is working and what is not. These case studies will discuss a systematic method of troubleshooting:

1. Listen, ask questions, and absorb as much information as possible from your dispatcher and the customer. At the job site, ask questions relative to what the problem is, what the system has been doing, and what may or may not have been done previous to this service call.
2. Survey the system as to equipment, installation, thermostat, and power supply(ies); read the wiring schematics if necessary; and list all **symptoms** of the problem.
3. What are the possible causes of the symptoms you have listed? List the **possible causes**.
4. With visual inspection, use the wiring schematic(s) and instrumentation (voltmeter, ohmmeter, velometer, leak detector, gauges/instruments, and necessary hand tools) to troubleshoot the system and repair or correct as needed.

Case Study One

For the first job of the day, the technician goes to a commercial assignment at the snack bar of a golf course. The air conditioning system was reported as not working, but no more details were gathered at the time the call was given to the dispatcher. When the technician arrived on the job, the snack bar manager stated that the system seemed to lose its cooling ability late in the afternoon. The snack bar is stuffy and humid, but the temperature is not overly warm because it is still cool outside.

Surveying the system, the technician locates the condensing unit outside and the indoor air handler in the mechanical room of this typical split air conditioning system. Locating the thermostat, he checks the setting and lowers the set point to ensure that the unit is calling for cooling. He feels for airflow in the return and supply grilles and determines that there is good airflow. When he checks to make sure the outdoor condensing unit is operating and rejecting heat, he finds that it has airflow but is not rejecting very much heat, so he is not sure if the compressor is operating or not.

Symptoms:

- Insufficient cooling
- Indoor system seems to be operating normally
- Condenser fan is operating on a call for cooling, but the compressor may or may not be

Possible causes:

- Compressor is not running or is short-cycling
- Refrigerant charge or restriction
- Inefficient (defective) compressor

Troubleshooting:

The technician removes the condensing unit panel and reviews the wiring schematic. By using his clamp-on ammeter, he confirms that the compressor is not running. With his voltmeter, he locates the problem: the compressor is not operating because the system has apparently been cycling on the low-pressure switch, shown in the previous Figure 7–2, which keeps the compressor from operating when low on refrigerant. This low-pressure safety is protection for the compressor. The technician installs his gauges to the high and low side of the system at the Schrader valves and confirms that the system does not have sufficient pressure and appears to be undercharged.

He shuts the system down and begins a visual inspection for an obvious leak. At the liquid line exiting the condensing unit housing, he notices oil on the liquid line. With his leak detector and soap bubbles, he finds that the leak is caused by the liquid line rubbing against the housing of the unit. The refrigerant is recovered, and the leak is repaired. The system is pressurized with nitrogen to be certain that no other leaks exist. This is a good practice because other minor leaks might be located and repaired.

No additional leaks were found, so the nitrogen pressure was released and a new liquid line filter drier installed. The drier was also leak checked after it was installed, and the system was evacuated and charged. System checks of the pressures, amp draw, superheat, and subcooling confirmed that the unit was operating correctly. The customer was satisfied because the job was completed before the lunch hour rush of hungry golfers.

Case Study Two

There is no cooling in a residential structure. The dispatcher stated that the customer called this morning and stated that the unit was not cooling and that he could not feel air coming from the duct system. The condensing unit was reported off.

Driving toward the service call, the technician starting thinking of what could cause the problem. Nothing is working. Usually that is an easy problem. Is it the thermostat? Is it a lack of voltage common to all parts of the system? Arriving at the job, the technician is told the same information as stated by the dispatcher.

Safety Tip

Each year in the United States, accidents involving ladders cause about 300 deaths and 130,000 injuries requiring emergency medical attention. Ladder accidents are usually caused by improper selection, care, or use, not by manufacturing defects. Common accidents include falls and electrical shock. Think about which ladder to use and how to use it safely.

Tech Tip

If you suspect an electrical problem, a quick-check method is to check the voltage supplied to the section that is not working and ohm out fuses (for infinity resistance reading) or reset breakers. A tripped circuit breaker can be reset by turning the breaker to the "off" position with force and then turning it to the "on" position. Always step to the side and look away when resetting a breaker or closing a disconnect. Wearing a glove is also suggested.

Checking the thermostat, lowering the set point, and moving the fan switch to on results in no change. Nothing operates.

Symptoms:

- Nothing comes on; no supply voltage to the indoor or outdoor section
- No 24-volt control voltage

Possible Causes:

- 24-volt transformer not producing the 24 volts; blown fuse or broken wire.
- No voltage to unit; blown fuse or circuit breaker

Troubleshooting:

With his voltmeter and hand tools, the technician checks the indoor unit first. This seems like a logical place to start because nothing is operating. Reviewing the wiring schematic on the unit cover, he starts with measuring voltage and finds 122 volts on the power coming into the unit and 122 volts on the primary of the transformer. Figure 7–9 shows a common transformer with primary and secondary leads. The secondary measured 24 volts, but the secondary voltage after the fuse measured zero volts. The fuse was checked, visually and with an ohmmeter, confirming that it was open and had infinite resistance (ohms).

The technician replaced the fuse, and everything began to operate as normal. The technician then checked the amperage draw of the line and low-voltage circuit. He then checked the evaporator differential temperature and airflow across the indoor coil. The system was cycled off and on several times. No problem was found, and the technician explained the solution to the customer. Sometimes fuses blow due to a fluctuation in voltage and sometimes due to an overloaded or short circuit. Because the system functioned properly after the fuse was changed, the technician told the owner that the fuse was open due to random and unknown reasons.

Case Study Three

There is no cooling from a commercial package unit, as seen in Figure 7–10. The dispatcher sent the technician to a city recreation center with a general complaint of no cooling from the recreation headquarters. One section of the building is warm.

Upon arriving at the site, the technician met the assistant manager, and she directed him to the area that was not cooling and turned the thermostat to cool. The technician could hear the blower come on, and air was blowing from the supply grille

Figure 7–9
Common transformer found in furnaces and air handlers. The yellow leads on the right are the 24-volt secondary. The three leads on the left are the primary or input voltage. Only two of three wires will be used. This is called a multitap transformer with the option of two possible voltages. In this case the input voltage would be 208 or 240 volts. (Courtesy of Bill Johnson)

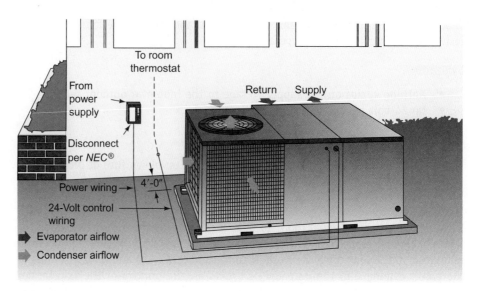

Figure 7–10
Package unit with duct work into the side of the building.

and returning through the duct system. The assistant manager said the package unit was on the south side of the building. After locating it, the technician notices that the condenser fan is operating but that the condenser is not rejecting heat.

Symptoms:

- No cooling
- Blower section seems to operate normally
- Condenser fan is operating, but no heat appears to be rejected at condenser

Possible causes:

- Compressor is not operating or is short-cycling.
- Mechanical or electrical failure with compressor
- Mechanical system problems related to refrigerant charge, restrictions, etc.

Troubleshooting:
The technician removes the package unit control and compressor panels, and feeling that the compressor is the same temperature as the ambient with no vibration, he determines that the compressor is not operating. The voltage at the disconnect and compressor contactor showed 240 volts. The technician disconnects the power to the unit and measures the resistance of the compressor motor windings from the contactor connection. The compressor is a three-phase motor, and the resistance reading on the three windings should be the same ohm reading. Two of the three wires did not measure resistance when checked from the contactor. What is the problem and the solution to this problem?

The technician diagnosed open compressor windings. He called the dispatcher and arranged for a replacement compressor and a crane to lift the replacement compressor over a high wall and remove the defective one. The technician installed the new compressor and new liquid line drier, leak checked the system, pulled a vacuum, and charged the system. When he turned on the system, the replacement compressor did not operate. The technician thought, "Not again. A defective compressor from the manufacturer!" Checking the resistance at the contactor revealed the same problem—open windings.

The technician called the supervisor, who discussed the troubleshooting procedure over the phone. The supervisor suggested checking the compressor winding at the compressor terminal, as illustrated in Figure 7–11. Doing so revealed that the

Figure 7-11
Terminals of a three-phase compressor. Resistance of all windings should be the same. (Photo by Susan Brubaker)

Tech Tip

Several clues will help the technician determine that a system is not charged properly without hooking up the gauges. The touch test should find the suction line cool and possibly condensing moisture if the humidity in the air is high. The discharge line should be hot. The hot discharge line will cause a burn, so to test it, tap it quickly with one wet finger. The liquid-line temperature will be near the ambient temperature. The amperage draw on the compressor should be no lower than 50% of normal rated amps. Next, go inside the building. Cold air should be coming from the register. Extremely cold or warm air could be a charging or airflow problem. To properly check the charge, the technician must hook up gauges and check pressures, superheat, and subcooling.

compressor was OK and that the windings were not open as previously thought. What is the new problem? What do you think?

After a more thorough check, the technician found that two of the wires from the contactor that fed power to the compressor had internal breaks and that the open wiring was preventing voltage from reaching the compressor terminals. The lesson from this service call is that motor resistance checks should be conducted at the closest point or at the terminal connections of the motor. The connecting wire may be open or strands burnt even if they appear OK.

Case Study Four

The customer relates to the dispatcher that there is poor cooling performance in the afternoon and that the house has a dust issue. The customer stated that her air conditioner cools well until about 3 or 4 p.m., and her housekeeper stated that there seemed to be more dust in the house in the past few months. The technician was not sure if the problems were related.

Arriving to the morning appointment at 8 a.m., the service technician conferred with the homeowner about the problem of cooling and dust. The technician began to check out the system by starting the cooling operation. The condenser coil seemed slightly dirty and would be cleaned before leaving, but the system was working. The liquid and suction line temperatures were normal, and the differential condenser air temperature appeared good. The air handler is located in the attic.

Symptoms:

- Insufficient cooling during the heat of the day
- Home appears to have more dust

Possible Causes:

- System or mechanical problems related to refrigerant, restrictions, or compressor efficiency.
- Insufficient airflow
- Compromised ductwork
- Dirty filters and or coils

Troubleshooting:

The technician shuts the unit down at the disconnect to check the air filters and coils for cleanliness. Because the condenser coil was dirty, he cleans the condenser and inspects the fan. The technician goes to the attic to check the air filter and changes it. The evaporator coil looks dirty, so he cleans it. While in the attic, the technician continues to investigate the possible duct air leakage problem. He begins to inspect the ductwork again and looks up the return air plenum. The duct leak is now obvious. The leak is on the side of the return air duct system opposite the technician's view when entering the attic. He tapes and seals the duct leak with the approved UL 181 sealing system and inspects the duct system for other leaks. He restarts the system, and the performance checks out fine.

The technician determined that the system performance was being hindered by pulling in hot attic air in the afternoon. The evaporator coil was dirty because the evaporator was ahead of the air filter, and the housekeeper's dust problem could be related to the duct leakage, which pulls in dusty attic air.

Before leaving the job, the technician discusses the solution with the homeowner. The technician also suggests a maintenance agreement and an upgrade to a better air filtration system. The homeowner seems interested in both options.

Case Study Five

When the technician reached the convenience store, the clerk restated that the air conditioning system that cooled the rear of the store was not working. She also said that air was not blowing from the ducts.

The technician went to the thermostat and verified that the system was in the cooling mode and that the thermostat was set several degrees below the room temperature. The technician also verified that there was no airflow from the supply registers. Setting the thermostat fan switch to the on position made no difference in the fan operation. The clerk notified the technician that the unit was on the roof. The technician climbed an interior ladder to the roof and found a package unit that services the problem area. The unit was silent.

Symptoms:

- Nothing is running; no supply voltage to the package unit
- No 24-volt control voltage

Possible Causes:

- 24-volt transformer not producing the 24 volts, bad transformer, blown fuse, or broken wire
- No voltage to unit; blown fuse or circuit breaker

Troubleshooting:

The technician went to the roof with her hand tools and volt/ohm/ammeter, opened the panel of the unit, and did not see any visual damage to the wiring or component parts. With the voltmeter, the technician found that there was no voltage being supplied to the system. She went downstairs and reset the tripped circuit breaker in the panel box. The breaker immediately tripped. A tripped breaker is shown in Figure 7–12. There appeared to be a short to ground causing the breaker to trip. How will the technician troubleshoot the unit if voltage cannot be supplied to it? Locating short-circuits can be a challenge.

Locate and solve the problem: One way to find a short circuit is to individually check the ohms of the major components and connecting wires of the following:

- Compressor
- Condenser fan motor
- Evaporator fan motor
- Control transformer
- Capacitors

Low resistance across any of these components will cause high-current draw and trip the breaker or blow a fuse. The technician disconnected the wires at the compressor contactor and, with the ohmmeter, found normal resistance of the start and run winding and no short to ground. She disconnected the condenser fan wires from the contactor and found that there was zero ohms of resistance on the winding as well as zero ohms of resistance from line to ground. The condenser fan motor had failed with a short to ground, and this is what caused the circuit breaker to trip. The condenser fan and capacitor were changed out, and the wiring was reconnected

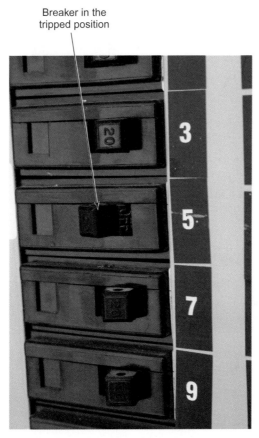

Figure 7–12
Tripped breaker caused by a shorted component. After resetting the breaker, it immediately tripped again because the short was not fixed.

and secured. The unit started, had normal amp draw, and began cooling the space. Problem solved, customer happy!

A short-circuit does not always result in zero ohms. A low resistance reading to ground or across a set of windings causes excessive current flow and opens the overcurrent protection. Figure 7-13 shows an example of a compressor motor winding with a 100 Ω resistance to ground. Motor resistance to ground should be infinite or a minimum of 1,000 ohms per applied volt.

Case Study Six

Case study six is an electric heating problem. Following the systematic method of troubleshooting, the technician will listen to the customer, list the symptoms, determine possible causes, and methodically troubleshoot and locate the problem. The customer states that the unit appears to heat OK during mild weather, but it runs continuously when the outdoor temperature starts getting below 10°F and won't maintain space temperature. The customer states that he normally keeps the thermostat set at 72 degrees, but during today's conditions (3°F outdoor ambient temp), he is only able to keep the space at 61°F and the unit runs continuously.

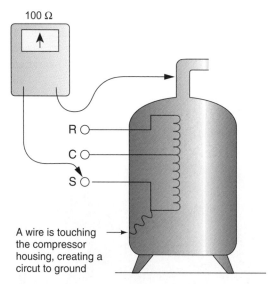

Figure 7-13
Low resistance to ground leads to excessive current draw that may trip the breaker or blow the fuse. Resistance from the motor winding to ground should be infinite or very high.

Tech Tip

Use the ohmmeter only when the power is disconnected and turned off. Remove the component from the circuit and then check it with the ohmmeter. Checking the resistance of a component wired in the circuit may read ohms through a different component wired in the same circuit. For example, if the circuit has two motors wired in parallel, the ohmmeter will read resistance through one motor winding even if the other motor has an open winding. When two loads are in a parallel circuit, the measured resistance will be much less than the resistance in the lower resistance load. This can be confusing, so it is best to disconnect the wire prior to measuring.

Symptoms:

- Electric furnace will not keep up during colder weather.
- Insufficient capacity.

Possible Causes:

- Undersized system for current load conditions
- Another condition causing increased load on the system such as open ductwork in attic or crawlspace
- System not running at full rated capacity
- Failed component

Troubleshooting:

The technician makes a quick check of the thermostat to verify that the setpoint is higher than the room temperature. He also checks to make sure that the thermostat is mounted in a good location, not in direct sunlight or near other sources of heat (for example, computers, copy machines, lamps, etc.). The thermostat looks normal, mounted on an inside wall near the cold air return. He then puts his hand in front of the baseboard supply diffuser and feels what appears to be normal airflow (see Figure 7–14), but it is only slightly warm. Next, the technician moves on to the electric furnace, which in this instance is located in the basement. He makes a quick visual inspection for open ductwork and missing panels, and all appears to be normal. He checks the unit rating plate and finds that there are several ratings based on what heater options are installed in the unit. Next, he shuts the unit disconnect off and removes the heater section panel to see what heaters are installed. He also checks the filter and blower wheel and makes a quick inspection for loose, discolored, or burned wiring. He determines that there are three 5 KW heating elements installed in this unit and marks the appropriate option on the unit rating tag. He then restores the unit to service and checks the individual heating elements using a clamp-on ammeter, and he determines that only two of the three heating elements are drawing current.

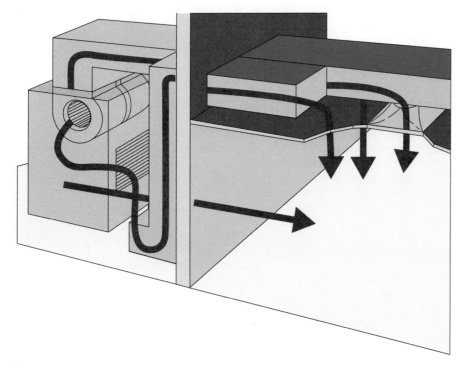

Figure 7–14
Checking the supply and return airflow and temperature difference. The temperature difference between the return and supply should be about 15°F to 20°F. (Courtesy of Trane)

Next, the technician thinks of the different things that might cause the third heater to not be working:

- An open circuit in the heating element
- An open circuit caused by limit switch
- An open circuit in the power supply to this heater (e.g., breaker or fuse open)
- An open in the control circuit (which could include thermostat staging, electric heat sequencers, or possibly an auxiliary thermostat that locks out the third stage of electric heat during warmer outdoor temperatures)

By checking the unit schematic, the technician determines that staging on this system is accomplished through the two-stage thermostat: the first-stage heating relay is energized by the first-stage thermostat contact, the second-stage heating relay is energized by the second-stage thermostat contacts, and the third-stage heating relay is supposed to be energized with the second-stage relay through a jumper, which the technician notices is not installed in this unit as illustrated on the unit wiring schematic. By tracing out the wiring, the technician finds that in place of the jumper is an outside temperature remote bulb thermostat mounted nearby on an outside wall and that the thermostat is set at −5°F. He then adjusts this to a setting of 15°F, hears a faint click, and goes back to the furnace to find that the third heating element is now drawing current. The technician reports back to the customer that he has corrected the problem. He shows him the location of the outdoor temperature thermostat and explains its purpose and operation. A faint burning smell is noted due to dust film burning off the third element of strip heat but quickly starts to dissipate. The technician then makes a quick check of the furnace temperature rise to make sure that it is within the limits stated on the rating plate and explains that the faint smell is normal for a heater that has not been energized in some time.

Prior to leaving, the technician makes a quick check of the thermostat to see that the space temperature is rising and that it is set to the customer's preference.

> **Safety Tip**
>
> Ladders have labels listing appropriate conditions for their use. A Type I Industrial ladder is considered heavy-duty with a load capacity of not more than 250 pounds. The Type II Commercial ladder is medium-duty with a load capacity of not more than 225 pounds and is probably not suited for HVACR work. The Type III household ladder is light-duty with a load capacity of 200 pounds and should not be used in our type of work.

> **Tech Tip**
>
> Personal ethics and professional conduct are characteristics that all technicians should strive to obtain. Following are some of the required traits:
> - Honesty implies a refusal to lie, steal, or deceive in any way.
> - "Dependability" means to be reliable, believable, and trustworthy.
> - A positive work ethic drives you to give a day's work for a day's pay.
> - Patience is the capacity to "stick it out" and continue without complaint. This is not easy to do with some customers.
> - "Empathy" means the ability to put yourself in the customer's position.
> - Finally, there is tolerance, which is the capacity to endure and "put up with" issues that come up with customers or the job.

SUMMARY

To be a successful technician, you must develop a problem-solving or troubleshooting skill set. Start by listening to the dispatcher and customer. After listening, make a list of the symptoms and possible causes to help you discover the problem. Start by checking the operation of blowers, the air filter, the thermostat (Figure 7–15), and the condensing unit. Check the voltage across the pressure switch. An open pressure switch will indicate loss of the control voltage.

Hooking up the gauges as shown in Figure 7–16 will determine if the system has pressure or a possible mechanical problem. Do not let the customer know you are having a problem finding what is wrong. If you need to call for help, excuse yourself from the customer and go to your vehicle to call for assistance. Do not waste time searching for something that you do not understand. The best person to contact for help is your supervisor. Technical support from a manufacturer is good, but many times those contacts are reserved for severe problems. Many manufacturers' representatives are very busy and have long telephone hold times.

At this point in your career, you should be able to isolate the problem. Look for the obvious problems like blown fuses, no power, or a burned wire. Replacing the fuse, restoring power, or replacing a burned wire may not solve the problem in all cases, but it is leading you to correcting the problem. Using a systematic approach will lead you to the problem and save time.

Most problems are electrical. Using an electrical diagram and the voltmeter in parallel to the load troubleshooting sequence will help you zero in on the faulty component. The topic of electrical problem solving will be covered in more detail in training found in years two and three. It is important to remember the function

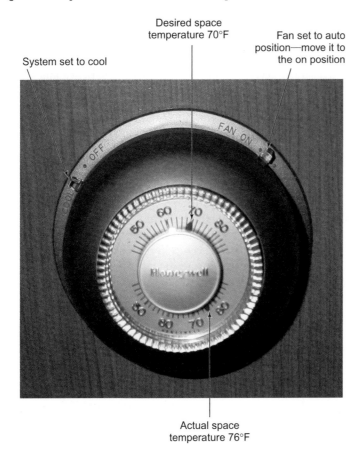

Figure 7–15
Set the thermostat to the fan "on" position to continuously operate the blower. Set the thermostat to cool and turn it down several degrees below the room temperature. (Courtesy of Eugene Silberstein)

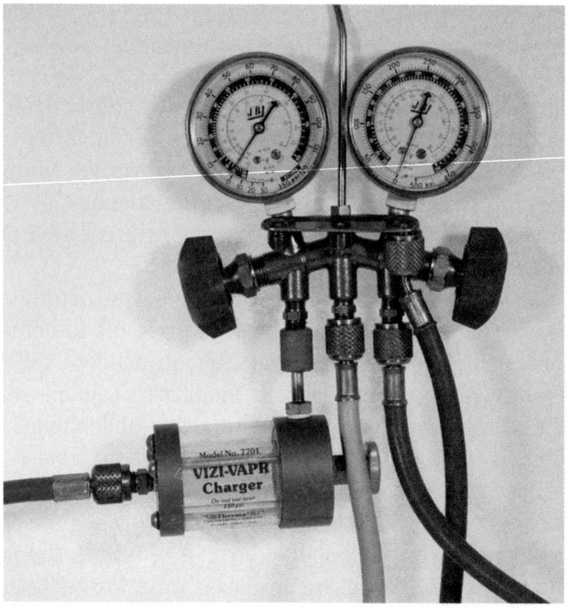

Figure 7-16
A charging device in the gauge line between the liquid refrigerant in the cylinder and the suction line in the system. Measuring pressure differences indicates that the compressor is functioning. The charge may not be correct or there may be some other problem. (Courtesy of John Tomczyk)

of each component in the system, as this will go a long way toward troubleshooting. You may not understand the function of a component in a refrigeration or electrical circuit, but if you understand the component itself, you will be able to check it for proper operation.

Problem solving is a learned skill, and the more you practice it, the better you will become as a service technician. Learn from your experience and the experience of others. Always ask how and why when you are unsure of something. Do not be afraid to ask for help. No technician knows everything.

FIELD EXERCISES

1. Part of what we learn here needs to apply to our everyday jobs. We can learn something new every day. Apply one of the principles you learned in this chapter to your problem solving skills. Be willing to discuss this with the class.
2. There is more than one way to successfully solve a problem. Look at an HVACR problem and figure out two or more ways of solving it.
3. Get someone else involved in your problem solving skills. When you have successfully solved a problem, review your solution with another tech or a supervisor. Select the right person because not everyone is willing to offer constructive criticism. If the person shares his experiences and skills, this is a good learning opportunity.

REVIEW QUESTIONS

1. Why is it important to review the problem with the customer prior to starting troubleshooting?
2. What is the voltage measured across the thermostat in Figure 7-1 when the light operates? What is the voltage across the thermostat when the light is burned out?

3. Describe the meter-voltage-reading troubleshooting technique used to troubleshoot an open compressor internal thermostat found in Figure 7-2.
4. Describe general steps you would take to locate a problem in a residential air conditioning system.
5. Why is productivity important to your company?
6. The technician is unable to figure out the problem after completing a few basic checks. The technician is stumped. What would you do if you were in this position?
7. What does a positive work ethic mean? What does honesty mean?
8. Why is it important to check motor winding resistance at the terminals rather than at the contactor?
9. On your preinspection of a no-cooling call, the compressor is not operating due to a burned terminal wire on the contactor switch. What would you do to repair this problem?
10. How do you determine if the compressor is operating prior to hooking up gauges?
11. Why is it important to use the right manifold gauge set on a condensing unit?
12. Why is it important to use the correct refrigerant gauges on a system?

CHAPTER 8

Basic Installation and Repair Methods

LEARNING OBJECTIVES

The student will:

- Describe the difference between brazing and soldering.
- List mechanical fasteners used in HVACR installations.
- Describe the importance of clearances around equipment.
- List five instruments used in the HVACR profession.
- List ten common hand tools used in installations.
- Describe the importance of insulation on the suction line and duct system.
- Describe how to install a condensate drain.
- List five important parts of an installation.

INTRODUCTION

This chapter relates to the practical aspects of installed HVACR equipment. We will discuss basic mechanical, brazing, and installation skills. Identifying and knowing the terms for fasteners are important for good communication with other technicians as well as the supply house. Fasteners are like tools—many names apply to the same item. General hand tools will also be discussed, and we will go over installation requirements that manufacturers have included with their pieces of equipment. In addition, mechanical code issues are important and will also be introduced in this chapter.

Along with these topics, other installation-related subjects such as piping, refrigerant, insulation, wiring, and condensate management are covered. The chapter will finish with discussions on a day in the life of a technician.

In general, an installation includes the following steps:

1. Set condensing unit and air handler/evaporator coil
2. Check ductwork
3. Run line set
4. Braze connections
5. Leak check the lines
6. Evacuate the system
7. Check/charge system
8. Check system airflow
9. Cycle system several times to check sequence of operation

If a technician is exposed to installation techniques, it will be easier to recognize problems for service and repair.

TYPICAL FIELD TOOLS

In order to install, service, or troubleshoot HVACR systems, the technician needs hand tools, gauges, and instruments. For example, to pull a deep vacuum, a micron gauge is required. The vacuum side of a compound or low-side gauge will not measure a deep vacuum because a deep vacuum is measured in microns.

Without a voltmeter, the technician cannot check voltage, which would limit troubleshooting because electrical problems typically outnumber mechanical problems. It is unsafe to touch a unit unless the voltage is zero. Using a voltmeter is the best way to determine this. The technician can turn off the power to a piece of equipment prior to working on it, but some equipment designs have more than one power source feeding the unit. Turning off one breaker or disconnect may not be enough to terminate the power going to a piece of equipment.

Most voltmeters are clamp-on or are part of a multimeter unit that has an ohmmeter and low-current measuring options. The ohmmeter is valuable for checking the complete circuit of a component and determining whether it is shorted to ground. Low-current measuring options are used to measure the operation of a gas furnace flame detector.

Gauges are important in measuring saturation pressures when troubleshooting or properly charging systems. Gauges are also needed to remove or add refrigerant and for general use during the evacuation process.

Hand tools, as discussed in this chapter, are important for equipment installation, access, and troubleshooting. Experienced technicians often have a second set of the more common tools. For example, nut drivers and screwdrivers are musts to access a piece of equipment.

BASIC INSTALLATION AND REPAIR METHODS

This section is an overview of basic installation requirements as seen in most air conditioning installations. Many of the components will be the same in refrigeration, although refrigeration components are specifically designed for their application. Ducts are not installed on refrigeration equipment.

The basic installation components of a split system, as illustrated in Figure 8–1, are the condensing unit, air handler, and evaporator coil. In this example, the evaporator coil is installed inside the air handler housing. There are interconnecting suction and liquid lines, and the condensing unit has an electrical disconnect. The refrigerant lines are sleeved with PVC as they penetrate the wall. This sleeve prevents damage to the vibrating refrigerant lines as well as any reaction between the copper and mortar. The air handler is upflow, sitting on top of an air filter assembly. The PVC drain removes condensate water from the coil and drains it to an approved place.

The commercial package unit has all the features of a split system. All that is required is power, a thermostat, and ductwork. The residential package unit is usually smaller than a commercial unit and may not have a fresh air intake as is required on the commercial installation.

Types of Air Handlers

There are three basic types of air handlers named for the direction of air movement:

- Upflow
- Downflow
- Horizontal

Figure 8–2 shows two examples of an upflow installation. One air handler is installed in the basement, and the other is in a closet with ductwork distributing airflow overhead. Some air handlers are designed for all three position options. These are called multipositional furnaces or air handlers. Minor changes in the equipment permit this flexibility.

Clearance

All equipment requires clearance around the unit. Clearance is required to work on the unit and to allow airflow so that heat is rejected. The condensing unit or package unit requires more clearance than the indoor section. Figure 8–3 shows general

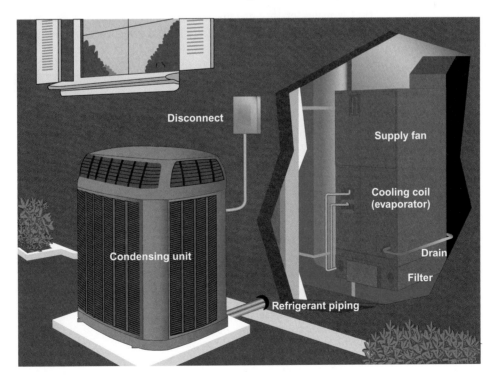

Figure 8–1
Split system with labeled components. The piping is straight, the equipment is level and square, the pipe penetration has a plastic sleeve, and there is a required condensing unit disconnect. (Courtesy of Trane)

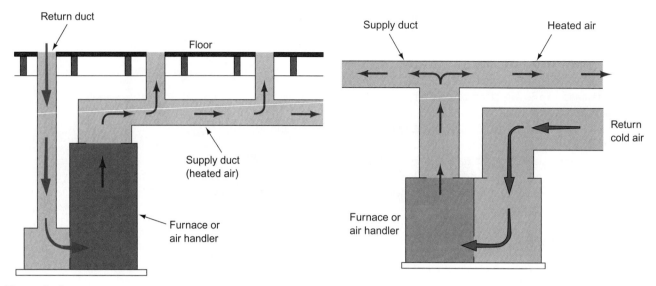

Figure 8-2
This is an example of two upflow heating systems. The air is returned from the base of the unit and discharged vertically. It is also called a vertical unit. Vertical systems are commonly installed in a basement or closet in a single story building with no basement. The air can be distributed through the floor or overhead ducts.

and specific clearance requirements. Some furnaces allow zero clearance on some sides, which means the equipment can be up against combustible construction, but the service panels and combustion air slots must not be restricted.

Manufacturers determine the clearances required around their equipment, as shown in Figure 8-3. The mechanical code generally determines the clearance around various heat-producing appliances, as seen in Table 8-1.

Outdoor Unit Placement

The outdoor unit should be placed on a 3-inch or higher concrete or plastic base that is made for this installation. The outdoor unit can be installed on a platform made of weather- and rot-resistant material. Heat pumps that operate in high

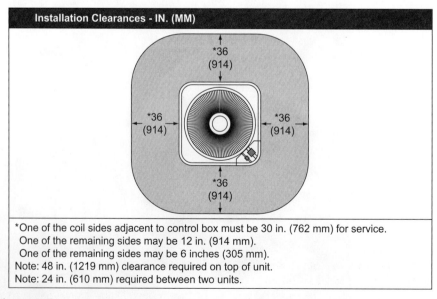

Figure 8-3
This is a picture from the installation manual of a condensing unit. The overhead view gives the installer the clearance requirements around all sides. The clearance is required for service and for air circulation to remove heat from the condenser. (Courtesy of Carrier Corporation)

Table 8-1 Standard Installation Clearances in Inches for Unlisted Heat-Producing Appliances

This is a partial sample of Table 3-1 from the *2006 Uniform Mechanical Code (UMC)* that has requirements for clearance around unlisted heat-producing appliances. The UMC has more clearance recommendations for commercial and industrial installations. Please refer to *2006 Uniform Mechanical Code* for completed table and special notations. (Courtesy of the International Association of Plumbing and Mechanical Officials)

See Section 304.0.   in × 25.4 = mm

RESIDENTIAL-TYPE APPLIANCES	FUEL	APPLIANCE ABOVE TOP OF CASING OR APPLIANCE	FROM TOP AND SIDES OF WARM-AIR BONNET OR PLENUM	FROM FRONT	FROM BACK	FROM SIDES
BOILERS[11] AND WATER HEATERS						
Steam Boilers—15 psi (103.4 kPa)	Automatic Oil or Comb.					
Water Boilers—250°F (121°C)	Gas-Oil	6		24	6	6
Water Heaters—200°F (93°C)	Automatic Gas	6		18	6	6
All Water Walled or Jacketed	Solid	6		48	6	6
FURNACES—CENTRAL;" OR HEATERS—ELECTRIC CENTRAL WARM-AIR FURNACES	Automatic Oil or Comb.					
	Gas-Oil	6^2	6^2	24	6	6
	Automatic Gas	6^2	6^2	18	6	6
Gravity, Upflow, Downflow,	Solid	18^3	18^3	48	18	18
Horizontal and Duct Warm-air —250°F (121°C) max.	Electric	6^2	6^2	18	6	6
FURNACES—FLOOR	Automatic Oil or Comb.					
For Mounting in	Gas-Oil	36		12	12	12
Combustible Floors	Automatic Gas	36		12	12	12
HEAT EXCHANGERS						
Steam—15 psi (103.4 kPa) max. Hot Water—250°F (121°C) max.		1	1	1	1	1
ROOM HEATERS[4]	Oil or Solid	36		24	12	12
Circulating Type	Gas	36		24	12	12
Radiant or Other Type	Oil or Solid	36		36	36	36
	Gas	36		36	18	18
	Gas with double metal or ceramic back	36		36	12	18
Fireplace Stove	Solid	48^6		54	48^6	48^5
RADIATORS						
Steam or Hot Water[4]		36		6	6	6

snowfall areas need to be installed a couple of feet or more above the expected annual snowfall level using snow legs that are usually either 6", 12" or 18" tall. Heat pumps installed in cold climates also need to be elevated to allow defrost water to run off and away from the outdoor unit. Various pads are illustrated in Figure 8-4.

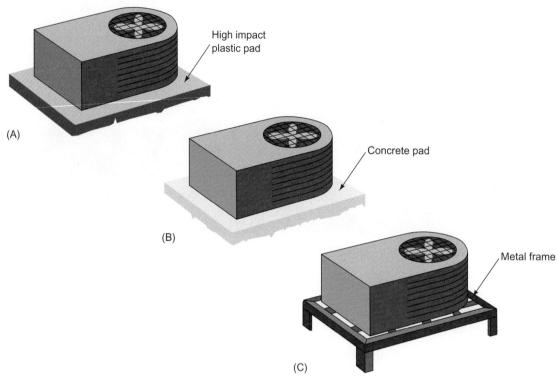

Figure 8-4
Condensing unit stands or pads come in various configurations. The condensing unit should be at least 3 inches from the ground to allow water to drain away.

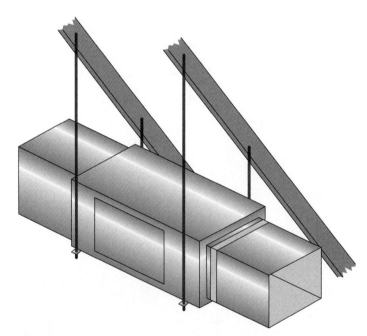

Figure 8-5
Air handler, furnaces, and ducts can be suspended. Suspending the equipment conserves floor space. Material used to suspend these items should not block the door or access to service the equipment.

Air Handler Installation

The proper installation of air handlers and furnaces is critical. The installation can be suspended as shown in Figure 8-5, installed vertically on a floor as shown in Figure 8-6, or fastened to a return air plenum box.

Hanging a unit reduces the vibrations that are transferred into the structure. Alternative ways of reducing blower vibration transfer are shown in Figure 8-7, Figure 8-8, and Figure 8-9. The system shown in Figure 8-6 is used on residential and commercial installations. Those shown in Figure 8-8 and Figure 8-9 are used in larger, commercial air moving equipment. Systems that have flexible duct or duct board are less likely to transmit vibrations in the ductwork.

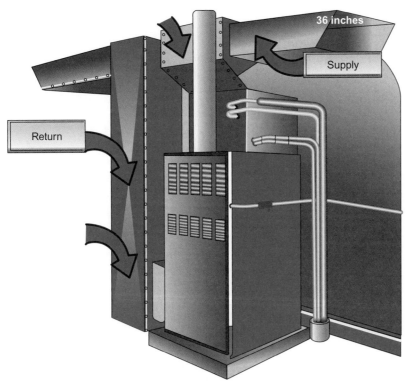

Figure 8-6
Vertical, upflow gas furnace is installed on a pad on the floor. Return air plenum comes into the side of the furnace. This is a basement installation. (Courtesy of Trane)

Figure 8-7
Canvas or flexible connector as seen from inside the duct system. The metal below the canvas connector is the air mover. The metal duct above the connector is the sheet metal duct. The purpose of the flexible connector is to prevent vibrations from being transmitted through the ductwork. (Courtesy of Carrier Corporation)

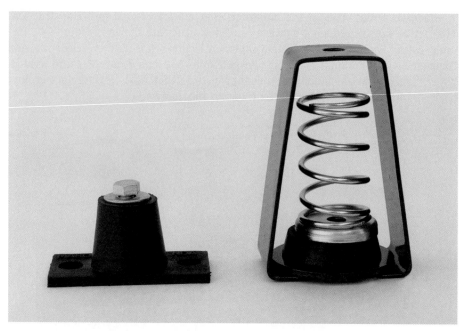

Figure 8-8
Two examples of vibration isolators used for mounting equipment. (Photo by Susan Brubaker)

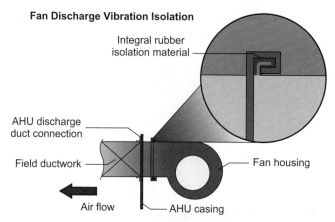

Figure 8-9
Fan discharge vibration isolation. The integral rubber isolation material is built into the duct connection. Blower vibrations are contained in the fan housing. (Courtesy of Carrier Corporation)

This air handling equipment needs to be secured so that it does not move. The panels on these units will not be blocked by the support design or by the structure itself. Structure members cannot be removed to satisfy these requirements. Within the local jurisdiction's mechanical code, the air handler or furnace generally needs 30 inches of clearance in front of the service panels as well as the evaporator section. These codes generally require a 30-inch platform on the service side of the indoor section. There are many other considerations depending upon the installation location. Always check your local code for clarification.

Basic Wiring

A package unit requires an electrical disconnect. A split system requires a disconnect for the air handler or furnace and a separate disconnect for the condensing unit. Figure 8-10 is a nameplate from a condensing unit and can be used to select

Figure 8-10
The nameplate reveals several pieces of electrical information required for condenser installation. The breaker size and circuit ampacity are two pieces of important information.

the wire size that feeds that component as discussed in Chapter 3. The installation instructions may also have a recommended wire size.

Disconnects are required to be within sight of the equipment. If working out of sight of the disconnect, the disconnect must be locked out and tagged to ensure that no one turns it on while someone is working on the equipment.

Installing the Condensing Unit or Package Unit

The following is a checklist that will help in planning the installation of a condensing unit or package unit.

1. Read the installation instructions.
2. Check local codes covering zoning, noise, platforms, clearances, etc.
3. If practical, avoid locations next to family rooms or bedroom windows. Install on the side or back of the building.
4. Avoid installations under roof overhangs without guttering because water draining from the roof onto the unit could produce excessive noise or cause recirculation of discharge air.
5. Place the unit in a well-drained area or where it is supported high enough so runoff will not enter the unit. Support pads are required.

Tech Tip

The nameplate may state "Maximum Overcurrent Protective Device Amps," which means that somewhere in the electrical circuit, there must be a time-delay fuse or an HACR circuit breaker, a term that is used in place of "HVACR." The current rating should not exceed this amperage. The time-delay fuse or circuit breaker is required because motors with low starting torque in the circuit draw very high current on start-up. This is known as locked rotor amps (LRA) or starting amps. Without the delay, the overcurrent protection may trip when the motor starts.

6. The soil conditions should be considered so the pad does not shift or settle excessively.
7. Do not locate where heat, lint, or exhaust fumes will be discharged onto the unit.
8. Do not install the unit in a recessed or confined area where recirculation of discharge air may occur. Watch for clearance around other units as per manufacturer installation instructions.
9. The site must be level and the platform or pad should be of permanent materials such as concrete, bricks, plastic, blocks, steel, pressure-treated timbers, or another type of permanent material. Whenever possible, avoid mounting units on pads that are rigidly attached to the structure due to vibration and noise concerns.
10. It is recommended that the soil be treated or the area be graveled to retard the growth of grasses and weeds.
11. Line sets should be installed keeping in mind that allowances should be made for unit settling.
12. Rooftop installations are acceptable providing the roof will support the unit and provisions are made for water drainage. Be alert to noise or vibration problems that may be transmitted through the structure.
13. Rooftop installations may not be recommended on wood frame structures due to weight distribution considerations.
14. If installation is on a flat roof, there are many considerations that must be made such as placing the unit over one or more load-bearing walls.
15. When using a restrictor or piston orifice metering device, check to make sure that the correct restrictor is being used for that particular condensing unit. The correct piston will be attached to the condensing unit.
16. Run the tubing by the most direct route and support with hangers as necessary.
17. Refrigerant lines must not exceed the length or vertical separation between indoor and outdoor units as per the manufacturer's recommendations.
18. Insulate the suction line to prevent sweating and possible water damage. The recommended thickness of insulation is $\frac{3}{8}''$ in conditioned spaces and $\frac{1}{2}''$ in an attic or high-humidity areas.
19. Check the data plate on the unit or the technical manual from the manufacturer for determining wire sizes and circuit protection.
20. Follow the National Electrical Code and local jurisdiction's codes or ordinances for permanently grounding the unit.
21. Make all outdoor electrical supply connections with rain-tight conduit and fittings.
22. Codes require that a disconnect switch be installed outdoors within sight of the unit.

Installing the Indoor Section

The following is a checklist that will help in planning the installation of the indoor section.

1. Read the installation instructions, including the warning labels in the instructions. This will assist you in following the manufacturer's recommendations and in making sure you are applying the equipment properly.
2. Provide adequate combustion air for gas or oil burning appliances.
3. The evaporator on an upflow furnace must be installed per the manufacturer's instructions.
4. When an uncased coil or multipositional coil is installed, ensure that installation does not allow air to bypass the coil.
5. Install the evaporator coil level for proper condensate drainage.

6. The furnace commonly has the evaporator coils located on the positive pressure side of the blower.
7. On a draw-through evaporator, a P-trap is required to ensure proper drainage. A P-trap is recommended on all installations.

START-UP AND FINAL CHECKOUT

The start-up checkout is done by the installation crew or a specific technician trained for this job. One technician is usually responsible for ensuring that the system works properly and briefing the customer on the operation. Survey the installation and be observant so that you spot any situations that can be corrected before they become a problem. Start-up and final checkout may include:

1. Make a visual inspection of connections, especially for tightness of connections and loose or disconnected wires.
2. Set up equipment components like dip switches and any jumpers required for the operation of the equipment.
3. Speed taps on multispeed motors need to be checked, as do adjustable pulleys on belt-driven blowers.
4. Check to see that the filters are clean and in place.
5. Check the supply voltage with the unit operating. It needs to be within ±10% of the equipment's data plate rating.
6. Check motor amp draw. This would need to be done on all motors: compressor, condenser fan, indoor blower, and any other amperage-drawing equipment.
7. Check fan rotation, especially on those motors that can be reversed. Remember, a blower that goes backward will deliver an insufficient volume of air.
8. Check for proper sequencing of the equipment.
9. Check that the condensate line is open and flowing.
10. Check for proper airflow. Remember, both the equipment and the customer need air, so we need to satisfy both the customer and the equipment.
11. Take temperature readings, including both the dry bulb and the wet bulb. The dry bulb will be an indication of the sensible heat removed, and the wet bulb will be an indication of the latent heat removed. You need to take both to determine how the equipment is operating.
12. Verify proper control voltage for installation.
13. Verify expansion device type and proper placement of sensing bulb if so equipped. Record pressures, temperatures, subcooling, superheat, and verify correct refrigerant charge per manufacturer's specifications.

CUTTING, SWAGING, SOLDERING, BRAZING, AND FLARING

The main methods used to connect refrigerant line and component parts are soldering, brazing, and flaring. Soldering and brazing use a filler metal to seal the gaps between tubing. Flaring compresses copper or aluminum between two metals, creating a tight gasket seal. Swaging is a way of expanding soft copper tubing to create a fitting that will allow tubing to slip inside it that is later sealed with solder or brazing filler material.

Cutting and Swaging

Before swaging, soldering, or brazing tubing, the technician must cut it to the proper size. A hacksaw should never be used because it develops metal shavings that can drop into the tubing. A tubing cutter, as shown in Figure 8–11, can be

Figure 8–11
Tubing cutters (not to scale) are used to cut copper or aluminum piping. The cutter on the left is for tubing sizes up to 2 inches. The cutter on the right is for smaller tubing and cutting in tight spaces. (Courtesy of Ritchie Engineering Co. - YELLOW JACKET Products Division)

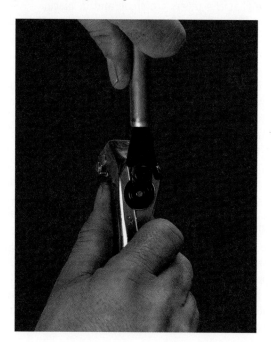

Figure 8–12
Removing copper burrs on the inner walls of copper tubing. A reaming tool is installed on the tubing cutter. The reamer is rotated several times inside the tubing. Do not allow shavings to drop into the piping.

used to accomplish this job. After the tubing is cut, it should be reamed to remove burrs created by the cutting process. The deburring procedure is shown in Figure 8–12.

Swaging is increasing the size of soft copper tubing enough to allow another piece of tubing of the same size to slip into it, as shown in Figure 8–13. Swaging reduces the need for a manufactured fitting. Common swaging tools, as shown in Figure 8–14, come in sizes ranging from $\frac{1''}{4}$ to $\frac{7''}{8}$. A stepped swaging tool, shown in Figure 8–14, in the right size can be used to replace several sizes, from $\frac{1''}{4}$ to $\frac{3''}{4}$. Figure 8–15 shows the swaging process. The tubing is extended a couple of inches

Figure 8-13
Swaging is a process of opening one end of soft copper tubing so that another piece can slip inside. This is a field-made fitting that requires sealing on one joint. (Courtesy of Bill Johnson)

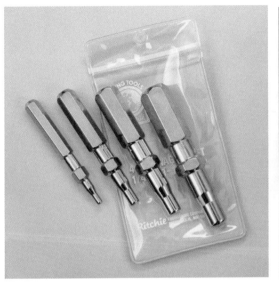 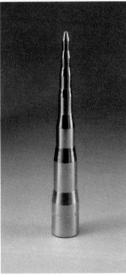

Figure 8-14
The set on the left are swaging tools for $\frac{1}{4}''$, $\frac{3}{8}''$, $\frac{1}{2}''$ and $\frac{5}{8}''$ soft tubing. The tool on the right is one swaging tool for the same sizes. (Courtesy of Ritchie Engineering Co. - YELLOW JACKET Products Division)

through a flaring block, which is clamped closed to prevent the tubing from slipping. The swaging tool is placed in the tubing, and the technician uses a hammer to strike the tool, driving it into the soft copper and expanding it. The depth of the swage is the diameter of the tubing. For example, $\frac{5}{8}''$ tubing will have a $\frac{5}{8}''$ swaged fitting. Extending the swaged fitting too far may split the tubing or make for a loose fitting. The final test is whether the piping fits snugly into the swaged fitting. A loose swage can be cut off and redone.

Soldering

Brazing and soldering are processes of joining metals by the use of a "filler" metal.

Brazing and soldering are performed at different temperatures. Soldering bonds metals at temperatures below 800°F, as illustrated in Figure 8-16. Brazing is the process of bonding metals at temperatures above 800°F, up to the melting temperature of the metal.

Air-acetylene, as seen in Figure 8-17, may be used for both low-temperature soldering and high-temperature brazing. Compared to oxygen-acetylene, the flame temperature of air-acetylene is lower even though the flame is much larger, as shown in Figure 8-18.

It will be difficult to use an air-acetylene flame to braze in low outdoor temperatures or when the wind is blowing across the brazing job. The cold temperature or wind cools the joint too rapidly to melt, penetrate, and seal the joint.

Safety Tip

The technician should use safety glasses and gloves when swaging tubing. The striking surface of the swaging tool may become mushroomed, allowing pieces to fly off when hit by a hammer. Do not use striking tools with mushroomed heads. The tool should be redressed using a grinder to remove the mushroomed metal. The glasses will protect against sparks or metal fragments that might fly off the swaging tool, and the gloves will offer minimal protection to the hand in case the hammer glances off the swaging tool.

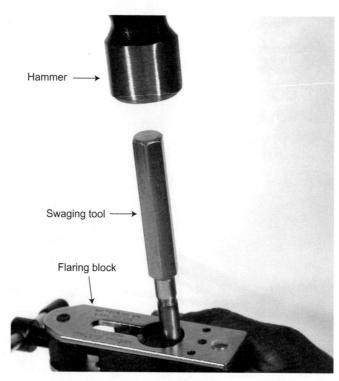

Figure 8-15
The swaging tool is placed in the soft tubing. A hammer is used to drive the tool into the tubing and open it enough so that another piece of tubing can slip inside. The flaring block holds the pipe so that it does not slip. Safety note: this technician should be wearing gloves. (Courtesy of Bill Johnson)

Figure 8-16
Soldering uses a larger flame that is at a lower temperature than brazing. A soldered joint requires flux to protect the joint while it is heated. The solder comes in a roll and is flexible. (Courtesy of Bill Johnson)

Figure 8-17
Air-acetylene torch outfit. The torch tip mixes acetylene with air. The flame temperature is lower than that of an oxygen-acetylene torch. All compressed gas cylinders should be chained to prevent accidental falls. (Courtesy of Bill Johnson)

Cutting, Swaging, Soldering, Brazing, and Flaring   **195**

Figure 8–18
The air-acetylene flame is larger but cooler than the flame made with an oxygen-acetylene torch.

Low-temperature soldering is done with silver-bearing filler metal and a matched flux. **Low-percent silver-bearing solder is used only in plumbing and condensate lines.**

The following materials are to be soldered or brazed:

- Copper to copper (or brass to brass)
- Copper to steel
- Copper to brass or bronze

Each of these materials can be soldered or brazed with either low-temperature or high-temperature filler metals. The difference will be in the strength of the joint and the ability to withstand piping vibrations. When in doubt, select brazing because it is less likely to leak due to vibrations. Low-temperature silver-bearing solder requires flux. The high-temperature silver filler metal requires a high-temperature flux in many cases. Phosphorous-bearing brazing alloys do not require flux when joining copper to copper fittings.

The technician will select an air-acetylene tip that is appropriate for the pipe size. The larger the pipe size is, the larger the tip required to heat the joint thoroughly.

Brazing

Oxyacetylene, the combination of oxygen and acetylene, is commonly used for high-temperature brazing. Common oxyacetylene tips are shown in Figure 8-19 and a common oxyacetylene rig is represented in Figure 8-20. High-temperature brazing can use silver- or phosphorous-bearing filler metal as shown in Figure 8-21. High silver content material, as shown in Figure 8-22, requires flux. Brazing rods with phosphorous does not require flux when brazing copper to copper, but flux is required when brazing copper to dissimilar metals.

Figure 8-23 shows examples of bad and good brazing and soldering practices. Generally, the brazing and soldering procedure requires that the pipe be heated first,

Tech Tip

The air-acetylene "swirl" tip, sometimes known as the high-velocity tip, will produce higher temperatures with the same air-gas mixture. It must be operated wide open at a high acetylene pressure of 15 psi. Otherwise, the tip will be destroyed in a short time.

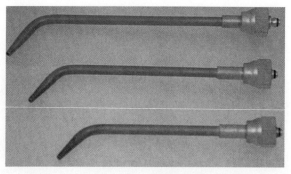

Figure 8–19
This illustration is of various sizes of oxyacetylene tips. (Courtesy of Bill Johnson)

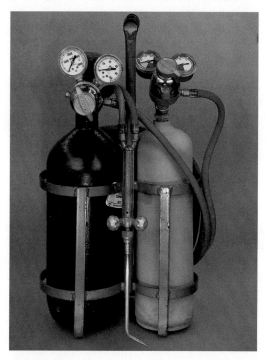

Figure 8–20
Oxyacetylene torch outfit. The oxygen cylinder is on the left and the acetylene cylinder is on the right. The regulators attached to the cylinder are used to control the pressure to the hoses and torch handle. (Courtesy of Bill Johnson)

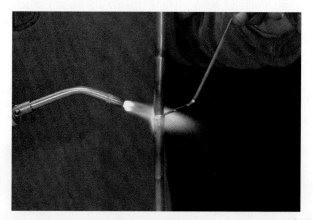

Figure 8–21
The phosphorous filler metal is used to seal the fitting. Brazing is high temperature and does not require a flux when sealing copper to copper, but does require a high-temperature flux when brazing copper to brass/bronze or to steel. Phosphorous in the brazing rod is considered the flux when mixed with silver and other alloys.

Figure 8–22
Some high-temperature silver-bearing brazing materials look like soft solder. This brazing material requires the flux designated for this material. This material is commonly used with dissimilar metals. (Photo by Susan Brubaker)

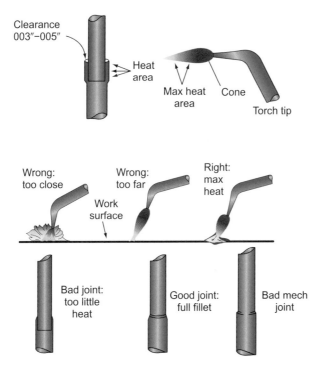

Figure 8–23
Various ways to produce a good or bad joint. Heat the pipe first, and then heat the joint and add the filler material. (Courtesy of Trane)

transferring heat deep into the fitting. Next, apply heat to the fitting to draw the filler material into the gap between the fitting and tubing. This drawing action is called capillary action and allows flux to flow evenly over the joint. It causes the molten brazing material to flow throughout the joint area. The brazing rod material will flow at a lower temperature, preventing burning of the fitting. This melting action causes the fitting to bond to the tube and fill the gap without the need of a third metal.

Safety Tip

Never use anything other than dry nitrogen to purge and pressure test a refrigeration system! Oxygen can create a reaction that will explode.

Table 8–2 Air-Acetylene Torch Tips

Tip chart that shows the tip size, gas flow at 14 psi, and tubing size. Acetylene pressure should never exceed more than 15 psi.

Tip No.	Tip Size in.	Tip Size mm	Gas Flow @ 14 psi ft³/hr	Gas Flow (0.9 Bar) m³/hr	Copper Tubing Size Capacity — Soft Solder in.	Soft Solder mm	Silver Solder in.	Silver Solder mm
A–2	$\frac{3}{16}$	4.8	2.0	.17	$\frac{1}{8}-\frac{1}{2}$	3–15	$\frac{1}{8}-\frac{1}{4}$	3–10
A–3	$\frac{1}{4}$	6.4	3.6	.31	$\frac{1}{4}-1$	5–25	$\frac{1}{8}-\frac{1}{2}$	3–12
A–5	$\frac{5}{16}$	7.9	5.7	.48	$\frac{3}{4}-1\frac{1}{2}$	20–40	$\frac{1}{4}-\frac{3}{4}$	10–20
A–8	$\frac{3}{8}$	9.5	8.3	.71	1–2	25–50	$\frac{1}{2}-1$	15–30
A–11	$\frac{7}{16}$	11.1	11.0	.94	$1\frac{1}{2}-3$	40–75	$\frac{7}{8}-1\frac{5}{8}$	20–40
A–14	$\frac{1}{2}$	12.7	14.5	1.23	$2-3\frac{1}{2}$	50–90	1–2	30–50
A–32*	$\frac{3}{4}$	19.0	33.2	2.82	4–6	100–150	$1\frac{1}{2}-4$	40–100
MSA–8	$\frac{3}{8}$	9.5	5.8	.50	$\frac{3}{4}-3$	20–40	$\frac{1}{4}-\frac{3}{4}$	10–20

*Use with large tank only.

NOTE: For air conditioning, add $\frac{1}{8}$ inch for type L tubing.

> **Safety Tip**
>
> Never use any petroleum products (oil or grease, etc.) on any oxygen component (gauges, regulators, fittings, etc.) due to the risk of spontaneous combustion.

> **Safety Tip**
>
> The following recommendations are provided for safe operation of soldering and brazing equipment:
>
> 1. Always use a pressure regulator on the nitrogen cylinder, and a relief valve is recommended when purging tubing.
> 2. Wear proper eye and skin protection.
> 3. Avoid breathing vapors from fluxes and filler metals.
> 4. Avoid prolonged skin contact with fluxes.
> 5. Under no circumstances should you ever pressurize a system with oxygen or compressor air. The mixture may become flammable and explode. Compressed air can also add moisture to the system.
> 6. Never look directly at the gauge on an oxygen, nitrogen, or acetylene pressure cylinder as you open the cylinder. If the gauge should "blow out," it will usually do so from the front. Stand to the side of the gauge faceplate when opening the valve in case it blows.
>
> *Continued*

Brazing requires a higher-temperature flame than soldering. An indicator that too much heat is applied to the soldered joint is that the solder will run off the pipe and drip onto the surface below. Brazing requires direct heat at the joint along with heat near the brazing rod feeding filler to seal the joint.

Like soldering, brazing has various tip sizes, with the larger tips required for large pipe. Larger tips are also required when brazing in cooler temperatures or in windy conditions. Table 8–2 is called a tip chart. It helps the technician decide what brazing tip to use based on gas pressure, copper tube size, and whether the filler material is soft solder or brazing silver solder.

Brazing or soldering components such as valves, flow controls, and sight glasses creates excessive heat that may damage these components. It is important to protect these components by purging with nitrogen while using a heat sink material around the valve. Also, use a heat shield to deflect heat away from anything that can be burned or scorched by the flames. Burns on the cover or component do not appear professional.

Proper ventilation is important for the technician and customer. Provide extra ventilation when working in confined areas. The by-products of fluxes, fillers, and residual refrigerant can be toxic. Do not ventilate smoke into the duct system. Care must be taken not to trip building fire alarm systems because this could lead to an embarrassing building evacuation with an automatic visit from the fire department. It does not take much smoke to sound an alarm. Contact the building management or owner when doing hot work inside, and have a fire extinguisher nearby when using a torch inside or outside a building.

> **Tech Tip**
>
> When brazing or soldering, always purge the tubing with an inert gas like nitrogen. This will prevent scale buildup inside the tubing. Scale is properly known as copper oxide and looks like carbon on the tubing. The scale or carbon may cause problems such as blockage at the metering devices and the compressor in the future.

Finally, do not exceed the 15 psi acetylene pressure. Above 15 psi, acetylene by itself becomes unstable and volatile. Acetone is placed in the tank of acetylene to keep it stable under higher pressures. As the acetylene is used, it separates from the acetone and flows through the regulator. Keep acetylene cylinders in the upright position to prevent acetone from draining from the tank.

Flaring

Flaring is the process of opening or fluting one end of soft copper tubing, making a metal gasket surface. A flare is pictured in Figure 8-24. The flare or gasket surface is squeezed between two other metal surfaces to make a sealed joint. Flares can be made only with soft copper tubing because hard-drawn copper is brittle and will crack if flared.

Flaring requires a flaring tool and a flaring yoke, as shown in Figure 8-25. The soft tubing is placed in the flaring tool, as shown in Figure 8-26. The tubing is tightened in the flaring block and should extend above the flaring block surface about the height of a quarter. The yoke is installed over the tubing, as seen in Figure 8-27, and the flaring block is then turned down to develop the flare.

The progress of the flare can be inspected anytime during the flaring process. An underflared fitting can sometimes be flared some more, but an oversized flared fitting will not fit inside the flare nut. The defective flare will have to be cut off and the process started over. Figure 8-28 shows another combined flaring tool and yoke mounted as one component.

The flare nut needs to be very tight. A backup wrench will be needed to prevent damage to the component being tightened for the flare joint. When a leaking flare is

7. Use only quality hoses for oxygen and acetylene use. Worn hoses are an accident waiting to happen.
8. Open the acetylene cylinder one-quarter to one-half turn, and leave the wrench on the valve so you can turn the tank off quickly.
9. Secure cylinders and rigs when transporting in a service vehicle. An accident can cause an explosion or fire.
10. Always turn the fuel gas (acetylene) on first and off last at the torch. This method will test both torch valves. A popping sound when the torch shuts down means that the oxygen torch valve is leaking. A small flame at the tip of the torch means the fuel gas valve is leaking. Finally, close the oxygen and acetylene cylinder valve and purge the pressure in the hoses.
11. If you must transport the oxygen or acetylene, use an approved hand cart or carrier with the cylinders secured and turned off at the tank. Purge all oxygen and acetylene from the hoses before storing or transporting, and remove all regulators from the cylinders before transporting. Transport in the upright position. Safety caps must be screwed onto cylinders.

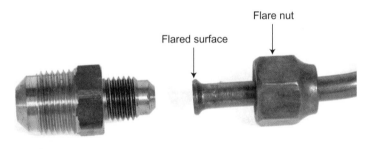

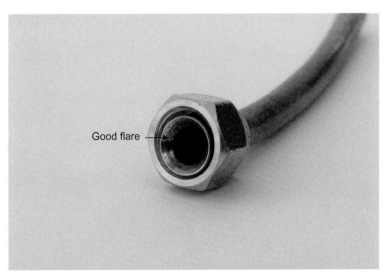

Figure 8-24
The upper image shows a common flare with a brass fitting and brass flare nut. The flare fits inside the nut snugly and will cover about 95% of the beveled edge of the mating surface. The flare nut should turn freely when seated fully over the flare. (Upper Photo Courtesy of Bill Johnson, Lower Photo by Susan Brubaker)

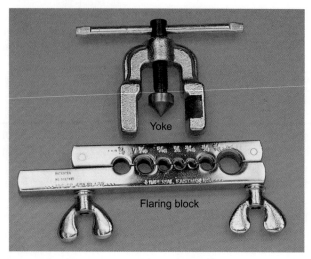

Figure 8–25
Common flaring tools. (Courtesy of Bill Johnson)

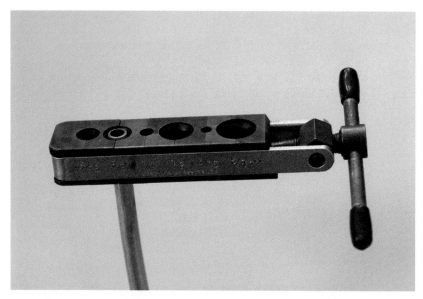

Figure 8–26
The copper tubing should extend above the surface of the block about approximately the thickness of a quarter. Extending the tubing too high will cause a flare to be too large. (Photo by Susan Brubaker)

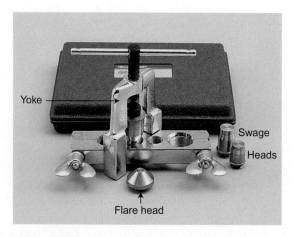

Figure 8–27
Yoke mounted on a flaring tool. This kit has optional sizes of flaring and swaging heads. This is a versatile kit that reduces the number of tools required. (Courtesy of Ritchie Engineering Co. - YELLOW JACKET Products Division)

Figure 8–28
This is a combination flaring tool and yoke. The yoke is mounted on the block for ease of flaring. (Courtesy of Ritchie Engineering, Co. - YELLOW JACKET Products Division)

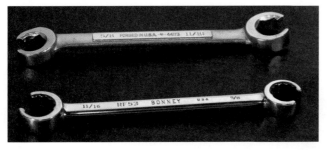

Figure 8–29
Flare nut wrench designed to slip over the tubing and gain access to the flare nut. (Courtesy of Bill Johnson)

found, the first thing to try is tightening the flare nut. In many cases, this solves the problem. If this does not stop the leak, cut the flare and make a new one.

A flare nut wrench like those in Figure 8–29 can be used to tighten a flare nut. This tool is designed with a notch to allow the wrench to fit over the tubing and then onto the flare nut.

FASTENERS

Fasteners are used to mechanically join materials. Examples of fasteners are screws, bolts, clamps, and nails. Electrical fasteners include components A through D in Figure 8–30. A and B are ring connectors, C is a female connector, and D is a wire butt connector used to join two pieces of wire. Item E is a plastic clamp that can be used to bundle wire or support piping. Items F through J are various fasteners used in many professions.

The color of the electrical connector designates the size of wire it handles. Red indicates wire size 22-16 AWG, blue 18-14 AWG, and yellow 12-10 AWG. The smaller the AWG size is, the larger the wire diameter.

Electrical connectors are used to complete an electrical wire connection. One type of connector is called a wire nut. Two stripped wires are slipped into the connector, and the connector is twisted until tight. This procedure creates a good bond between the wires in the connection. The color of the connector determines the size of wire it can handle.

A cable tie, commonly known as a nylon tie, is used to bundle wire and strap flexible duct to starting collars. Ties come in various lengths from a few inches long to several feet long. The longer ties are usually thicker and wider and have greater strength capacity. Ties can be connected in series to increase their length.

Safety Tip

The first-aid kit and Material Safety Data Sheet (MSDS) are two important parts of a service technician's life. HVAC work exposes the technician to cuts, scrapes, and sharp edges. The first-aid kit, Figure 8–31, comes in handy to treat minor injuries. The MSDS, seen in Figure 8–32, is used to determine any dangers in handling the various chemicals, fluids, other treatments, and other materials. The MSDS material must be available in a binder in the vehicle or at the job site.

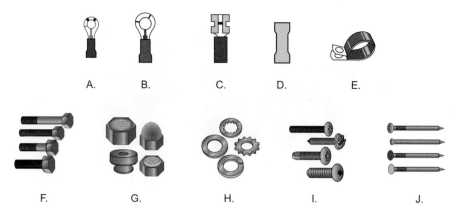

Figure 8–30
Electrical connectors are shown in pictures A through D. A and B are ring connectors, C is a quick female connector, and D is a wire butt connector. Item E is a clamp that is used to secure and support piping or wiring bundles. Items F are bolts. Items G are various types of nuts. Items H are washers. Items I are screws, and Items K are assorted nails.

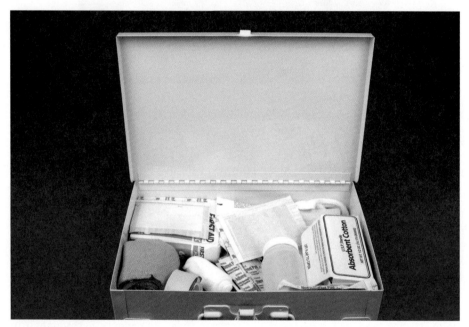

Figure 8–31
A first aid kit should be kept on a service vehicle or at an installation location. (Courtesy of iStock Photo)

CODES

When we say the word "codes," we mean a host of different rules and regulations that govern our industry. The codes are developed by many different professional officials. Even though they are written by different code organizations, they have many of the same requirements. Some cities, counties, and states modify these codes and adopt revised versions of them. Local jurisdiction codes always prevail.

Codes generally have a 3-year cycle. Most of the changes made in the 3-year cycle are minor and not a burden on our industry. The purpose of the code is to provide a minimum standard for the health, safety and welfare of building occupants.

Figure 8-32
Material Safety Data Sheet.

MANUFACTURERS' REQUIREMENTS

Manufacturers provide installation instructions with their equipment. Manufacturers' installation information varies among companies. Manufacturers' instructions include clearance around the equipment and general installation requirements such as the following:

- Charging recommendations
- Pipe sizing
- Wire size and circuit protection requirements
- Condensate trap installation
- Safety practices
- Instructions for venting gas or oil furnaces
- Combustion air requirements

TOOLS AND INSTRUMENTS

This section is a checklist for most of the tools and instruments required in the HVAC trade. It seems that when a technician thinks she has all the tools needed to complete a job, another new, helpful invention comes along. This section will not cover every tool and instrument required in the trade but rather the most common items needed to complete the common day-to-day jobs. Having the right tool for the job shows professionalism.

Manifold Gauge Set

The manifold gauge set, normally referred to as gauges, is the sign of an air conditioning and refrigeration technician. The gauges are not typically used in heating service unless dealing with a heat pump. The gauges can be seen in Figure 8-33. A repair kit like the one in Figure 8-34 can be purchased to maintain and extend the life of gauge set.

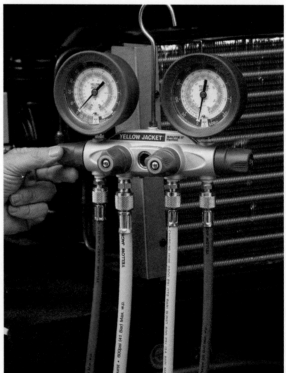

Figure 8-33
A manifold gauge set is used to check the high and low side pressures. (Courtesy of Ritchie Engineering, Co. - YELLOW JACKET Products Division)

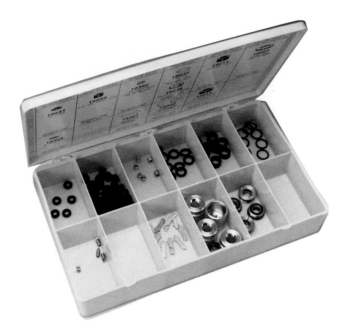

Figure 8-34
This repair kit has specific parts for one brand of manifold gauge set. If a tech has two or more gauge sets by the same manufacturer, she can make minor repairs out of the same kit.
(Courtesy of Ritchie Engineering, Co. - YELLOW JACKET Products Division)

Valve Core Removal Tool

The valve core removal tool, as seen in Figure 8-35, can be used in a few ways. It can remove the Schrader valve with minimal loss of charge. A leaking valve core can be replaced without having to recover the refrigerant charge, and the core can be removed to speed evacuation and the charging process. This is a time-saving tool.

Temperature Testers

Temperature testers are required when checking air, water, or refrigerant line temperatures. Normally, an experienced technician will require two or more types of testers, each being used for a specific purpose. For example, the pocket thermometer is good for measuring air or water temperatures but not refrigerant line temperatures. A surface probe or contact probe will be needed to check the refrigerant lines in order to measure superheat and subcooling.

Digital Psychrometer

The digital psychrometer, as shown in Figure 8-36, is used to measure dry bulb temperature, wet bulb temperature, and relative humidity. These readings may be necessary to properly charge an air conditioning system. The sling psychrometer is also used in the study of properties of air and the psychrometric chart.

Power Tools

A variety of power tools are used to make the job easier and faster to finish. Figure 8-37 shows one of the most useful tools, and that is the cordless drill.

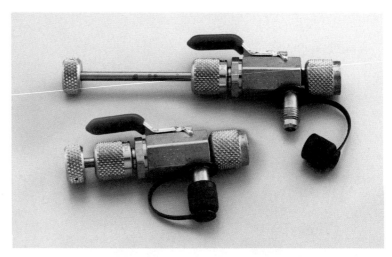

Figure 8–35
Valve core removal tool. The valve core can be removed or replaced with minimal loss of refrigerant. (Courtesy of Ritchie Engineering, Co. - YELLOW JACKET Products Division)

Figure 8–36
Digital psychrometer that can measure dry bulb, wet bulb, and relative humidity. (Courtesy of Ritchie Engineering, Co. - YELLOW JACKET Products Division)

Figure 8–37
Power tools are useful in removing and reinstalling the various fasteners used in units you install or service. They can also be used to drill holes. (Courtesy of iStock Photo)

Figure 8-38
Assorted vacuum pumps used to remove air and moisture from a refrigeration system. (Courtesy of Ritchie Engineering, Co. - YELLOW JACKET Products Division)

Cordless drills with higher voltage ratings will maintain greater torque for a longer period of time. Always have a battery charger and a spare charged battery when working on the job. Other useful cordless power tools include the circular saw and reciprocating saw.

Vacuum Pumps

Vacuum pumps, as shown Figure 8-38, are used to remove noncondensibles from the refrigeration system. The pumps come in various sizes, and which is appropriate to use depends on the job. The larger vacuum pumps are used on larger capacity systems. Two smaller vacuum pumps can be used in place of one large pump. If this is done, the pumps will be placed in two different parts of the system, such as one on the suction side and one on the high side. Changing the oil in the vacuum pump after each use will ensure a good deep vacuum and a long vacuum pump life because oil traps moisture and contaminants that would damage the pump.

Airflow Measuring Devices

There are many types of airflow measuring devices, as shown in Figure 8-39 and Figure 8-40. Airflow is generally expressed in cubic feet per minute (CFM). An air conditioning system needs approximately 400 CFM per ton of air conditioning. A 10-ton system will need 4,000 CFM of airflow. Some of the air measuring devices shown measure CFM directly, while others measure airflow in velocity as feet per minute (FPM), which is a measure of the speed of air moving by a point. FPM can be converted to CFM by multiplying and using the follow simple formula: CFM = FPM $\times$ Area of the duct in square feet.

Airflow will be discussed in more detail in years two and three of your training. At this time, it is important that you know that airflow is significant and that there are tools to measure it.

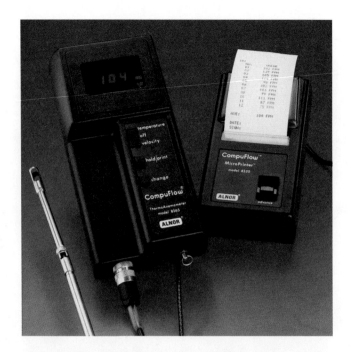

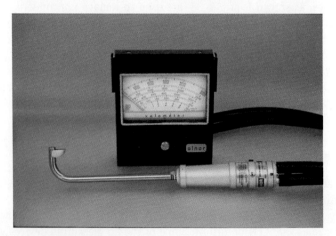

Figure 8-39
Airflow measuring devices. Airflow measurement is important to determine the proper operation of the system. (Photos by Susan Brubaker, Alnor Instrument, and Bill Johnson)

Figure 8-40
The airflow hold can be used to measure supply air or return air. The readout will give a direct reading of the airflow in cubic feet per minute (CFM). (Courtesy of Alnor Instrument Company)

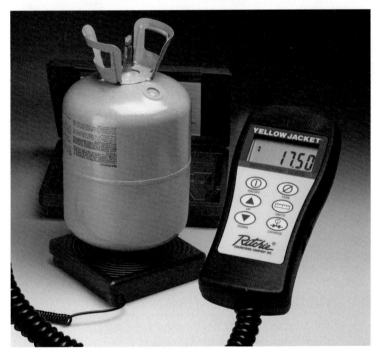

Figure 8-41
A digital scale has many uses in the HVACR trade. It can be used to weigh in a complete refrigerant charge. It can be used to weigh a recovery cylinder to make sure that it is not beyond 80% of its capacity. Finally, it can be used to weigh in a few pounds of refrigerant and ensure an accurate charge per pound to the customer. (Courtesy of Ritchie Engineering, Co. - YELLOW JACKET Products Division)

Digital Scale

A digital scale, as shown in Figure 8-41, is used to weigh in a total charge, which will allow an accurate invoice to the customer because customers are charged for the amount of refrigerant used. Some systems are field-charged because the total charge will depend on the installation. A digital scale should be used to determine the proper charge for these systems and the total charge should be recorded and noted on the unit label for future reference.

Figure 8-42
This is an example of a four-size refrigeration ratchet. This side has $\frac{1}{4}''$ and $\frac{3}{8}''$, turning over the wrench exposes sizes $\frac{3}{16}''$ and $\frac{5}{16}''$. (Courtesy of Ritchie Engineering, Co. - YELLOW JACKET Products Division)

Figure 8-43
The refrigeration wrench can be used to open a service valve or adjust the superheat on a TXV. (Photo by Joe Moravek)

Volt, Ohm, and Milliammeter

The volt, ohm, and milliammeter or multimeters were discussed in Chapter 3. These tools are vital in diagnosing electrical problems, and a backup meter should be readily available for each of these instruments. These tools should periodically be checked for proper calibration.

Hand Tools

The HVACR technician needs a variety of hand tools in a tool pouch, tool bucket, or small toolbox. This section will discuss some of these tools.

Refrigeration Wrench

The refrigeration wrench or ratchet is often used to regulate the position of service valves or adjust flow controls. The wrench and its uses are illustrated in Figure 8-42 and Figure 8-43. The wrench will fit on the square head stem of a service valve, and some TXVs and flow controls. The refrigeration wrench is one of the handiest tools in the technician's toolbox.

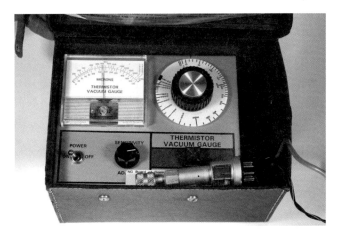

Figure 8-44
The vacuum gauge is hooked to the refrigeration system so that it continues to read vacuum in microns with the vacuum pump isolated and turned off. The vacuum should hold 500 microns or less. (Courtesy of Bill Johnson)

Micron Gauge

The micron gauge is an instrument used to measure a vacuum in microns. Figure 8-44 shows a digital micron gauge. The micron gauge is superior to the vacuum side of a compound gauge. The lower the micron reading is, the deeper the vacuum. A level of 500 microns or lower is considered a deep vacuum. The micron gauge should be hooked up in parallel with the vacuum pump. The micron gauge reads the vacuum when the vacuum pump is isolated and removed from the system. A rising micron level means that there is a leak in the system, possible moisture contamination exists, or additional refrigerant is still boiling out of oil.

Various Hand Tools

Figure 8-45 shows various screwdrivers and nut drivers that are used every day in the HVACR profession. The first picture shows a Phillips or crosspoint screwdriver and a flat-blade screwdriver. The second picture shows common nut drivers. Nut driver handles are color coded to indicate their size. A yellow-handled driver is $\frac{5}{16}''$, and a red-handled driver is $\frac{1}{4}''$.

Figures 8-46 through 8-50 are samples of pliers, which can be used for removing fasteners, cutting, crimping, mashing, wire stripping, duct shearing, and line piercing. The tool in Figure 8-51 is unique in that it is able to pierce a refrigerant line, allowing the technician to hook up a refrigerant hose and recover refrigerant.

A measuring tool is required in many aspects of the job for measuring pipe size, duct size, or room size. A retractable tape measure is shown in Figure 8-52. The utility cutter in Figure 8-53 is used for general cutting and for cutting flexible duct and duct board. Spare blades should be stored in the tool handle.

A variety of files and hammers as seen in Figure 8-54 and Figure 8-55 are required for numerous special jobs.

Several sizes of adjustable wrenches are required. An adjustable wrench is shown in Figure 8-56. Fixed open-end and box-end wrenches are better options, but the adjustable wrench can be used in place of several sizes of the fixed wrenches. Care must be taken when using adjustable wrenches because a loosely adjusted wrench can round off a bolt head. The fixed wrench is the best choice.

The pipe wrench, as seen in Figure 8-57, is used primarily on galvanized or black iron pipe. The technician will use this tool when working on gas lines or water lines. A pipe wrench is not recommended for use on a bolt head because it may round off the head.

Metal and wood saws are used mainly in installation jobs. The hacksaw in Figure 8-58 is used to cut metal. It is not advisable to cut copper tubing with a saw

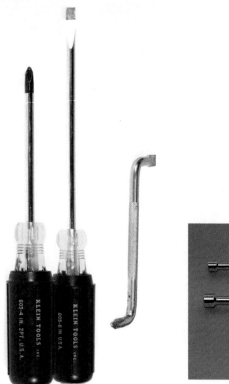
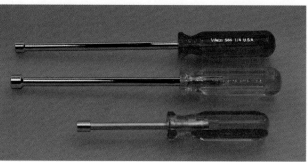

Figure 8–45
Various screw and nut drivers are used everyday in the HVACR profession. Tool 1 is a Phillips or crosspoint screwdriver. Tool 2 is a flat blade screwdriver. Tool 3 is an offset screwdriver. The second picture shows three common nut drivers. The yellow handle driver is $\frac{5}{16}''$ and the red handle driver is $\frac{1}{4}''$. Nut driver handles are color coded to the size of the hex head it will fit. (Courtesy of Bill Johnson)

Figure 8–46
Common slip joint pliers sometimes called pump pliers or by the brand name Channel Locks. (Courtesty of Bill Johnson)

Figure 8–47
Needle-nose pliers can be used for cutting or bending wires. Tools with insulated handles are not necessarily rated against electrical shock. (Courtesy of Bill Johnson)

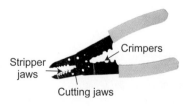

Figure 8–48
Wire stripper and wire cutter pliers. This model has a crimping feature. (Courtesy of Ideal Industries, Inc.)

Figure 8–49
Lineman pliers used to cut and bend wire. Can cut heavy-gauge wire. (Courtesy of Ideal Industries, Inc.)

Figure 8–50
Various types of sheet-metal cutters. The red-handled cutter will cut to the left or counterclockwise. The yellow will make a straight cut, and the green-handled cuts to the right or clockwise. (Photo by Susan Brubaker)

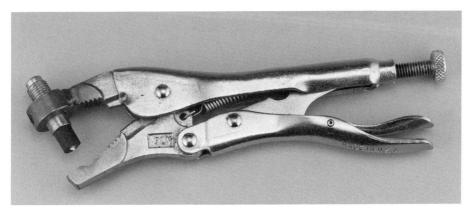

Figure 8–51
Refrigerant-line-piercing locking pliers also known as piercing vise grips. The refrigerant hose can be attached to the Schrader valve to measure pressure. The refrigerant needs to be removed before this tool is removed because a hole will be left in the pipe. (Courtesy of Ritchie Engineering, Co. - YELLOW JACKET Products Division)

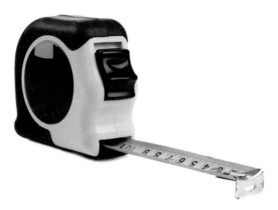

Figure 8–52
This is a retractable tape measure used to measure inches and feet. (Courtesy of Ideal Industries, Inc.)

Figure 8–53
Utility knife used to cut insulation, cardboard, and general use. Also known as a box cutter. (Courtesy of Ideal Industries, Inc.)

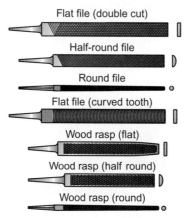

Figure 8–54
Assorted metal and wood files. The rasp is used to shave wood.

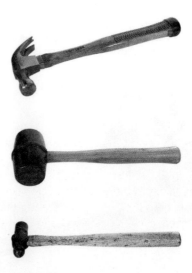

Figure 8–55
Assorted hammers. From top to bottom they are the claw hammer, soft blow hammer, and ball peen hammer. (Courtesy of Bill Johnson)

Figure 8-56
Typical adjustable wrench. (Photo by Susan Brubaker)

Figure 8-57
Pipe wrench used on galvanized or black steel pipe. Not recommended on copper tubing because the sharp teeth will dig into the pipe, creating a weak spot. (Photo by Susan Brubaker)

Figure 8-58
Hacksaw used to cut pipe or metal. (Photo by Susan Brubaker)

because some of copper shavings will drop into the tubing. The shavings circulate with the oil and can plug metering devices or get into the compressor motor, creating electrical or bearing damage.

The level in Figure 8-59 is used in leveling air handlers, evaporators, and thermostats. All HVACR equipment should be level and plumb. The air handler may include an evaporator that needs to be level to drain properly. Mechanical thermostats must be level for correct cycling operation. Digital thermostats do not need to be level to operate, but the appearance of a level device exhibits a professional installation with pride in workmanship.

Next is a fish tape in Figure 8-60. It is a long, semirigid wire or fiberglass material that is rolled up inside the case shown in this figure and used to "fish" wire through conduit. The tape is extended through conduit until it exits from the other end. The

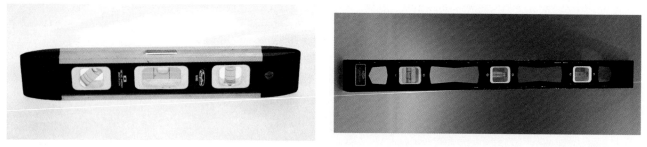

Figure 8-59
The level on the left can be used to measure vertical, horizontal, and 45° angles. The level to the right is only 4 inches long and will make a good thermostat level. (Courtesy of Bill Johnson)

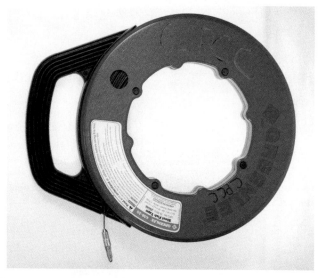

Figure 8-60
Fish tape used to pull wire through conduit. The tape is a semi-rigid wire that is pushed through conduit. Wire is attached to the loop and pulled back through the conduit. (Courtesy of Bill Johnson)

electrical wire, control, or thermostat wire is secured to the looped end of the tape, and the tape is then rolled back into the case as it pulls the wire to the point where it is needed.

The hex or Allen wrenches in Figure 8-61 are used to tighten set screws on pulleys and to open or close service valves. The ones shown here are encased in something similar to a pocketknife and are many different sizes. A hex wrench can be an individual tool, as shown in Figure 8-61. Longer individual wrenches are useful in hard-to-reach installations such as squirrel cage fans.

A wire brush is used to clean copper tubing before it is brazed or soldered. The brush in Figure 8-62 is used to clean inside the tubing or fitting. An alternative way to clean tubing is with a cleaning pad such as a Scotch-Brite® pad, which does not contain soap. Sandpaper is not recommended; the sand grains can get into the tiny crevices made by the sanding action, and the sand will melt when heated and turn to glass, which may break over time and cause a leak.

There are two types of tubing benders. Only soft copper tubing can be bent. Soft copper can be hand bent, but this incurs the risk of kinks. Figure 8-63 and Figure 8-64 are examples of the spring bender used to bend up to 90 degrees. Another type of bender is the lever-type, which can create up to a 180° angle, shown in Figure 8-65. Tubing benders reduce the number of joints, thus reducing the number of potential leaks as well as the time and material needed to install fittings.

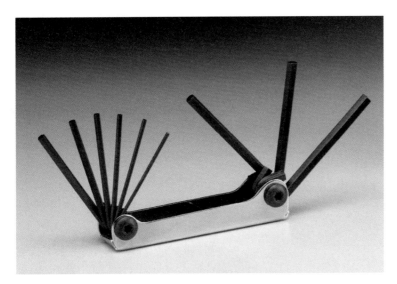

Figure 8-61
Hex wrenches are used in six-sided holes found in pulleys and some service valves. (Courtesy of Ritchie Engineering, Co. - YELLOW JACKET Products Division)

Figure 8-62
Wire brushes used to clean out copper tubing before brazing or soldering. (Photo by Susan Brubaker)

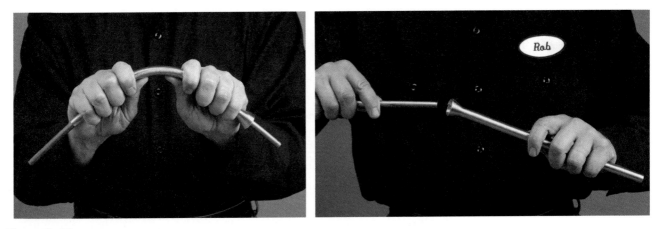

Figure 8-63
Spring tube bender. The soft copper tubing can be bent to an angle of 90° without creating a kink. (Courtesy of Ritchie Engineering, Co. - YELLOW JACKET Products Division)

Figure 8-64
This is a set of used spring benders. Spring benders come in several sizes from $\frac{1}{4}"$ to $\frac{7}{8}"$. (Courtesy of Bill Johnson)

Figure 8-65
Level type tubing bender. The bender can make up to a 180° bend. Each bender is designed for one pipe size. The bender prevents kinks and restrictions in the tubing. (Courtesy of Ritchie Engineering, Co. - YELLOW JACKET Products Division)

It is important to know what tools are available and the proper name for those tools, especially when a special application requires a tool that you do not have in your inventory. You will need to have the right tools to complete the job.

PIPING MATERIALS AND METHODS

Following are some common piping materials used for HVACR installations:

- Soft copper tubing
- Rigid copper tubing
- Black steel pipe
- Galvanized steel pipe
- PVC plastic pipe
- ABS plastic pipe

Soft and rigid copper are used in the refrigerant circuit, and rigid copper is sometimes used in water lines. Black pipe is used to join gas piping, galvanized pipe can be used for water lines, PVC is used for water or drain lines, and ABS pipe is used for drainage. For our purposes here, we will focus on copper tubing.

Types and Sizes of Copper Tubing

Copper tubing comes in hard-drawn and soft copper rolls, as shown in Figure 8-66 and Figure 8-67. The hard-drawn copper comes in 20-foot lengths and is purged with nitrogen and capped to keep it clean until installed on the job. The soft copper comes in coils of various lengths and is also purged and capped.

Figure 8-66
Hard-drawn copper usually comes in 20 foot lengths, capped at both ends to prevent contamination. (Photo by Joe Moravek)

Figure 8-67
Soft copper comes in a 50-foot roll. Once soft copper is extended, it becomes hardened and is difficult to roll up to its original size. It is a good practice to only unroll what is needed for the job. (Courtesy of Bill Johnson)

The bottom picture in Figure 8-67 shows the proper way of unrolling soft copper tubing. Only unroll what is needed for the job because excess tubing will not roll up to the original size. Unroll on a firm, flat surface in order to maintain a straight piece of tubing. Any kinks need to be cut out of copper tubing.

Refrigerant line is measured in outside diameter, and tubing used for plumbing is measured by the approximate inside diameter. Figure 8-68 is an example of $\frac{5}{8}''$ tubing. The actual usable area for a $\frac{5}{8}''$ tube is $\frac{1}{2}''$. The usable inside diameter is dependent on the wall thickness of the copper. Type L is the most common wall thickness used in the refrigerant industry. Type M has a very thin wall and is never used to handle refrigerant but could be used for nonpressurized drain line. Type K has a thicker wall than type L and is rarely used unless specified by the installation. Some government and specialty contracts specify type K copper.

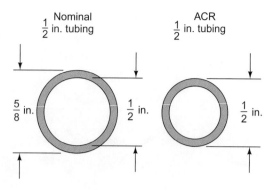

Figure 8-68
Copper tubing used for plumbing and heating is sized by its inside diameter, as shown in the left drawing. ACR tubing is measured by its outside diameter. (Courtesy of Silberstein)

There is a difference in the diameter of copper tubing used in plumbing and that used in HVAC. Air conditioning piping is generally called refrigeration piping. The diameter of copper used in plumbing is measured by its internal diameter, or ID. The diameter of air conditioner copper tubing is measured by the outside diameter, or OD. For example, in Figure 8-68, the piping on the left would be $\frac{1}{2}''$ plumbing tubing or $\frac{5}{8}''$ refrigeration tubing. The pipe on the right is $\frac{1}{2}''$ air conditioner tubing, which is less than $\frac{1}{2}''$ internally.

Copper Fittings

Copper fittings are used to connect soft or hard-drawn copper tubing. Fittings include elbows, couplings, tees, caps, plugs, and adapters as seen in Figure 8-69.

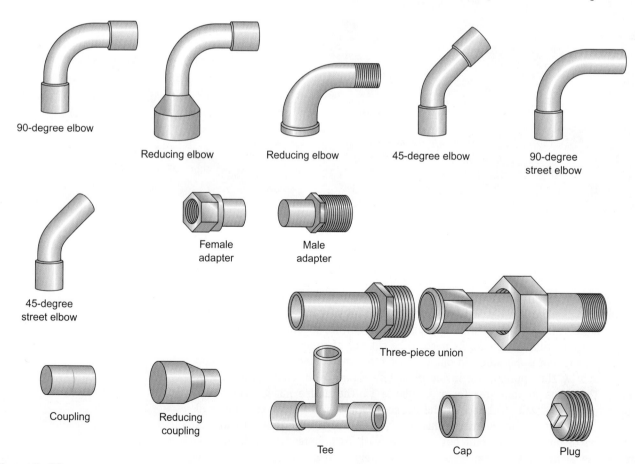

Figure 8-69
Pipe fittings. Fittings include elbows, couplings, tees, caps, plugs, and adapters.

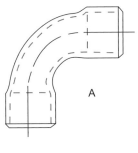

A

90° Standard radius

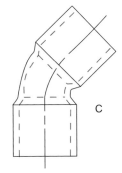

C

45° Standard radius

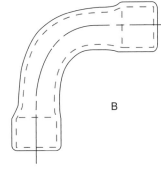

B

90° Long radius

Tube Bend Losses
Fitting losses in equivalent ft

Tube size OD	Reference diagram		
(in.)	A	B	C
$\frac{5}{8}$	1.6	1.0	0.8
$\frac{3}{4}$	1.8	1.2	0.9
$\frac{7}{8}$	2.0	1.4	1.0
$1\frac{1}{8}$	3.6	1.7	1.3

Figure 8-70
Each fitting has a different fitting loss rating. Comparing standard and long radius elbows, the long radius has a lower pressure drop and a lower fitting loss equivalent. Compare the three fittings A, B, and C. (Courtesy of Carrier Corporation)

Figure 8-70 shows the difference between the standard radius elbow and the long radius elbow. The long radius elbow has a long sweep; therefore, there is less pressure drop in this type of fitting. The figure also illustrates a 45° fitting, which is used when less than a 90° turn is needed. With a short piece of copper, two 45° fittings can be brazed together to make an elbow.

Refrigeration Trap Requirements

The purpose of the P-trap is to return oil moving through the refrigerant piping back to the compressor, where it is intended to be used to lubricate the compressor. Anytime the evaporator is mounted at a lower elevation than the compressor, manufacturer's instructions should be consulted. A P-trap may be required to assist in returning oil to the compressor.

More elaborate traps, such as the double suction riser in Figure 8-71, are required in commercial systems with varying loads. When a system is working at less than full capacity, the velocity of the refrigerant is lower and the oil drops out and pools in the piping. In this case, the double suction riser fills the lower trap with oil and forces the refrigerant through the smaller riser. The smaller riser has a greater velocity and thus carries the oil back to the compressor. When the refrigerant capacity increases, it will increase the velocity in the piping. This increase in velocity will cause the oil trap to open, allowing refrigerant to pass through it and pick up the trapped oil to be returned to the compressor.

Any reduction in velocity allows the oil to drop out of the refrigerant and collect in the refrigerant tubing. This will cause less oil return to the compressor and can cause the compressor to seize due to lack of lubrication.

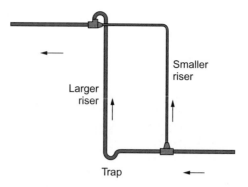

Figure 8-71
The double suction riser is used with some commercial systems that have varying capacity. At full capacity both risers are used to return refrigerant and oil. At reduced capacity, the trap fills with oil and blocks the trap. This forces the refrigerant and oil up the smaller riser, which creates a restriction and higher velocity, thus carrying the oil with it. (Courtesy of Trane)

INSULATION

Insulation is used in three major places when installing air conditioning, refrigeration, and heating equipment:

- Refrigerant pipe insulation
- Ductwork insulation
- Condensate drain line insulation

Refrigerant Pipe Insulation

The suction line is insulated to prevent condensation on the cold pipe, add to system efficiency, and reduce superheat that would cause high compressor discharge temperatures. Pipe insulation reduces the amount of superheat added to the refrigerant as it travels from the evaporator to the compressor. The liquid line is sometimes insulated, depending on equipment and application. Common piping insulation thicknesses are $\frac{3}{8}''$ and $\frac{1}{2}''$. Some energy codes require thicker wall insulation.

Ductwork Insulation

Ductwork must be insulated when used for cooling when running through an unconditioned space such as an attic, basement, crawlspace, or outside. The insulation stops condensation and heat transfer through the duct system. Common R-values for duct insulations are R-4, R-6, and R-8. R-6 and R-8 indicate a better insulating material and may be required by your local code.

The duct is insulated in several different ways. With metal duct, the insulation can be fastened on the inside of the duct or wrapped outside the duct, as shown in Figure 8-72. Figure 8-73 shows flexible duct, which has insulation between the inner liner and the outer vapor barrier. The inner liner is pulled over a metal starting collar and secured with UL 181B duct tape and a nylon cable tie. The insulation is pulled over the inner liner and starting collar, and the vapor barrier is pulled over the insulation and sealed.

Duct board, shown in Figure 8-74, has insulation on the inside of the duct. The outer duct is a foil vapor barrier. The specific size of duct board duct needed can be manufactured at the installation site.

Condensate Drain Line Insulation

The condensate drain line should be insulated in situations where such condensation would cause a problem. The drain line should be sloped away from the evaporator drain pan. Insulating the drain line will prevent condensation from forming and creating water damage on the structure.

Figure 8-72
The duct fastener is being hammered to keep the section of duct together. The duct joint will be sealed with mastic or UL 181 duct tape. Duct wrap will be installed over metal duct.

Figure 8-73
The inner duct liner is secured with a nylon cable tie or UL 181B duct tape. The insulation is pulled over the inner liner, and the outer vapor barrier is slipped over the insulation and sealed again. No metal should be exposed because this will cause condensation.

CONDENSATE DRAIN

Coil condensate is collected and drain to a disposal location. This section will cover the following information that relates to condensate:

- Drain line size
- Drain line material
- Pitching of the line
- Traps
- Auxiliary drain pans
- Condensate pump
- Safety float switch

The size of the condensate line should not be smaller than the fitting leaving the condensate drain pan.

Figure 8-74
Duct board can be cut to size at the site of installation. The insulation is located on the inside of the duct. A foil vapor barrier is outside the duct and is designed to give it strength as well as reflect heat and moisture. (Courtesy of Knauf Insulation GmbH)

The drain line material is usually PVC (polyvinyl chloride) or ABS (acrylontrile butadiene styrene) piping. Other materials such as galvanized or copper tubing can be used.

The pitch of the condensate line should be a minimum of $\frac{1}{8}''$ per foot of run. The greater the slope is, the less likely a plugged drain line will occur.

TYPICAL INSTALLATION PROBLEMS AND FAILURES

If you were to poll many manufacturers' technical support groups, you would find that the biggest problems in a new installation are the refrigerant charge and airflow issues. The system is either over- or undercharged enough to affect performance and capacity, or the airflow is usually too low. The supply or return or both are too small to move the right amount of air to develop the rated capacity. Sometimes, ducts might be the correct size but are restricted by the installation process. This is especially true if flexible duct is not installed according to manufacturer's recommendations.

Tech Tip

Additional installation practices will be covered in future training. For example, combustion air is required when installing gas or oil burning heating systems. There are specific requirements for the amount of outside air that must be brought into the room surrounding these systems.

Commercial installations require that outside ventilation be brought into the building to break up indoor air pollution. This is usually done by installing the appropriately sized duct on the return side of a duct system.

Some systems require a certain amount of air balancing, which ensures an even distribution of air throughout the building. Balancing dampers need to be installed for this process to be accomplished.

These are just a few of the advanced installation techniques you will learn on the job and in training programs like this one.

SUMMARY

This chapter covered many aspects of the installation job. Your supervisor will guide you through the correct activities involved in installation procedures. Installation instructions must always be read and followed.

It is important to know the different options for joining refrigerant piping. Brazing, soldering, or flaring can be used to seal the tubing. Fasteners are used to seal equipment and ductwork and to connect electrical circuits.

Manufacturers' installation requirements will guide the installation process. These instructions include clearances, wiring and charging information, and other requirements.

The chapter gave an overview of common instruments and tools used in our profession. A technician will collect a large inventory of useful items over a short period of time.

The common piping materials are soft or hard-drawn copper tubing. The tubing, fittings, and valves must be appropriately sized for good refrigerant velocity and oil return. Undersized piping and fittings create capacity loss.

Insulation is required on any cold surface to prevent condensation and water pooling on the structure. Ductwork, suction lines, and condensate lines can be cold and need to be insulated. Insulation needs to be completely sealed with a vapor barrier to prevent condensation from occurring in the material.

Finally, the condensate drain size was discussed. The chapter ended with typical installation problems and various optional components that make maintenance and service more attractive.

FIELD EXERCISES

1. Develop a list of tools that will be required when doing an installation. Do you have all of these tools?
2. You are going to install a split system. What materials are required for this installation?
3. Review an air conditioning installation. How would you have improved this installation? How would you rate this installation?
4. Make a list of additional tools you will need to do a complete installation.

REVIEW QUESTIONS

1. What are four reasons that clearance is required around installed equipment?
2. What is the advantage of suspending an air handler?
3. What size platform is required on the service side of a furnace?
4. Why is it not a good idea to install a condensing unit under a roof overhang that does not have a gutter?
5. What are the main reasons to insulate the suction line?
6. What is the difference between soldering and brazing?
7. What is swaging?
8. Why is it important to wear safety glasses and gloves when grinding?
9. What is the advantage of using a larger air-acetylene tip?
10. When is flux not required when brazing copper-to-copper piping?
11. What are five pieces of information that you might find in the manufacturer's condensing unit installation instructions?
12. What are two reasons to use a valve core removal tool?
13. What are three uses of temperature testers?

14. What maintenance can a technician do to increase the life of the vacuum pump?
15. Airflow is stated in _____.
16. What is the approximate airflow requirement for a 5-ton system?
17. What are two purposes of a digital scale?
18. What two tools could you use to bend soft copper tubing?
19. Why is it important not to uncoil excess soft copper tubing?
20. What are five reasons to correctly size refrigerant lines?
21. List the three types of copper tubing in order from thinnest to thickest wall thickness.
22. What is the purpose of a refrigerant trap?
23. What are the purposes of duct insulation?

CHAPTER 9

Energy Efficient Installations

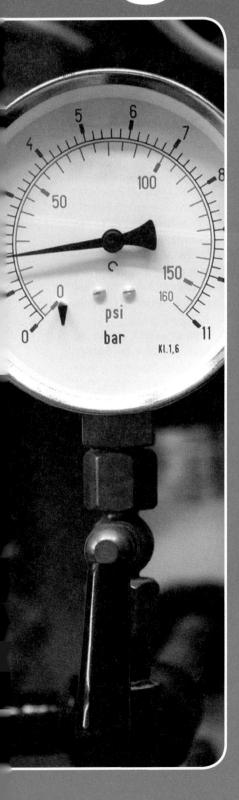

LEARNING OBJECTIVES

The student will:
- Define "EER."
- Define "SEER."
- Define "COP."
- Define "HSPF."
- Define the BTU output of a watt and a cubic foot of gas.
- Describe the human comfort zone.
- List four examples of alternate energy.
- Describe the purpose of the Energy Star program.
- List 10 good installation practices.
- Describe the effects of a lack of maintenance on an air conditioning system.

INTRODUCTION

This chapter will address the terms that are used to identify energy efficiency levels and the features of a good, efficient installation. As you will see, there are many different ways to use the word "efficiency." It is necessary to understand the term and how to use it with each type of equipment. For example, the Seasonal Energy Efficiency Ratio (SEER) and the Energy Efficiency Ratio (EER) can be used to compare the efficiency of air conditioners. SEER is more appropriate because it is used as an industry standard for comparing equipment. The higher the SEER rating is, the lower the electric bill. A 21 SEER system will save more energy than a 16 SEER system. In 2006, the federal government mandated a minimum 13 SEER rating for residential equipment. Prior to that, the minimum SEER rating was 10. Commercial-equipment SEER ratings are being developed.

What does going "green" mean? The term "green" is understood to mean using energy efficient products and appliances. It refers to not only energy efficiency but also to using the least amount of energy necessary to provide comfort, a safe, healthy environment and transportation, as well as to produce an item using recyclable materials where possible. The energy used to transport an item to market is taken into consideration.

Green HVACR equipment and buildings will use less energy than standard models. Equipment that uses less energy uses less electricity or fossil fuels and therefore generates less greenhouse gases, which are thought to contribute to global climate change. Many customers want to save energy and protect the environment. Offer customers energy-saving products and remind them that these products are environmentally friendly; it's a good selling point if they care about these issues.

The most efficient gas furnace can be 90% to 95% efficient at the site of use. The 5% to 10% loss is attributed to venting. Gas heating is often less expensive because it does not have the generating and transmission losses. Comparing fuel costs is tricky and confusing because the price of the fuel is governed by supply and demand.

This chapter will discuss efficiency and the many forms in which it can be expressed. The technician needs to understand these terms in order to convey proper information to the customer. In this case, the home builder consultant confused energy efficiency with energy cost. Electric strip heat costs a lot more than gas heat or electric heat provided by a heat pump system.

FIRST COST, MAINTENANCE COST, AND OPERATING COST

First cost is defined as the total cost of the installation, which includes major components:

- Condensing unit
- Furnace
- Evaporator
- Refrigerant piping
- Duct system material
- Drain piping
- Vent piping
- Thermostat
- Installation
- Insulation materials
- Electrical, wiring, disconnect

Secondary components on a new installation include power supply to operate the system, gas piping, and venting for a gas furnace. Electrical work and plumbing are usually completed by a different skilled professional. There are important minor considerations such as support material, thermostat wiring, supplies to fasten the ducts, and piping and pipe insulation.

In many cases, first cost is most important to the customer because energy efficient equipment costs more. If the customer plans to live in the home or occupy an office space for 5 years or more, he should consider energy-efficient options. In many cases, the customer who purchases energy-efficient equipment ends up with a better quality product and a payback through utility savings.

You will be able to sell more efficient equipment by focusing on the features and benefits of the system. Ask the customer about his comfort concerns, which include high utility bills; indoor air quality (dust, allergies, proper humidity, and fresh air); old, leaky, or dirty duct systems; safety features of the equipment; and extended warranties and priority service.

Maintenance Cost

Maintenance cost means how much money it costs to maintain the equipment, including regularly scheduled maintenance expenses for both heating and cooling and repair costs after the warranty period.

Operating Cost

Operating cost includes the electricity and fuels used to operate the system. Operating costs are higher in the summer and winter months when these systems are likely to be used the most. The customer must remember that part of the monthly utility bill is for water heating, lights, cooking, and the operation of appliances. Some utilities have varying rates and charge more for a utility when it is used the most. For example, electric rates may be higher in the summer when it is demanded the most, and gas rates may increase in the winter when heating consumption is greatest.

The overall operating cost of a HVAC system may be higher than its initial cost. The average life of a system is about 15 years. With this in mind, it is easy to understand the operating cost exceeding the initial investment on a system; therefore, keeping the operating cost low will make an impact over the life of the equipment. Remember, in most locations, the utility rates increase over time, making the investment in energy-efficient equipment worth the extra up-front investment.

Discuss payback with the customer. Payback is found by dividing the additional cost of the equipment by the annual energy savings in dollars.

$$\text{Payback} = \frac{\text{Additional cost of equipment}}{\text{Annual energy savings}}$$

Customers may be eager to invest in an energy-saving product. Most customers expect a 5-year payback period for their extra energy investment. The higher the SEER or annual fuel utilization efficiency (AFUE) rating, the less energy will be used.

WATTS, KILOWATTS, AND COST OF ENERGY

The watt is a measurement of power. One watt equals the product of 1 volt and 1 amp. Watts are stated in the formula as Watts = Volts × Amps.

Most electric utilities bill their customers in kilowatt hours (KWH); "kilo" means "1,000," so a kilowatt is 1,000 watts. A kilowatt hour would be 1,000 watts used in 1 hour. For example, ten 100-watt incandescent light bulbs burning for 1 hour would use 1 KW, expressed as 10 standard bulbs × 100 watts = 1,000 watts, or 1 KWH.

KWH costs vary widely, from 5 cents a KW to as much as 25 cents a KW. Areas that use hydroelectric generators generally have the lowest electric rates, and areas that must have fuel shipped to operate generators have the highest rates.

Most residential rates are based on simple KWH usage with charges for fuel generating, state and local taxes, and several other miscellaneous fees. The easy way to calculate KW usage is to divide the dollar amount by the total KW. For example, the total electric bill for the month is $250 when using 2,000 KWH.

$$\text{Cost per KW} = \frac{\text{Total bill}}{\text{KW}} = \frac{250}{2000} = 0.125, \text{ or } 12.5 \text{ cents per kilowatt hour}$$

Most utilities advertise a rate prior to adding taxes and miscellaneous charges, which can add a cent or two to each KW used.

Some commercial and industrial users are billed by KWH and demand charges. Demand charges are fees charged to a user that has peaks or spikes in its kilowatt usage during high-demand periods. Electric companies prefer customers with relatively flat KW usage rates because they will not need to create additional power by starting another generator or purchasing power from another company in the electrical grid. Fees in the form of demand charges are billed to customers who have fluctuations in their power usage. A percentage of the highest peak may be billed throughout the year. There are also standard KW charges along with the demand charges.

Electric and natural gas rates vary throughout the United States. It is best to have a general understanding of how these rates are established. Specific rate information can be obtained from the local energy provider.

Residential natural gas charges are billed by 100 cubic feet of gas, stated as CCF. The letter "C" means 100 and "CF" is cubic feet; therefore, 1 CCF is equal to 100 cubic feet of gas. Total charges for gas, taxes, and miscellaneous fees can be as high as $2 per CCF. The heat capacity of 1 cubic foot of gas is approximately 1,000 BTUs, and some gas utilities bill in "therms." One therm is equal to 100,000 BTUs, or 100 CCF.

CONVERSION AND COMPARISON OF FUELS

Before we begin this section, you must understand that heat is measured in BTUs. A BTU is a BTU no matter whether it comes from electricity, natural gas, or a burning log.

Common forms of energy like electricity and natural gas can be converted into BTUs per hour, or BTUH. There are 3.4 BTUs of heat per watt of electricity. To be exact, it is 3.416 BTUs, but 3.4 will suffice in most calculations. One KW equals about 3,400, BTUs, calculated as follows:

$$\text{BTUH} = \text{Kilowatts} \times 3.4 \text{ BTUs per watt}$$
$$3,400 \text{ BTUH} = 1,000 \text{ W} \times 3.4$$

One cubic foot of natural gas is approximately 1,000 BTUs. The quantity of the BTU heat content depends on the purity of the gas. Some gas supplies provide as low as 950 BTUs, and others provide as much as 1,060 BTUs to their customers, so 1,000 BTUs is a convenient average for common calculations. If a specific heat capacity is required, check with the local natural gas supplier.

Natural gas furnaces are rated in BTU input, not heat output. For example, a 100,000 BTU gas furnace with an efficiency rating of 80% will deliver 80,000 BTUs of heat as calculated by the following formula. First, convert 80% to a decimal by moving a decimal point two digits to the left to make it 0.80:

$$\text{BTU output} = \text{BTU input} \times .80$$
$$80,000 \text{ BTUH output} = 100,000 \text{ BTUH input} \times 0.80$$

Fuel oil and propane can also be converted to BTUs. The efficiency of these furnaces will determine how much heat is delivered to the space being heated.

MEASURES OF EFFICIENCY: EER, SEER, COP, AND HSPF

The air conditioning industry and the Department of Energy have developed standards for measuring energy efficiency. The terms are EER, SEER, Coefficient of Performance (COP), and Heating Season Performance Factor (HSPF). Let's review what these terms mean.

Energy Efficiency Ratio

The Energy Efficiency Ratio, or EER, is the ratio of the cooling capacity divided by the watts at full-load design conditions. For example, the EER of a 3-ton system at 95°F is calculated by dividing the measured total heat-removing output of 36,000 BTUH by the measured 3,000 watts.

$$\text{EER} = \frac{\text{BTUH}}{\text{W}} = \frac{36,000}{3,000} = 12.00 \text{ EER}$$

EER is measured at the design temperature, which is 95°F. The problem with EER is that it does not consider the complete operating temperatures of an air conditioning system. In many climates, 95°F is the high temperature for a few hours on a hot day. The system operates at a lower temperature the other 20 or so hours of the day. What is the efficiency at those operating conditions? In some cases, the equipment is more efficient in cooler temperatures. SEER was developed to obtain a more realistic view of the energy use over a range of operating conditions. EER was replaced by SEER in the 1980s to reflect this more accurate formula. The SEER rating will be higher than the EER rating because it includes temperatures where the system operates more efficiently.

Seasonal Energy Efficiency Ratio

The SEER calculation includes a more complex formula and procedure. A simplified definition of SEER is the total amount of heat removed in a cooling season divided by the total watts used during the cooling period. This is represented by the following formula:

$$\text{SEER} = \frac{\text{Total Cooling Season BTUs}}{\text{Total Watts Consumed}}$$

This is a more realistic measure of performance because capacity and power consumption vary at different outdoor temperatures. Most systems have reduced capacity at warmer temperatures of 95°F and greater. Table 9-1 shows a table that illustrates capacity changes of a 3-ton system at 10°F increments. The temperature of the outdoor air entering the condenser is found in the left column of Table 9-1 under the heading "O.D. D.B." The table also shows the sensible heat capacity at various return air temperatures.

For example, with an outdoor dry bulb temperature of 85 degrees and an indoor wet bulb temperature of 63 degrees, total capacity of the system is 36,000 BTUH and the Total KW usage is 2.82. If the outdoor temperature increases to 95°F and the indoor wet bulb temperature holds at 63°F, the capacity drops to 34,500 BTUH and the KW usage increases to 3.17. Increases in the outdoor temperature reduce the system capacity while increasing the KW usage, which reduces the EER of the system. After reviewing the higher outdoor temperature, you will come to the same conclusion of lower capacity and lower efficiency.

Coefficient of Performance

Coefficient of performance, or COP, is used to determine the efficiency of electric heat and heat pumps. *COP is the ratio of total heat output divided by watts input*

Table 9-1 This manufacturer's table illustrates that as the temperature of the outdoor increases, the capacity drops and the KW usage increases, which leads to reduced efficiency at high temperatures. This is a realistic example but should not be used to determine the capacity of any system. Use the manufacturer's data to determine capacity.

O.D.D.B.	I.D.W.B.	TOT. CAP. (× 1,000)	SENS. CAP. AT ENTERING D.B. TEMP.				TOTAL KW
			72	75	78	80	
85	59	34.6	27.7	31.1	34.4	34.6	2.81
	63	36.0	22.8	26.2	29.7	31.9	2.82
	67	38.5	18.0	21.6	25.1	27.4	2.86
	71	41.3	13.2	16.7	20.2	22.6	2.87
95	59	33.3	27.3	30.6	33.3	33.3	3.15
	63	34.5	22.3	25.8	29.2	31.4	3.17
	67	37.0	17.5	21.1	24.6	26.9	3.21
	71	39.7	12.7	16.2	19.7	22.1	3.22
105	59	31.5	26.6	30.0	31.5	31.5	3.54
	63	32.7	21.7	25.2	28.5	30.8	3.56
	67	35.1	16.9	20.4	23.9	26.2	3.60
	71	37.6	12.0	15.5	19.1	21.5	3.62
115	59	29.3	25.8	29.2	29.3	29.3	3.98
	63	30.4	20.9	24.3	27.7	30.0	4.00
	67	32.6	16.0	19.6	23.1	25.4	4.05
	71	35.0	11.1	14.7	18.3	20.7	4.06

converted to heat. COP for a heat pump excludes any supplementary resistance (electric) heat. Because the fan motor adds heat to the air, it is included in this rating. COP is represented by the following formula:

$$COP = \frac{BTU \text{ heat output}}{\text{Electric heat equivalent (Watts} \times 3.4)}$$

Question: What is the COP of a 5 KW electric heat strip?
Solution: The technician measures 16,000 BTUs from the heat strip and fan motor. The technician can calculate the heat output of a system by measuring the airflow and temperature difference between the supply and return air. The airflow is measured by an airflow device like a flow hood, and the temperature is measured with two thermometers. This information is plugged into the following formula:
BTUH = CFM × (supply air temp − return air temp × 1.08)
Using a multimeter, the technician measures 245 volts and 20 amps. The technician calculates the heat output to be 16,000 BTUH. Next, the technician uses the COP formula to calculate the efficiency.

$$COP = \frac{BTUH \text{ heat output}}{\text{Electric heat equivalent (Watts} \times 3.4)}$$

$$COP = \frac{16,000 \text{ BTUH}}{245 \text{ V} \times 20 \text{ A} \times 3.4}$$

$$COP = 0.96$$

The COP of electric heat is close to 1.0 because all of the electrical energy supplied to the heating unit (or device) is converted to heat. The operation is 100% efficient but very expensive to use.

The next example is a heat pump. A heat pump's cooling efficiency is measured by the SEER, like a standard air conditioning system. The heating efficiency is measured as COP or, as we will see later, as HSPF.

Question: What is the COP of a 3-ton heat pump?

Solution: The technician measures 33,000 BTUH output of the heat pump system. Using a multimeter, the technician measures 240 volts and 15 amps on the running outdoor unit. The technician uses the COP formula to calculate the COP.

$$\text{COP} = \frac{\text{BTUH heat output}}{\text{Electric heat equivalent (Watts} \times 3.4)}$$

$$\text{COP} = \frac{33{,}000 \text{ BTUH}}{240 \text{ V} \times 15 \text{ A} \times 3.4}$$

$$\text{COP} = 2.7$$

The COP is 2.7 for the conditions measured. As the outdoor temperature drops, there is less heat in the air and the COP decreases. In this case, the heat pump is 2.7 times more efficient than standard electric heat.

COP Comparison Summary

Here is a comparison summary using the COP of electric heat and a heat pump:

- 1 KW of electric heat will produce about 3,400 BTUs of heat (1,000 × 3.4 = 3,400). The electric heat output does not vary with the outdoor temperature.
- 1 KW input to a heat pump operating at a COP of 2.5 will produce 8,500 BTUs of heat (1,000 × 3.4 × 2.5 = 8,500 BTUs). This is 2.5 times as much heat per KW than electric heat strips.
- The COP rating for an air source heat pump is variable. The heat output will depend on the outdoor temperature. The cooler the outdoor temperature is, the less heat is in the air and the less heat is absorbed by the heat pump. As the temperature drops, the COP drops. (Note: geothermal/earth heat pumps will have higher COP, SEER and HSPF.)
- The higher the COP is, the more efficient the operation and the less it costs to operate.

Heating Season Performance Factor

The Heating Season Performance Factor, or HSPF, was developed to give the customer and the contractor a better tool to evaluate the efficiency of a heat pump. To explain it in its simplest form, *HSPF is the total heat output of a heat pump (including supplementary electric heat) during the heating season divided by the total electric power in watts*. It could be stated as the following formula:

$$\text{HSPF} = \frac{\text{Total BTU heating season output}}{\text{Total heating season watts}}$$

HUMAN COMFORT

Human comfort depends on age, health, and physical activity. An older person may like a warmer environment, but a sick person may want a colder or warmer condition depending on the illness and body temperature. People involved in physical activities require cooler and drier conditions to be comfortable.

Human comfort, as illustrated by the psychrometric chart in Figure 9–1, *is a combination of temperature, moisture, and airflow conditions that will create comfort for 80% of the people in a conditioned space*. Human comfort can be condensed to the following conditions:

- 68°F–80°F
- 25%–60% relative humidity (RH)
- air velocity of less than 15 feet per minute

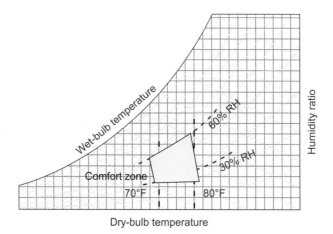

Figure 9–1
The human comfort zone can be illustrated by the standard psychrometric chart. The center yellow box represents the range in which 80% of people will feel comfortable. The range of comfort is approximately 68°F to 80°F and 25% to 60% relative humidity. (Courtesy of Trane)

A temperature of 68°F–80°F is acceptable to most people. Human comfort relative humidity ranges from 25% to 60%. Lower than 25% RH is too dry for most people because dry conditions produce dry nasal membranes and dry mouths and can possibly be irritating to the lungs. The skin will evaporate moisture quickly, cooling the body more quickly. An RH above 60% is comfortable for the skin and nasal passages. High humidity reduces evaporation on the skin, thereby reducing heat transfer from the body, but high-moisture conditions can create mold and mildew problems.

Figure 9–2 illustrates that moisture, as indicated by a percentage of relative humidity, is a key to the number of indoor air quality problems that can develop. Keeping the relative humidity between 40% and 60% will create a healthy environment.

ALTERNATIVE ENERGY SYSTEMS

"Alternative energy systems" can have several meanings. *The most common meaning of "alternative energy" is the use of energy sources other than fossil fuels.* Two common examples of alternative energy systems are solar and wind energy. Site-generated energy is usually more expensive and only used as backup power when the commercial electric grid is down. Our discussion will stay with the more common meanings of solar energy and wind energy.

Solar Energy

Solar energy uses the sun to generate electricity or heat water. Photovoltaic electricity is captured solar energy that develops a direct current (DC) voltage from solar cells. The DC is used to charge batteries that can be used to operate DC

Tech Tip

Wintertime heating at 68°F and 60% relative humidity will provide the same comfort as 72°F and 30% relative humidity. Lower utilities will result if the air is humidified during the heating season. A humidifier may be required to raise the moisture content of the air, but excessive moisture content in the air may cause moisture problems on the windows and walls.

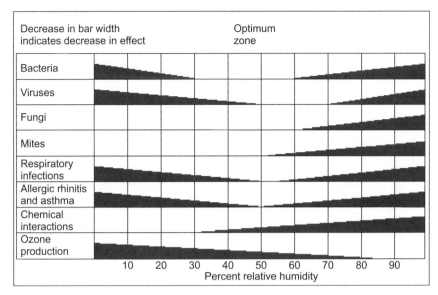

Figure 9-2
A relative humidity between 40% and 60% reduces the number of health problems. The air conditioner or dehumidifier can reduce the moisture content in the air, and a humidifier adds moisture to the air and may be used in the heating season. (Courtesy of ACCA Comfortools)

equipment. DC can be converted from direct current to alternating current (AC), which is used in many common appliances. Figure 9-3 is an array of photovoltaic cells used to provide part or all of the energy for a house.

Solar energy can also be used for heating domestic water or warming water to be used in space heating. A picture of this kind of installation is in Figure 9-4, and Figure 9-5 is a diagram of the process.

Another way to use solar energy is by the design of a structure. With proper design, a building can be oriented to pick up more heat in the winter for "free heating." Totally exposed windows facing south are washed with the sun many hours of a winter day. The sun in the winter is lower on the horizon than in the summertime, when the sun travels directly overhead. The winter sun angle allows it to penetrate into the glass, heating up the space in the building. Figure 9-6 illustrates this occurrence. For a south-facing structure, the summer solar heat is

Figure 9-3
An array of photovoltaic (PV) collectors. The sun's energy is converted to direct current voltage. DC can be used to power direct current loads or charge batteries and can be converted to alternating current for use with household appliances. (Courtesy of iStock Photo)

Figure 9-4
Flat-plate solar collectors on a tile roof of a new house to be used for water heating. (Courtesy of iStock Photo)

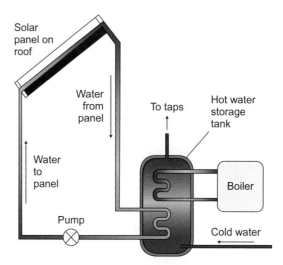

Figure 9-5
The water is heated by the sun in the solar panel. The heated water is piped to a heat exchanger and is used for domestic hot water and preheated boiler water that is used for space heating.

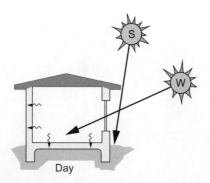

Figure 9-6
The properly designed overhang will shade a southern wall in the summer and allow sunlight to strike the wall in the winter to heat the building. The sun with the "S" shows the high angle of the solar rays in the summer that can be blocked by the correctly designed overhang. As represented by the sun with the "W," the same solar rays will go under the overhang in the winter for free heating.

reduced due to the angle at which the structure receives sunlight. Overhangs can be added to further reduce the solar gain.

Wind Energy

Wind energy uses large "windmills" to turn generators that develop DC electricity, which can be used to charge a battery pack or converted to AC electricity. Wind technology is shown in Figure 9-7. Wind energy can be used to supplement energy at a site. Wind farms generate thousands of kilowatts of energy and feed a nearby electrical grid. The initial investment for wind energy equipment is high; therefore, wind-developed energy is found only in areas of the country that have a fairly constant wind with enough velocity to develop electricity on a daily basis.

Geothermal Systems

"Geothermal heat" can be defined as heat extracted from or added to the earth. Here, "earth" is defined in a broad sense and includes the ground, water, and steam. There are geothermal heat pumps that reject or absorb heat from dependable water sources such as wells, rivers, or lakes, as seen in Figure 9-8. The geothermal heat pump may have hundreds of feet of tubing buried vertically or horizontally in the earth, as shown in Figure 9-9. Heat is rejected or absorbed into the circulating fluid (water and glycol) and then into the surrounding earth. The earth's temperature is usually stable and lower than the outdoor ambient temperature in the summer and higher than the winter ambient temperature. This creates a more efficient operating heat

Figure 9-7
Single wind generator on the left. Wind farm on the right. The wind farms have multiple wind generators that can place electricity on the utility grid. (Courtesy of iStock Photo)

Figure 9-8
Open-loop system, using a water well to remove or add heat to the refrigerant. The pond is used to collect the pumped water. The water can be pumped down to a different water level to keep the local aquifers changed with water.

pump in the extreme conditions of summer and winter. For this reason, the geothermal heat pump has the highest efficiency rating when compared with other types of equipment. Its SEER ratings are over 20, and its HSPF ratings are over 8.

Geothermal energy also involves taking hot water or steam from the earth to turn turbines and generate electricity. Geothermal water or steam can be taken from the earth and used to heat buildings or water. In this case, the hot geothermal source needs to be close to the building using this energy source.

Future Fuel Cells

Hydrogen fuel cells are expected to play a major role in the nation's energy future. After helping develop and introduce the first fuel cells into the market, industry and

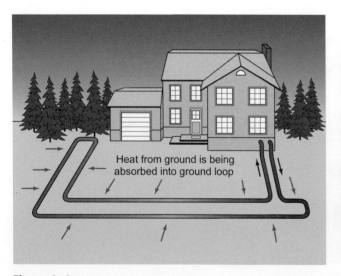
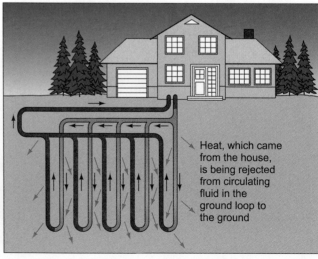

Figure 9-9
Geothermal heat pumps can use a reliable source of water to reject or absorb heat. In this illustration, the heat is carried though hundreds of feet of plastic tubing that is buried horizontally or vertically in the ground.

government are now working on ways to dramatically lower the cells' costs and expand their use in power generation, military applications, and transportation.

Fuel cells are an energy user's dream: an efficient, combustionless, virtually pollution-free power source capable of being located in downtown urban areas or in remote regions and that runs almost silently, having few moving parts.

Using an electrochemical process discovered more than 150 years ago, fuel cells began supplying electric power for spacecraft in the 1960s. Today, they are being developed for more down-to-earth applications: to provide on-site power (and waste heat in some cases) for military bases, banks, police stations, and office buildings from natural gas. Fuel cells can also convert the energy in waste gases from water treatment plants to electricity.

Fuel cells operate much like a battery, using electrodes and an electrolyte to generate electricity. Unlike a battery, however, fuel cells never lose their charge. As long as there is a constant fuel source, fuel cells will generate electricity.

Green Buildings

"Green buildings" are structures that use fewer resources than the average building. The United States Green Building Council is one organization that promotes green buildings, thereby reducing energy use, conserving water resources, and lowering greenhouse gases by selecting building materials that are durable and require less maintenance. To have their project classified as a green building, the designer and builder must improve the thermal quality of the structure by adding more wall and ceiling insulation and using high-efficiency windows. The building should be oriented with minimal exposure on the east and west walls, which will reduce heat gain in the summer and improve it in the winter. Windows on the east and west sides of the structure should be minimized or eliminated. The southern orientation of the building should have the most window exposure and an overhang or external summer window shading. The northern window exposure should be less than the southern exposure and more than the east or west sides. The amount of northern window exposure should be governed by the amount of natural light needed by that side of the structure. Natural day lighting is also part of a green building's design. The builder should design window orientation to minimize the use of lights when sunlight is available.

A green building uses energy-efficient light and HVAC systems in addition to building materials that have a long life cycle and require less upkeep. Finally, green buildings use efficient appliances, solar energy, and low-flow plumbing fixtures and incorporate ways to recycle rain and gray water to be used in landscape watering.

"Sustainability" is another term used in the discussion of green building. The term implies a preference for systems that can be productive indefinitely. A building is referred to as sustainable along with all it components, one of which is the HVAC system. If a green building were to have a sustainable HVAC system, it would need to be designed to last longer than the normal system and be energy efficient. An example of this would be a water-cooled condenser. These systems have a long life if properly maintained, and their utility costs are lower than those of air-cooled condensers. Rainwater could be trapped, filtered, and stored for use in the cooling tower. The idea is to spend more money up front in the various construction components to save on operating expenses and to select durable equipment that will last for the life of the building.

EFFECTS OF MAINTENANCE ON EFFICIENCY

The effects of annual maintenance create a system that operates at its rated performance and at the lowest utility cost. Even when systems are maintained properly, losses occur because not all the dirt can be removed from the indoor and outdoor coils and the compressor experiences mechanical wear, which reduces pumping capacity.

It is difficult to determine the capacity and efficiency losses in a system that is not maintained. Some systems are so dirty that customers must call for service when the system is not cooling because air cannot flow through the filter, condenser, or evaporator coils. As for heating systems, the filter can be dirty and cause the furnace to cut off at the high-temperature limit switch.

There is no doubt that scheduled maintenance reduces operating cost, maintains system efficiency, and increases dependability. Regularly scheduled maintenance extends the life of the equipment because the system has minimum operating hours. Finally, another benefit of scheduled maintenance is to head off problems found in the inspection process. The technician may find and correct a problem before it requires an unscheduled service call, which usually occurs in the heat of the summer or cold of winter.

ENERGY CODE

Minimum energy codes or standards can be adopted from various governmental entities or other organizations. When locally adopted, the governing authority usually modifies the code to meet local needs and requirements. In some cases, the code is modified to require more energy-efficient standards or requirements due to climate conditions in the area. One code cannot possibly cover all local climate requirements.

The energy code enables the effective use of energy in new building construction. It regulates the design and selection of the following:

- Building envelope
- Mechanical systems
- Electrical systems
- Service water heating systems
- When does the energy code apply?
- New construction
- Newly conditioned space
- New construction in existing buildings
- Alterations to existing spaces
- Additions
- Mixed-use buildings
- Change in occupancy
- Make-up air

GOVERNMENT AND UTILITY INCENTIVES

Local, state, and federal energy-efficiency incentives vary widely over the years. Sometimes there are incentives such as tax credits, which may be limited to 1 to 3 years. Governmental incentives come in the form of tax deductions or tax credits. Tax deductions reduce the amount of taxable income, while tax credits are a direct credit against taxes owed to the government entity. Tax credits are more like a direct rebate for part of an energy investment.

Government incentives are directed to residential or commercial business users. Private incentives from utility companies are a different story in that they are directed to the purchaser, the contractor, or the home builder to promote conservation of electricity or natural gas. The incentives are granted after a spot inspection to ensure compliance with the utility energy program. Examples of features that earn incentives include the installation of heat pumps or high-efficiency air conditioning equipment, adding insulation, and using high-efficiency windows. The incentives are given in the form of a rebate per ton of equipment installed. For example, the utility company may offer a $100-per-ton rebate to the contractor who installs high-efficiency heat pumps. Incentive programs change from year to year.

ENERGY STAR PROGRAM

Energy Star is a joint program of the United States Environmental Protection Agency (EPA) and the United States Department of Energy (DOE) designed to help all customers save energy and protect the environment through the use of energy-efficient products and practices.

Results are already adding up. Americans, with the help of Energy Star, saved enough energy in 2007 alone to avoid greenhouse gas emissions equivalent to those from 25 million cars while also saving $14 billion on their utility bills.

More information can be obtained from the agencies' Web sites at http://www.epa.gov and http://www.energystar.gov.

Energy Star for the Home

For the home, energy-efficient choices can save families about a third of their normal energy bills, with similar savings of greenhouse gas emissions, without sacrificing features, style, or comfort. The Energy Star program helps the contractor and customer make energy-efficient choices. Figure 9–10 shows that 17% of greenhouse gases come from residential energy use.

If looking for new household products, many customers look for the ones that have earned the Energy Star seal. They meet strict energy-efficiency guidelines set by the EPA and DOE. If looking for a new home, many customers will look for one that has earned the Energy Star seal.

If customers are looking to make larger improvements in their homes, the EPA offers tools and resources to help plan and undertake projects to reduce energy bills and improve home comfort.

ENERGY-EFFICIENT INSTALLATION: BEST PRACTICES

This section will discuss what goes into an energy-efficient installation. Load calculations, equipment selection and placement, duct sizing, and refrigerant piping sizing are a few of the important considerations when installing an energy-efficient system.

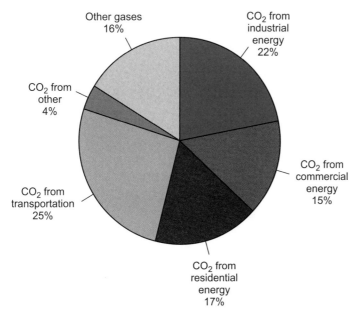

Figure 9–10
Greenhouse gases are created by several major sectors in our country. Transportation is first with 25%, industry is second with 22%, and residential use is third with 17%. Energy Star appliances can reduce residential-generated greenhouse gases.

"Best practices" means doing all the right things in an ideal installation. It is vital to strive for the best practices goal.

Cooling and Heating Load Calculations

Prior to installation, a load calculation will determine the necessary size of the heating and cooling system. This will be calculated by an estimator, designer, or in large commercial jobs a mechanical engineer. Oversizing a cooling system will result in short-cycling because the system will be too large for the structure. An oversized system cools well but does not dehumidify the air and remove moisture, an excess of which leads to high relative humidity with the possibility of mold and mildew formation. Short-cycling due to oversizing does not allow the system to reach its operating efficiency, as illustrated in Figure 9–11.

An undersized cooling system will work fine until the afternoon, when the weather gets hot. The system will operate continuously without satisfying the set indoor temperature. It will finally cool the structure when the outside temperature drops.

Oversizing a heating system may cause failure in the gas furnace heat exchanger and vent system because the furnace will not operate long enough to thoroughly warm the heat exchanger and venting system. An undersized heating system will not provide the required heat on cold days.

The Air Conditioning Contractors of America's (ACCA) Manual J will serve as a standard tool for determining the heating and cooling requirements for a residential structure. Manual N can be used to calculate the loads on commercial buildings. The covers of these CDs and manuals are seen in Figure 9–12.

Manual J and Manual N have recommended winter and summer design conditions based on weather data for hundreds of small and large cities in the United States. Figure 9–13 is a map of the continental United States showing general climate zones. Zone 1 is the warmest, and Zone 7 the coldest. There are pockets of higher and lower temperature zones within these areas.

Once the equipment is selected based on an accurate load calculation and the ductwork sized, the contractor should pull a permit from the agency responsible for enforcing the mechanical code. The agency may not require that formal plans

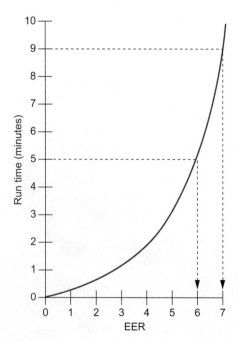

Figure 9–11
This chart shows the relationship between efficiency and air conditioning run time. All systems need to run for a period of time before they reach maximum efficiency. The amount of time varies depending on equipment and conditions.

Figure 9-12
The ACCA Manual J is the standard for residential load calculation. (Courtesy of ACCA)

be approved prior to residential installations, but it will require an approved set of plans prior to approving permits for new commercial installations.

Equipment Sizing, Selection, and Efficiency

The ACCA Manual S is the American National Standards Institute (ANSI) standard for equipment selection. A cooling system should not be oversized by more than 15% because this will cause short-cycling and a lack of runtime to dehumidify the space.

Selecting a two-stage condensing unit can give the customer the option of oversizing the system because the air conditioner can operate on low-capacity cooling until extra cooling is necessary. Extra cooling may be required when record-breaking summer temperatures are experienced or when a large load is added to the building, such as when entertaining a large group of people.

Tech Tip

Many residential air conditioning systems are too large for the homes they serve. Short-cycling due to oversized equipment or system problems creates the following problems:

- Reduced equipment life
- Reduced efficiency (SEER)
- Poor dehumidification
- Reduced filter effectiveness

244   CHAPTER 9

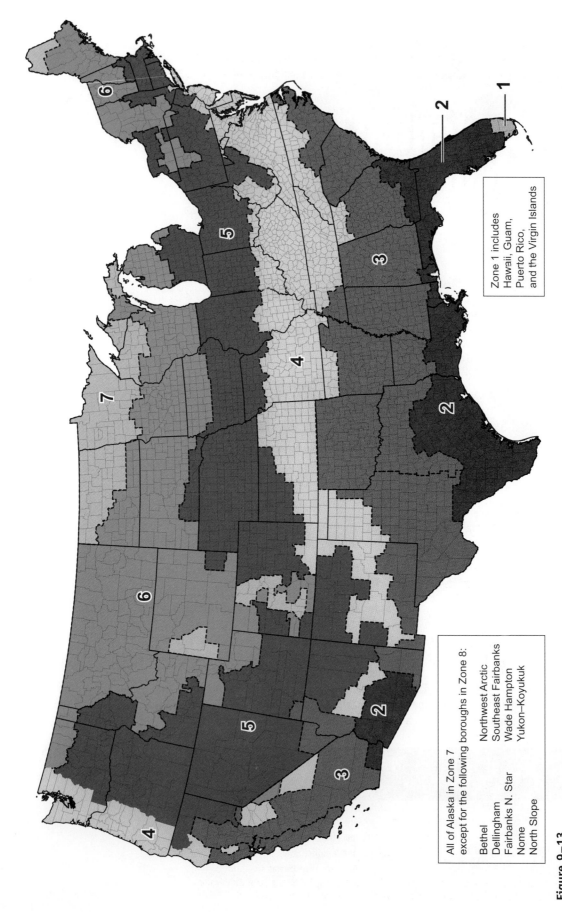

Figure 9–13
Climate zone map of the continental United States. The lower zones are warmer and will require more cooling. The higher zone numbers will require larger heating systems.

When installing new HVAC equipment or changing out a portion of the system, it is important to match the major components (condensing unit, evaporator, and furnace) to achieve the required capacity and efficiency rating. Installing a 16 SEER condenser will not ensure energy savings if an unmatched evaporator coil, metering device, or blower is installed in the system. Changing out the condensing unit requires changing the evaporator coil to achieve correct capacity and efficiency.

Duct Sizing, Support, and Sealing

The ACCA Manual D is the ANSI standard for duct design. The ductwork will be appropriately sized to distribute the correct amount of air to each conditioned space. The design will create a "throw" pattern that pushes the air across 75% of the room at a termination velocity of 50 feet per minute, as seen in Figure 9-14. Most duct systems are designed to move approximately 400 cubic feet of air per minute (CFM) per ton; therefore, a 3-ton system will move a total of 1,200 CFM (400 CFM/ton × 3 tons). Airflow for stand-alone gas heating systems is usually lower, but airflow for electric heat and heat pump equals or exceeds that of a standard air conditioning system.

The duct system should be designed to supply and return the same amount of airflow. If 1,200 CFM is supplied to the space through the duct system, 1,200 CFM should be returned through a properly sized grille, filter, and duct system.

The installer will be given instructions on what size duct to install in which room. The size of the duct is determined by the size of the room and the required air conditioning capacity. The estimator or designer will also develop a grille schedule that tells the installer what size and type of grille is required for each room. The return air should be installed in a central location to return air evenly from all rooms. ACCA has duct calculators and other manuals that can be used to size duct systems. For example, ACCA Manual Q (Figure 9-15) is used for commercial low pressure, low velocity duct system design.

Ducts need to be sealed with proper material. Tapes, mastics, and heat-sealing tapes used with fiberglass ducts are listed with UL 181A. Tapes and mastics used with flexible air ducts and metal ducts are listed with UL 181B. Tapes and mastics rated for UL 181 A or B can be used on any duct system.

Refrigerant Pipe Sizing, Support, and Insulation

When installing a matched system, the installer has a guide for selecting the suction and liquid lines. Match the size of the suction line with the larger fittings on the condenser and evaporator. The liquid line size is determined by the fittings on the condenser and evaporator coil. This match is usually good if the refrigerant line is shorter than 50 feet, with a liquid line rise of no more than 15 feet and fewer than five 90° elbows. Long refrigerant lines with a high liquid line lift and numerous elbows create excessive pressure drop that affects subcooling and could cause

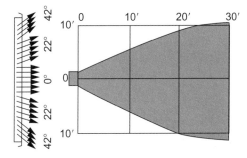

Figure 9-14
The air-throw velocity must be great enough to sweep the air across the area being conditioned. The velocity of the air must not be so great as to make noise as the duct passes through the register louvers. In this example, the airflow pattern will "throw" the air 30 feet before reaching its terminal velocity, which is an airflow of 50 feet per minute (FPM).

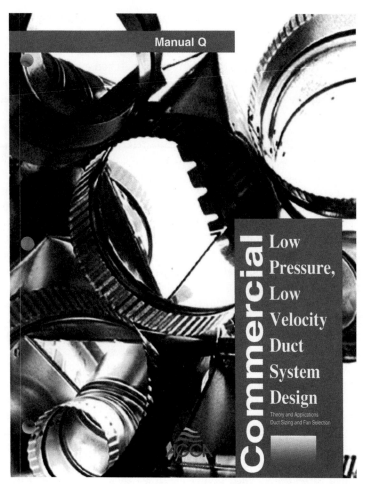

Figure 9–15
ACCA Manual Q is used for commercial low pressure, low velocity duct system design. (Courtesy of ACCA)

> **Tech Tip**
>
> Duct leakage may increase energy consumption by 40% or more. Supply air leaks to the outside create negative pressures in a building, resulting in unconditioned moist and dirty air being drawn into the building. Return duct leaks may bring unwanted, unconditioned air from the attic or crawlspace, and/or bypass the filter grille and may cause problems such as dust, human discomfort, insufficient system capacity, higher utility bills, and mold or other allergens. Factory-made air ducts such as flexible duct and duct board have specific installation instructions from the manufacturer. Use these instructions to ensure a tight duct system that meets mechanical code requirements.

liquid refrigerant to flash. To compensate for these losses, the installer may need to use larger refrigerant piping. Use manufacturers' recommendations or engineering piping charts to determine correct refrigerant pipe sizes.

Pipe support is governed by the mechanical code, which has recommendations based on the size of the piping and whether the piping is installed vertically or horizontally. Support is essential to keep the piping from bending. Support also reduces vibrations that may cause leaks or transmit noise to the structure.

The cool suction line must be insulated to prevent condensation from damaging the structure and unwanted increases in superheat.

Thermostat Installation

The thermostat should be installed in a central location so it can sample the average air temperature in the structure. It is best to install the thermostat about 5 feet high, near a central return air grille, out of direct sunlight, and on an interior wall. Modern-day thermostats use digital technology that does not require leveling, but a professional installer will make this device level and plumb. Some mechanical thermostats such as mercury-bulb thermostats must be level in order to operate properly.

Many digital thermostats require a simple setup program unless they are designed for a specific piece of equipment. Some setup options may include selecting the type of heat, choosing the number of operating cycles per hour, and programming setback temperatures to save energy in the heating and cooling modes, as shown in Figure 9–16. Always read all instructions.

Miscellaneous Components

Other components that might be installed are high-efficiency air filters and humidifiers. A humidifier is used to add moisture to dry air and is controlled by a humidistat, as demonstrated in Figure 9–17. An ultraviolet (UV) light may assist in killing viruses, reducing mold, and keeping coil surfaces free of organic growth.

Dampers should be installed in the take-off fittings of the branch runs. The dampers will allow balancing of the airflow through the duct system. Even a well-designed duct system may require airflow balancing. Chilled water for cooling and hydronic piping for heating require water-flow balancing, and are part of a best practices installation.

Finally, heating systems that burn fossil fuels like gas or oil must have outside combustion air, which is required for proper burning of the fuel as well as venting of the flue gases. Local mechanical codes specify various methods of calculating the amount of outside air required for combustion air. For example, small fossil fuel furnaces in large open areas may not require outside combustion air.

Figure 9–16
This is a commercial digital thermostat that can be programmed to alter the temperature during unoccupied periods. (Courtesy of iStock Photo)

Figure 9-17
This digital humidistat is used to operate a humidifier that will add moisture to air during the heating cycle. (Courtesy of Carrier Corporation)

The National Fuel Gas Code and other mechanical codes have guidelines for the amount of combustion air requirements. The standard method of calculating combustion air is a minimum room of volume of 50 cubic feet per 1,000 BTUH of the gas input. No outside air is required if the volume meets these requirements, but most gas installations do not meet these space requirements. The gas code has several options. For example, vertical ducts to the outside must be sized for 1 square inch of opening per 4,000 BTUH of input rating. A horizontal duct requires 1 square inch per 2,000 BTUH. An upper and lower duct must be supplied to the appliance enclosure.

Replacing the Condensing Unit

In most cases, replacing the outdoor section requires more thought and planning than just installing a new condensing unit. In most change-outs, the condensing unit is over 10 years old. Installing a new, high-efficiency condensing unit requires installing a matching evaporator, a feed device (like a TXV), and sometimes different size refrigerant lines. Without the correct equipment match, the condensing unit will not produce the capacity and efficiency the customer paid for. Here are more problems that result when only the condensing unit is replaced:

1. Reduced capacity in cooling, meaning the system may not be able to keep up with the thermostat setting in the hottest part of the day.
2. For heat pumps, reduced capacity in heating, resulting in the strip heat coming on earlier and running up the operating cost.
3. System charging, especially on critically charged systems with fixed bore metering devices.
4. Inadequate moisture removal.

All of these situations result in poor comfort conditions and extra service calls, and most likely the only way to resolve the issue will be to replace the indoor coil with a matched coil.

> **Safety Tip**
>
> Correct lifting methods are important to prevent back injury. Figure 9-18 shows the steps in good lifting procedures.

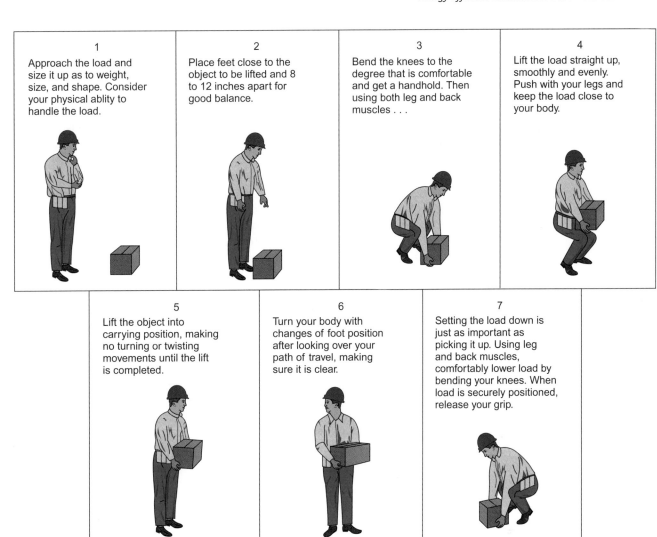

Figure 9–18
Correct lifting procedure.

SUMMARY

This chapter discussed the issues of first cost, maintenance cost, and operating cost. It is helpful to know the terminology and how it is properly used in conversation with a customer. We learned that electricity is measured in kilowatt hours and natural gas in cubic feet. Electricity and gas can be converted into BTUs, which is a measure of heat. Converting these sources of energy makes it easier to compare them. One KW equals about 3,400 BTUs (3.4 × 1,000). One hundred cubic feet of gas, CCF, equals about 100,000 BTUs (100 cubic feet × 1,000 BTUs per cubic foot of gas). Other fuels like propane, butane, and fuel oil can be converted into BTUs. Converting a fuel to BTUs makes it easier to compare its value against others' in heating or air conditioning. Another comparison that must be made is the cost of fuel per BTU.

The chapter discussed four measurements of efficiency, including EER and SEER for cooling and COP and HSPF for heating.

The range of human comfort means that 80% of people will prefer a certain range of temperatures, moisture conditions, airflow, and radiation affects. Our HVAC systems are designed to satisfy that audience.

Finally, we learned about government and utility promotions of energy and resource conservation efforts. The EPA and DOE, both government agencies, have joined forces on the Energy Star program for this purpose. The Air Conditioning Contractors of America (ACCA), Plumbing-Heating-Cooling Contractors Association (PHCC), and Refrigeration Service Engineers Society (RSES) are organizations that educate the community in energy and resource conservation planning.

FIELD EXERCISES

1. On an installation you are working or visiting, list 10 things that you see that are considered "best practices."
2. Look for the yellow energy label on a piece of heating or air conditioning equipment. What does it say? How much energy does it use compared with other models?
3. Review an HVAC system and structure for energy-saving options. What can be done to save more energy? Is this reasonable, or can the initial cost be recovered in a reasonable amount of time?

REVIEW QUESTIONS

1. What is the difference between first cost, maintenance cost, and operating cost?
2. How many watts are in 10 KW?
3. How many BTUH are in 5 KW?
4. How many BTUs are in 1 cubic foot of natural gas? How many BTUs are in 1 CCF?
5. What is EER? How is it calculated?
6. What is SEER? How is it calculated?
7. What is COP? How is it calculated?
8. What is HSPF? How is it calculated?
9. What rating is used to describe the efficiency of an air conditioner?
10. What rating is used to describe the efficiency of a heat pump?
11. List the ranges of the human comfort zone.
12. What are two ways that solar energy is used?
13. What is wind energy?
14. What is the purpose of an energy code?

15. What is the purpose of the Energy Star program?
16. What type of material is used to seal metal ductwork seams?
17. What type of material is used to seal fiberglass duct board?
18. As the outdoor temperature rises above design ambient, the cooling capacity of an air conditioner _____.
19. The total on a residential electrical bill is $300. The customer used 2,000 KWH in this billing period. What is the cost per KWH?
20. The technician wants to determine the BTUH output of electric strip heat. The technician measures 40 amps at 240 volts. What is the BTUH output?
21. According to the human comfort chart, as the relative humidity in the heating season decreases, the human comfort level _____.
22. Describe the effects of relative humidity on viruses and bacteria.
23. Why is a heating and cooling load calculation important?
24. A 4-ton system is going to be installed in a small commercial business. Approximately how many CFM will the air handler need to move?

CHAPTER 10

Selling and Customer Service

LEARNING OBJECTIVES

The student will:

- Define what is meant by "customer satisfaction."
- List the steps of listening.
- Discuss the value added from the customer's perspective.
- Describe how to "value sell."
- List the reasons for discussing repair or replacement with the customer.
- Discuss the role of a technician as a salesperson.
- Describe flat-rate pricing and how to use it.
- Discuss why it is essential that the technician watch her language, dress, and habits around customers and in public.
- List the reasons for obtaining industry certification.
- Name the organizations that promote HVACR education.

INTRODUCTION

This chapter will cover aspects of selling and customer service that will help you become a better technician. Customer service is "job one" for all of us who work in this profession. Installing or repairing equipment is a service to a customer, and selling is an art that should be developed and enjoyed.

Replacing or repairing equipment will place the technician in direct contact with the customer. The contact may be just a short greeting, or it may be continual contact throughout the installation or repair process.

Topics to be covered in this chapter are customer satisfaction, listening skills, understanding the value you add to your product, and value selling. The chapter will also discuss the decision of whether to replace or repair equipment and the value of investment. You will learn the importance of the energy-saving capabilities, quality, and performance of a system as perceived by a customer.

Additional subjects include flat-rate pricing, professional certification, and the importance of a clean appearance, appropriate language on the job, and customer interaction. Look at this chapter as an opportunity to learn something that is not technical but still significant in your advancement as well as the promotion your company. In the eyes of the customer, you are the company. This is vital to remember.

Tech Tip

It is important to remind yourself that your job is customer service. Customer service is not only getting the equipment operating but also making your visit a positive experience for the customer. One way to do this is to leave the job site in better condition than you found it. Remove trash and old pieces of abandoned parts or material from the job. If the trash is in a space that the customer does not see, show him that you are removing materials that were left behind from a previous job. If the job site is clean of debris, think of other ways to leave the job better than you found it and make sure the customer knows of the steps you take beyond the call of duty.

Field Problem

On one job, the technician found that the old flexduct system was in bad shape. The technician showed the customer air leaks and the deteriorating vapor barrier. The outer, gray vapor barrier had split open, and most of the fiberglass insulation was exposed. The technician recommended that the duct be replaced because it was going to turn into a real problem once the insulation started becoming saturated with water. The technician asked the homeowner if the duct had ever been cleaned, and she stated that in the 11 years she had lived in the house, it had not cleaned. The technician stated that the internal liner of the duct was probably dirty, and replacing the duct had the advantage of having a new as well as a clean air duct. The technician also stated that the new sealing requirements and better quality vapor barriers would ensure a longer duct life. The technician also explained the advantage of sheet-metal ducts should the customer want to go with that option.

The homeowner realized that replacing the duct system needed to be done soon and contracted with the technician's company to get the work done.

> **Tech Tip**
>
> Empathize. "Empathy" means to put yourself in the shoes of the customer. If you were the customer and this happened to you, how would you feel?
>
> When you empathize with the customer by letting her know that you understand how she feels, you release tension out of the situation. Empathy is expressed by stating that you genuinely understand why the customer feels a certain way and that you want to help improve the situation.

> **Tech Tip**
>
> Don't take it personally. The customer is not mad at you. The customer may be angry at your company or something else beyond your control. You just happen to be the representative of the company at the time the customer is expressing her dissatisfaction. When you think that you are to blame, just tell yourself that it is not your fault. In some instances, personalities clash and the customer will never like you or you may never like the customer. Let your supervisor know of this situation.

CUSTOMER SATISFACTION

What is customer satisfaction? *"Customer satisfaction" means keeping the user happy.* In most cases, this is fairly easy because the customer will be satisfied if the system is operating within the human comfort zone. Other factors not found in the comfort zone are temperature, humidity, airflow and radiation effects, as well as sound. These are all subject to customer satisfaction or complaints.

Sometimes, customers are sold expectations. What does this mean? The person selling the job discusses the features and benefits with the customer. The customer expects to be comfortable as well as experience whatever else is being promoted, such as the equipment being quieter, running more efficiently, filtering better, or lower utility bills. In this example, the customer may expect a system that does not make any sound, that her utility bill will drop by 50%, and that she will not need to dust again. The sales associate may have given inaccurate information, or the customer may have misinterpreted what was being said. The sales associate should review the job with the customer to prevent the wrong expectations being formed.

Finally, some companies have a customer satisfaction survey. After the customer is shown how to operate the system, the technician reviews a brief customer service checklist with the customer. Review the operation of the thermostat, ask the customer if she is comfortable, and check for quiet operation. The system will not be totally silent, but the sound should be minimal. The approval of acceptable sound levels should include the indoor and outdoor sections, as well as air noise from the grilles and registers. No vibrations or vibration-related sounds should be noticed. The walk-through inspection will help head off any post-installation problems because an experienced technician will be there to answer customer questions and address concerns. Finally, it is best to have all adults responsible for the structure at the walk-through inspection. This will prevent a return call if everyone is not satisfied.

Part of customer satisfaction is explaining to the customer what work is going to be completed prior to starting. State what will be done and how much it will cost. It is important that the customer understand all the conditions of the work; therefore, after a question-and-answer period, the customer should sign a simple contract authorizing the work. Tell the customer when the work will start and finish. (Courtesy of iStock Photo)

LISTEN

How can you handle customer service complaints by listening?

1. Listen: Allow the customer to talk without interruption. This is sometimes difficult to do. Let the customer explain the whole problem, and take brief notes on the important parts of the communication or you may forget something.
2. Repeat: By repeating the complaint, you can ensure that you understand the problem. It is critical to understand the complaint in order to solve the problem.
3. Empathize: Place yourself in the customer's shoes. The customer has a problem she wants solved. It may not bother you or seem to be a problem for you, but she is willing to pay to have something corrected. The vast majority of complaints are legitimate.
4. Apologize: State that you are sorry to hear that the customer is having a problem but that you are here to solve the problem. The problem may not be related to anything your company did on the job. It may not even be the equipment your company installs. The customer must believe that you are concerned.
5. Assure: Tell the customer that you understand the problem and you will investigate to determine the cause.
6. Solve: Repair the problem. If you cannot solve the problem, call for help. This is usually done by excusing yourself and going to the service vehicle to contact a supervisor for advice. No one technician knows everything. Ask for help! The customer does not need to know that you do not know the solution.
7. Check: When a repair is completed, make sure that everything works properly. For example, after replacing a condenser fan motor, check the system performance. If the charge is incorrect, the customer will think you did not fix the problem. Do not give the customer a reason to complain and

call you back a day later. She is not going to want to pay another service charge for something that should have been repaired on the previous trip.

8. Review: Go over the invoice with the customer. Explain what was done, what parts were replaced, and the charges. Ask the customer if she has any questions.
9. Appreciate: Thank the customer for her business and ask her to call your company if she has any more problems.

These are all important listening and action skills. To develop good listening skills, you must spend more time listening and less time talking.

It is very important to listen to the customer. Having both members of a couple present is not important when it comes to scheduled maintenance or minor repairs, but it is extremely important to have both present when making a presentation on a big-ticket item like a condensing unit, furnace, or replacement system. If both are not present, reschedule the presentation for when both are available. Discourage meetings at dinner time or when the customer is in a rush. Try to schedule the meeting within that day so that you do not miss the sale. (Courtesy of iStock Photo)

UNDERSTANDING ADDED VALUE FROM THE CUSTOMER'S PERSPECTIVE

When equipment is replaced or repaired, the customer has no idea of the value added to his system. If the system needs a minor repair such as the replacement of a thermostat, the customer does not expect any extra value. He expects that the system will operate like it did prior to the replacement. He understands the value of minor repairs. However, replacing a mechanical thermostat with a digital

Tech Tip

Ask. Always ask your customer about the problem you are sent out to repair. This is an opportunity to find out the problem as well as break the ice with the customer. This will instill a sense of trust in the customer as well as clarify the problem in your mind. Do not worry about asking the wrong question. Almost any question is going to have a positive impact on the customer.

one may add value to the operation of the system. If the customer does not know this, he may be pleased to see that he can see the temperature readout easier and that the system seems to control the temperature better than the previous thermostat. This is a case of adding value beyond the customer's expectation. The goal is to always exceed the customer's expectations.

From the beginning of your conversation, you must ensure that the customer understand the value you mention. In addition, you should document the added value in writing in the form of a contract or a brochure. When reviewing a value item on a brochure, circle or highlight the information to emphasize its importance. Do not overemphasize a value or emphasize one that cannot be delivered. Review the value, but do not oversell it because the customer will not want to deal with a company that does not deliver what he perceives as a promised value.

VALUE SELLING

"Value selling" means explaining features and benefits that will meet the customer's needs and stay within her budget. Many home and business owners do not realize that there is more to HVAC systems than just heating and cooling. Most customers realize that there are different levels of efficiency and that efficiency carries a higher price, but many customers are unaware of the extra values and features that a system can offer.

Value starts with a determination of what the customer wants. A short survey will help the customer determine his comfort needs. Ask the customer how much he is concerned about these items:

- Comfort
- Energy bills
- Clean air
- Repairs
- Maintenance
- The environment
- Other problems relating to HVAC

The purpose of the survey is to get the customer to think about what he wants. Once you determine his priorities, you can make several suggestions. Some customers will want the basics because they are moving soon or do not want to take the time to learn that value can be added to a standard HVAC system. Try to identify this customer at the outset of the discussion so as not to waste their time. Let's examine the value of comfort concerns.

Comfort Value

Talk to the customer and find out what the comfort concerns are. Ask questions:

1. Is the temperature satisfactory?
2. What about humidity in the summer and winter?
3. Are there cold or warm rooms?

Tech Tip

The first thing to do when you arrive on the job is to check in with the customer. Introduce yourself, your company and what you are there to do. It is a common courtesy and shows professionalism. The customer may give you additional pertinent information that will help you solve the problem. It also helps to establish a relationship with the customer, and if done correctly, helps to ensure additional business when the customer has additional needs in the future. Most customers tend to give business to people they know and like.

A customer survey is important. The survey will reveal what comfort needs are requested by the customer. Items on the survey will include temperature and humidity problems and the desire for a quiet system that produces clean and filtered air that will reduce the utility bills. (Courtesy of iStock Photo)

Tech Tip

Positive word-of-mouth is the most effective and least expensive form of advertising. A dissatisfied customer will tell people about a bad customer experience the very day it happens. Within a week, many people may know about it and become reluctant to do business with your company. Do everything you can to keep the customer satisfied.

4. Is the temperature uniform throughout the structure?
5. Do you feel the air is clean?
6. What do you think about your energy bills?

This is a good way to get the customer to open up and start talking. Make notes on your survey to confirm what the customer values. Quickly review the customer's comfort concerns.

Energy Bill Value

The value of energy savings relates to how a customer views his utility bills. A customer's concern about high electric bills is an opportunity to discuss the value of a setback thermostat. If the customer is away from the home 8 hours or more a day, tell him that a setback thermostat can be programmed to change the heating and

Tech Tip

Apologize. An apology is a minimal response to an upset customer. If your customer has been wronged or even just thinks he has, it is the least you can do to try to save the day. You can tell the customer that you are sorry for what has happened, which is an apology for the situation without necessarily accepting fault.

cooling setpoints a few degrees in the summer and winter to obtain energy savings by reducing load during unoccupied periods. The thermostat should be set to the temperature setting the customer desires. Some customers are uncomfortable with using digital thermostats, which gives you the opportunity to offer value to your service by programming it to their specifications and demonstrating its basic operation.

If the customer values lower utility bills, suggest high-efficiency equipment as the best way to reduce utility cost.

Indoor Air Quality (IAQ)

IAQ must be discussed from two angles, the most obvious being the air filtering system. Is the customer aware of high-efficiency air filtering systems? Does she notice excessive dust in the building? This refers to airborne and settled dust, not dirt tracked in on shoes. Does anyone in the family have allergies? Good air filtration systems may not totally stop dust or allergens, but can reduce them. Make the customer aware that a good filtration system keeps the coil cleaner and discuss options for ventilation and fresh air.

The second point to discuss is duct cleaning. Has the duct system ever been cleaned? Cleaning a duct system will remove most of the dirt and dust but not all of it; only installing a new duct system can do that. The supply and return plenum can be changed or cleaned as part of this recommendation.

Scheduled Maintenance Value

Scheduled maintenance plans, service agreements, and the benefits and costs of those plans, vary from company to company and may include the following:

- First year no-cost inspection
- Scheduled maintenance inspection to ensure that systems operate at peak performance and to help catch minor problems before they become serious
- Parts discounts included with service agreements
- Priority service if a problem should arise

These items add a real value for the customer as well as for your company because they help provide work during slow periods.

Scheduled maintenance agreements are an important part of keeping a system working at peak performance and forecasting future problems with a thorough inspection. Review the maintenance check points with the customer. Give a copy to the customer and file a second copy away at the office or a dry place on the job. (Courtesy of iStock Photo)

Other Problems Relating to HVAC

At the end of the customer discussion, ask about other problems related to HVAC. The customer may have other issues that a replacement system could address. Do not be afraid to ask an open-ended question that may generate more comfort concerns that will add value in the eyes of the customer and profit to the job.

WHY REPLACE VERSUS REPAIR

Sometimes the question comes up of whether to repair or replace a major component part or possibly the whole system. Several factors must be considered, the most important of which is the age of the components. Has a major component like the condensing unit, furnace, or duct system served its useful life? Does it need to be repaired or replaced? The average life of a compressor is 15 years, but replacement compressors tend to have a much shorter life because the unsolved problem that caused the compressor failure the first time may not be corrected. The compressor can be replaced and will operate a few more years. Replacement compressors are usually given a one-year warranty unless the condensing unit is still in its warranty period, in which case the compressor warranty ends at the termination of its original warranty as part of the unit.

What other considerations go into the decision to repair or replace? An owner planning to leave the property within one year will probably not want a replacement unit because he may not benefit from long-term use. A replacement cooling or heating system may save the customer some money, reduce equipment downtime and repair cost, and may come with a new warranty on all parts and labor for a specified time period.

Systems that have a history of repairs should be replaced if they are over 10 years old. The average life of a compressor is 15 years. To replace an expensive compressor in an old system may be a poor investment as other components are likely to fail. Explain to the customer that with an old system, it is best to replace everything so that she can have matched components for the maximum comfort and lower utility operation. New equipment comes with 5 and 10 year warranties, which will be another attractive feature and encourage system replacement. (Courtesy of iStock Photo)

The Environment

Environmentally-friendly green consumers can be an easy sell for high efficiency systems if you explain the relationship between carbon dioxide and our enviroment.

Value of Investment

The value of an HVAC investment includes first cost, utility savings, payback, and human comfort. The customer must understand and be willing to pay for his comfort concerns. The greater the number of comfort concerns, the greater the first cost. For example, to have a high-efficiency filtration system costs more money than a less efficient one. Is this important enough for the customer to increase the first cost?

A system with a high SEER rating will have a higher first cost and lower operating cost. The payback will be utility savings. The price of electricity will increase over the life of the system, making the payback even better.

As mentioned earlier, the value of investment includes comfort. Do you measure or notice comfort unless there is a problem? As long as the customer is comfortable, he takes his human comfort for granted. The customer must be aware that he is purchasing comfort and should be given the options of comfort benefits. Knowing the features and benefits of a system is important to selling a good comfort control system. It also makes the customer feel good about his investment.

> **Safety Tip**
>
> A small fire extinguisher is important when checking out a fossil fuel furnace or when brazing or soldering. The fire extinguisher should be nearby when working on or with any heat-producing activities. Have this device within a few feet of your working location.

Safety Tip: A small fire extinguisher is important when checking out a fossil fuel furnace or when brazing or soldering. The fire extinguisher should be nearby when working on or with any heat-producing activities. A fire extinguisher left in a service vehicle will do little good to stop a fire once it is started. Have this device within a few feet of your working location. (Courtesy of iStock Photo)

ROLE OF THE TECHNICIAN AS A SALESPERSON

The role of the technician as a salesperson is significant, and the technician must realize its importance. Even if the technician does not offer an item for sale, the customer sees the technician as a salesperson for comfort. Repairing or replacing a component or system sells comfort.

Because you are a technician, your customer expects you to be clean and well-groomed. Carry an extra uniform in case you get dirty early in the day. The customer does not expect you to be perfectly clean, but she will be impressed with a sharp-looking technician. (Courtesy of iStock Photo)

The technician is the number one salesperson because he is on the job representing the company and has the opportunity to earn the trust of the customer.

One approach to total system replacement is to determine what the customer would like in the new system. For example, does the customer want a better air filtration system? Does the customer want high-efficiency equipment? What are the comfort concerns? Offer the customer good, better, and best options. Start with the best option first. Explain the options in the best package, followed by the total price including taxes. If the customer has objections to the price, review the comfort options to make certain he understands every option being offered. If price is still an objection, the customer must decide which option to take off the table to reduce the price and value to the "better" category. Every time the price is reduced, the value of the original price quote must be dropped. Let the customer choose what he wants to eliminate because he is the one making the financial decision. You do not want to be the one making the decision for him.

> **Tech Tip**
>
> No one should feel uncomfortable about selling. Offer something to the customer that will truly add value to his system. In many cases, the customer will not accept your recommendation, but the time may come that he will reconsider it. That time may be your next visit or at the time the equipment is replaced.

Finally, you can offer the customer one improvement suggestion every time you make a service or maintenance call. You may notice something important that needs to be replaced. If it needs a little patch job, repair it. If it is something important, it should be mentioned at every visit. Programmable thermostats may be promoted one visit and a better air filtration system the next. Heating or cooling systems in need of replacement should be the number one priority even if they are functioning or patched from year to year. When a heating or cooling system is changed, customers are more likely to go with some of the recommended upgrades. When doing repairs or scheduled maintenance, write at least one recommendation upgrade on the invoice after finding what the customer needs so the customer will see that what you say is worthy enough to be documented.

FLAT-RATE PRICING

Flat-rate pricing is a way to bill a customer based on an industry standard. When you take your vehicle to get the oil changed, the oil-changing station gives you a price before they get started. This is flat-rate pricing—knowing the price of a repair or service before it is started.

Flat-rate pricing establishes a price for a service call. The flat rate establishes a price to bring the technician to the job and troubleshoot the problem. The distance the technician needs to travel should be considered when establishing these flat-rate service call rates.

Flat-rate pricing develops a price for time and materials for the most common service procedures. How much does it cost to replace a Brand ABC compressor model number 124A? The flat-rate pricing book will have the total price, which will include the price of the compressor, a drier, and the time it takes to recover refrigerant, replace the compressor, pull a vacuum, and recharge. Flat-rate pricing covers time and material, and sometimes it includes sales tax. The customer will know the total bill before the work starts. From the flat-rate pricing book, the technician shows the customer what the service costs, so the technician does not need to decide what to charge the customer. The flat-rate pricing program does it for them.

In summary, flat-rate pricing allows the customer to know what the charge will be prior to the start of a repair job.

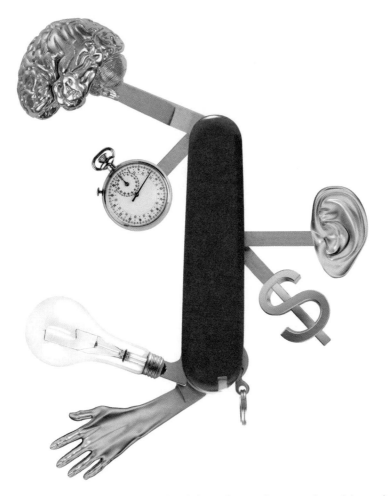

Everyone needs help in our industry, whether it be assistance from a tool or advice on how to handle a sale. Be willing to ask questions and request help when you are in doubt about how to handle a situation. (Courtesy iStock Photo)

VALUE OF ENERGY SAVINGS

As mentioned in the section on value of energy efficiency, efficiency translates into energy savings for the customer. Utility bills seem to always be increasing; therefore, energy-saving systems are more important. Energy savings increase as the price of energy increases.

The price rise of energy is compounded. What does this mean? Here is an illustration of compounding. If the price of a utility increases 10% a year, the price will more than double in 7.5 years. Ten percent is multiplied by the 10% from the previous year and then on to the next years, as shown. Let's start with a $1,000 annual utility bill rising at 10% a year:

- Beginning of year one, the annual utility bill is $1,000
- End of year one is 110% = 1.10 × $1,000 = $1,100
- End of year two is 110% = 1.10 × $1,100 = $1,210
- End of year three 110% = 1.11 × $1,210 = $1,331
- End of year four 110% = 1.10 × 1,331 = $1,464
- End of year five 110% = 1.10 × 1,464 = $1,610
- End of year six 110% = 1.10 × 1,610 = $1,771
- End of year seven 110% = 1.10 × 1,771 = $1,948

Utility costs may not rise by 10% every year, but whatever the rise, a customer saves more money because of compounding.

UNIFORMS, LANGUAGE, AND CUSTOMER INTERACTION

Uniforms not only standardize the appearance of the company's employees, but they also make the technicians part of a team. Sports teams have uniforms to show their unity and separate them from the competition. The customer recognizes the uniform, associates its appearance with your company, and feels more comfortable when the technician arrives at the door with the familiar uniform. A uniform can be as simple as a shirt with the company name and logo, or it can be detailed, including the individual's name on the shirt, matching pants, a belt, and safety boots. Some uniform materials meet safety standards for fire-retardant treatment. These may be required by employers when working in plants that produce flammable components.

A clean, neat uniform should be a standard. You are trying to impress customers. It is a good plan to carry an extra uniform if you going to more than one location in a day. Should the uniform get dirty at the first stop, you can change it prior to arriving at the following stop. Customers do not expect you to stay clean, but they do expect you to arrive clean even if it is late in the day.

Language

Slang, profanity, cursing, and other inappropriate terms make most customers feel uncomfortable even if they do not say anything about it. Putting down one's wife, husband, or friend in front of a customer creates a barrier of mistrust and doubt. Be careful about what you say and where you say it. You may think a customer cannot hear you because you are in the attic or basement, but that is not always true. Language should be guarded when on a customer's property. Not using questionable language is always the best policy.

We all like to tell and listen to jokes, and many are very funny, but off-color jokes or stories can be another source of trouble. Even the most seemingly innocent joke may offend someone. Keep all jokes and questionable stories away from the job site.

Do not criticize your supervisor, company, or competition when on the customer's property. Even if this does not get back to the company, the customer will have less respect for you. It makes the customer uncomfortable to hear about "dirty laundry" that relates to your business.

Customer Interaction

A certain amount of customer interaction is desirable. The first goal is to make your customer comfortable. When you approach the door, knock or ring the doorbell and then step back about three steps. Only one technician should approach the house. Stand so that the customer can easily see you through the peephole, door window, or side window. Park the company vehicle in front of the house on the street so that it is visible to the customer. If you have a clip-on name tag, have it visibly located on your shirt or jacket. Offer a business card or business brochure.

Before leaving the job, even for lunch or to get parts, the technician should alert the customer in case she needs to leave and lock up the property. When the job is completed, one technician should review the work and invoice with the customer. This is important to ensure that the customer is satisfied. If the customer is not satisfied, it is time to call the office and get the situation straightened out. Some HVAC companies follow up with a phone call to determine if the customer is pleased and find out how they can improve their service.

QUALITY AND CLEANLINESS

The customer expects quality and cleanliness from HVAC technicians. The customer feels she is paying for a service that will bring her heating or cooling system up to peak performance. The customer trusts the technician and the company to do what they are paid to do.

> **Tech Tip**
>
> Do not blame. No one cares who is at fault. Blaming someone else for the customer's problem makes you look bad and makes the company look bad. Blame is the first response of a small person. Do not show yourself to be in that league by immediately jumping to blame someone. Instead, make a point to simply acknowledge what the problem is and state how you're going to fix it.

The customer wants to see a clean service vehicle and a relatively clean technician.

A fresh shave or well-trimmed beard and hair create a positive first impression. Long hair should be tucked up under a cap.

When the technician enters a home or business, she should remove her shoes or place covers over the shoes per the company's policy. When changing pieces of equipment in the home, the technician should lay out runners to protect the floor covering.

PROFESSIONAL AFFILIATION AND CERTIFICATION

Professional affiliation and certification show that a technician has the potential to become a valuable employee. Persons pursuing the HVAC profession can obtain education by taking courses in select high schools, attending HVAC courses at a community college or university, or as many new technicians do, learning the profession in the field supplemented by attending apprentice and manufacturers' training classes. The technician may be able to obtain a certificate or associate's degree from an apprenticeship training program or local community college, and some private technician schools offer certificates in HVAC. In addition, some university and four-year colleges offer bachelor's degrees in HVAC.

Even if a technician attends formal education, continuing education is needed to upgrade skills in a changing profession. No one school or university can teach it all. Just as you want your doctor to learn the latest medical skills to practice medicine, you will need to do the same in order to work on the doctor's home and office HVAC systems.

Here is a list of air conditioning organizations that will help you in your professional education and certification:

- Air Conditioning Contractors of America (ACCA)
- Plumbing-Heating-Cooling Contractors – National Association (PHCC)
- Refrigeration Service Engineers Society (RSES)
- North American Technician Excellence (NATE)
- HVAC Excellence
- UA STAR

The role of each of these organizations is discussed next.

> **Tech Tip**
>
> Remember to leave the job better than you found it. Yes, you will leave it repaired, which is better than you found, but this adage refers to cleaning up the site. Clean the site or do something that the customer will notice as a general improvement. With the customer's permission, trim grass, plants, or branches away from the condensing unit. Some companies actually wax the condensing unit as a finishing touch on a clean-and-check maintenance job.

Air Conditioning Contractors of America (ACCA)

ACCA is the largest HVAC contractor organization in the world, with members in all 50 states as well as more than a dozen foreign countries. ACCA is a contractor-led organization that exists for the purpose of improving the industry and helping its members succeed.

The ACCA Web site is http://www.acca.org.

Plumbing-Heating-Cooling Contractors—National Association (PHCC)

PHCC, a national association, is the oldest trade association in the construction industry and a premier organization for the plumbing, heating, and cooling profession. Since 1883, PHCC has been the leader in promoting, advancing, educating, and training plumbing and HVAC professionals. Today, PHCC has more than 4,100 contractor members from open and union shops who work in the residential, commercial, new construction, industrial, and service and repair industry segments.

Learn more about PHCC by visiting its Web site at http://www.phccweb.org.

Refrigeration Service Engineers Society (RSES)

RSES is the world's leading education, training, and certification association for HVACR professionals. For more than 70 years, RSES has provided opportunities for advancement in the HVACR industry. RSES membership allows access to training and education, which is increasingly important for succeeding in today's HVACR industry. Visit the RSES Web site at http://www.rses.org.

North American Technician Excellence (NATE)

NATE is an independent, third-party certification body for HVAC technicians. NATE tests technicians while others like ACCA, PHCC, and RSES train. Testing validates the technician's knowledge and a training program's instruction. NATE-approved testing organizations throughout the United States and Canada are open for technicians to take the NATE exams.

Review the NATE Web site at http://www.natex.org for more information on how to become NATE certified.

HVAC Excellence

The mission of HVAC Excellence is to improve the future and current technical workforce's competency through quality education. To fulfill this goal, HVAC Excellence developed its "Five Elements to Achieve a Competent Workforce." These five elements are programmatic accreditation, educator credentialing, employment-ready certification, professional-level written technician certification exams, and master specialist hands-on technician certification exams.

More information can be obtained by going to http://www.HVACExcellence.org.

UA STAR

The United Association of Journeymen and Apprentices of the Plumbing and Pipe-fitting Industry of United States and Canada (UA) offers an HVACR certification exam and the opportunity to earn 30 college credits toward an Associate Degree in HVACR or Construction Supervision. This certification is available only for union-affiliated technicians.

See the UA STAR Web site at http://www.uastar.info for more information on how to become STAR certified.

PERFORMANCE OF A SYSTEM AS PERCEIVED BY THE CUSTOMER

Understanding how a system works is critical to understanding a problem that a customer may have. A customer must never be degraded or laughed at if he does not know how to operate his thermostat, locate the air filter, or understand how his system performs. The next section covers how to explain what a customer should know without overwhelming him with too much information.

Performance of a System: What Do I Say to Mrs. Smith?

If the system is heating or cooling well, tell Mrs. Smith everything is working according to manufacturer's standards. If it is not working properly, the customer should be told so. Some customers do not realize that the heating or cooling system may have a problem because it may be operating at a reduced capacity. The problem could be long run times, yet the system could be keeping the customer at a comfortable temperature. Some customers do not know what to expect of their systems and may not realize that wide temperature swings between heating or cooling cycles is not normal. It is up to the technician to report any problems and give the customer options to correct them. If the customer is satisfied with the operation, at least she was advised that the system is not performing as designed. The customer will appreciate that you located the problem and allowed her to make the decision on whether to correct the problem at this time.

Right System for the Right Building at the Right Price

The right system for the right building means that the heating, cooling, and duct systems are designed for the home or business where they are installed. A heating and cooling load calculation needs to be completed to determine these HVAC needs. The formal load calculation, equipment selection, and duct design are recommended to legally cover this critical aspect of the job.

The "right price" is an arbitrary topic. Everyone likes to receive the lowest price for the greatest value and comfort, but the price will depend on the equipment's features and benefits. HVAC companies, like all businesses, are not created equal. It is up to the HVAC company to educate the customer and to point out ways their service or installation is superior. This is best done by a brochure listing and showing the customer the value of the service she is receiving. Point it out to her.

Finally, as was discussed earlier, it is vital to determine what the customer wants and then develop good, better, and best options with prices that match. If price is an objection, ask the customer which options she wants to remove to reduce the price. Customers want comfort and are willing to pay extra for it. It is essential that the customer be made aware of what is available so that she can make what she considers good choices. Help your customers make the correct decisions.

Unit and Component Capacity

Capacity issues are more common on replacement condensing units. In most cases it is recommended that with condensing unit replacements, the evaporator should also be replaced. This matched replacement maintains the rated efficiency and capacity of the cooling system. The refrigerant lines will need to be checked. Many R-410A systems require smaller suction lines than R-22 systems. A different evaporator and an expansion valve are also required along with new refrigerant lines when changing to a R-410A unit.

SUMMARY

This chapter discussed selling and customer service skills. The technician "sells" service in the form of installing, repairing, or replacing HVAC equipment. It may not seem like selling, but the technician is selling a service. The service is to have the equipment operating within the needs of the human comfort zone. Selling also includes simply helping the customer realize that other options are available to make him more comfortable with possibly more efficiency. Let the customer decide what options work best for him.

Customer service is part of our everyday dealings with all customers. The customer wants to know who you are and what you are going to do and wants to be updated on the progress of your work on the problem. Finally, the customer wants an explanation of what was done. This is a final opportunity to make a good impression. Complete the customer invoice and discuss all details on it, give the customer a valuable recommendation for improvement on his system, and write the recommendation on the invoice. The customer may not take advantage of it now, but the suggestion may pay off in a return visit.

This chapter discussed the various professional organizations that a technician should consider working with to keep abreast of the changes in the field. ACCA, PHCC, and RSES are all professional organizations that can fulfill this continuing education role. HVAC Excellence, NATE, RSES, and UA STAR are testing and certifying agencies that are recognized nationally. Passing one of these certification exams is not easy, so having a certification is truly a benefit to the technician, the company that hires him, and the customer.

FIELD EXERCISES

1. Go to the Air-Conditioning, Heating, and Refrigeration Institute Web site at http://www.ahri.org. Under "Specifiers/Contractors," click on "Certified HVACR Equipment." Accept the terms of use, and this will take you to the AHRI Unitary Directory of Certified Products page. Print out this page. What information will you need in order to obtain system information? What information can this Web site give you?
2. Go to the ACCA Web site at http://www.acca.org and list five benefits you can derive from being an ACCA member.
3. Go to the PHCC Web site at http://www.phccweb.org and list five benefits you can derive from being a PHCC member.
4. Go to the RSES Web site at http://www.rses.org and list five benefits you can derive from being an RSES member.
5. Go to the NATE Web site at http://www.natex.org and find the KATEs (Knowledge Areas of Technician Expertise) section (http://www.natex.org/data_kates.htm). KATEs are used as a study guide to help prepare for the NATE exam.

REVIEW QUESTIONS

1. What is customer service?
2. List five features of good listening skills.
3. How is value added when dealing with a customer?
4. What is value selling?
5. When should a system be repaired? When should a system be replaced?
6. What is the role of a technician as a salesperson?

7. Describe flat-rate pricing. Include charges for the service call and parts/equipment.
8. What is the purpose of a company uniform?
9. Why is it important to watch your language on a job and at the office?
10. Why is customer interaction important?
11. Who is the target audience of ACCA?
12. PHCC supports what professions?
13. What are two main goals of RSES?
14. How does NATE help the HVAC industry?
15. How does HVAC Excellence help the HVAC industry?

Answer Key

Chapter 1

1. What percentage of growth for HVACR employment is expected by the United States Bureau of Labor Statistics from 2004 to 2014? *18–26%*
2. What does the acronym HVACR mean? *Heating, ventilation, air conditioning, and refrigeration*
3. Describe five different career opportunities found in our profession. *Open-ended answers such as installer, technician, sales, owner/operator, inspector, estimator, and other related positions are acceptable.*
4. Discuss your goals in our profession. How do you expect to accomplish these goals? *Open-ended answer. Evaluate answer based on class discussion on this topic.*
5. What does a technician do at a job site? *Repairs, installs, and/or maintains HVACR equipment and practices good customer service.*
6. What required certification will you need to obtain as a technician in order to legally install gauges on a refrigeration system? *EPA Refrigerant Handling Certification*
7. Name five general locations where you might be working while doing HVACR work. *Here are some answer options: construction, homes, commercial and industrial complexes, health care facilities, transportation, manufacturing, educational complexes, military, refineries, chemical plants, and other sites.*
8. Why is record keeping a very important part of the technician's job? *Gives the customer detailed records of the work completed along with the cost involved, and records are used to bill for time and materials on the job.*

Chapter 2

1. Draw the refrigeration cycle by connecting the refrigerant lines in Figure 2–38. *The compressor will be connected to the condenser; the condenser to the metering device; the metering device to the evaporator; and the evaporator back to the compressor.*
2. Label the refrigerant lines in Figure 2–38. *The line between the compressor and condenser is the discharge or hot vapor line. The line between the condenser and metering device is the liquid line. The line between the evaporator and compressor is the suction line.*
3. Which component in the refrigeration cycle rejects heat? *Condenser*
4. Which component in the refrigeration cycle absorbs heat? *Evaporator*
5. Which components change the pressure in the refrigeration cycle? *Compressor and metering device*
6. What are the major components of the gas furnace? *Controls section, blower section, burner and heat exchanger section, and venting section*
7. What is hydronic heating? *Circulating heated water or steam to a heat exchanger for space heating*
8. Draw the refrigeration cycle for the heat pump in the cooling mode. *Use Figure 2–32 as a reference.*
9. Which component in the heat pump heats the indoor air? *Indoor coil (or condenser)*
10. What are two sources of heat for a geothermal heat pump? *Water from a lake, well, river, or waterway, or buried piping in the ground with water circulating through it*
11. What is the difference between an open-loop and a closed-loop geothermal heat pump system? *An open-loop system pulls water from a well or open water source. A closed-loop system circulates water through piping that is buried in the earth.*
12. Give two examples of active solar energy applications. *Photovoltaic cells that generate voltage, or direct water heating applications*
13. What is the difference between ACR tubing and water piping? *ACR tubing is cleaned, purged with nitrogen, and capped.*
14. What are three common duct materials? *Sheet metal, flexduct, and duct board*
15. How many BTUH are produced by electric heat strips that equal 10 KW? *10,000 × 3.4 = 34,000 BTUH*

Chapter 3

1. Define "volt." *Electrical pressure that pushes electrons through a circuit*
2. Define "ampere." *The number of electrons flowing in a circuit*
3. Define "ohm." *Opposition to current flow*
4. Define "watt." *In power, a measure of work being done*
5. How many volts will be measured in an electric heat strip if the resistance of a heater is 25 Ω and it measures 10 amperes? *250 volts*
6. How many amps are measured in a heating circuit that has 240 volts applied with a resistance of 10 Ω? *24 amps*
7. How many ohms of resistance will be in a heat strip that draws 20 amps at 240 volts? *12 ohms*

8. How many watts will a set of electric heat strips draw if the total measured amperage is 45 A and they have an applied voltage of 250 V? *11,250 watts*
9. Convert the answer to question 8 to KW. *11.250 KW*
10. What are two precautions you must observe when measuring resistance? *Remove power from the circuit and remove the component from the circuit*
11. What is the difference between an analog meter and a digital meter? *The analog meter uses a needle movement to indicate a reading. The digital meter uses a number display to indicate a reading.*
12. Name five safety features that you must require when purchasing a multimeter. *Select five of the seven listed*:
    - *Fuse protected*
    - *Tested by independent laboratories*
    - *600 or 1000 volt CAT III or 600 volt CAT IV rating*
    - *Ohms circuit protected to the same level as the voltage test circuit*
    - *Measures true RMS voltage*
    - *No broken case, worn test leads, or faded digital display*
    - *Has the features or options that are required for your job*
13. Draw and label twenty electrical symbols used in the HVACR industry. *See symbol drawings in the chapter.*
14. What are the components of a complete circuit? *Power, load, and interconnecting wires*
15. Draw an air conditioning circuit that includes the compressor, condenser, indoor fan, thermostat, and control circuit. Label the components in abbreviated form. *See diagram in the chapter as a guide.*
16. What factors determine the resistance of a conductor? *Resistance in a wire is determined by its length, diameter, type of material, and surrounding temperature.*
17. How many amps will a #12 AWG-TW (and UF) copper conductor handle in an air conditioning circuit? *25 amps*
18. How many amps will a #10 AWG-THHN copper conductor handle in a refrigeration circuit? *40 amps*
19. How do you measure low amperage using a clamp-on ammeter? *Wrap the wire being measured ten times around the jaws of the clamp-on meter, and then divide the amperage reading by 10*
20. Name three types of wiring methods used in HVACR. What is the construction of each type?
    - *NM is a nonmetallic casing enclosing the wire.*
    - *MC is a metal-clad casing.*
    - *EMT is electrical-metallic tubing.*

Chapter 4

1. How many BTUs will it take to heat a pound of water from 50°F to 100°F? *50 BTUs*
2. How many BTUs will it take to heat 10 pounds of water from 50°F to 100°F? *500 BTUs*
3. How many BTUs will it take to cool 10 pounds of water from 100°F to 50°F? *500 BTUs*
4. Give an example using Boyle's law. *Refrigerant compressed in a cylinder*
5. Give an example using Charles's law. *The pressure in a cylinder increasing as the temperature around it increases*
6. Give an example using Dalton's law. *The pressure of mixing two refrigerants with different pressures will be additive.*
7. What is the difference between sensible heat and latent heat? *Sensible heat can be measured with a thermometer and is associated with a change in temperature. Latent heat is associated with a change of state without a change in temperature.*
8. Give an example of sensible heat and latent heat. *An example of sensible heat will be anything that changes the temperature of an object. An example of latent heat will be changing water to ice or ice to water or water to steam or steam to water.*
9. What is specific heat? Where is specific heat applied? *Specific heat is the amount of heat required to raise 1 pound of a material 1 degree Fahrenheit. It can be applied to air, water, or commercial refrigeration loads such as fruit, vegetables, or meats.*
10. Describe three methods of heat transfer. Give an example of each method. *Conduction: Heat moving in a solid object by molecule-to-molecule transfer; Convection: Heat transferred by slow movement of a fluid such as warm air or water rising as it is heated; Radiation: Heat transferred through space without heating the space such as sunrays going through a window and heating the inside of a building*
11. What is pressure? *Force over unit area*
12. What is the difference between gauge pressure and absolute pressure? *Gauge pressure is typically considered to be zero psig under normal atmospheric conditions, and absolute pressure is the pressure of the air above us added to gauge pressure.*
13. Convert a 300 psig pressure reading to psia. *314.7 psia*
14. What is thermodynamics? *The study of heat transfer*
15. What is a ton of air conditioning? How many BTUH will a 10-ton air conditioning system remove? *A ton of air conditioning is 12,000 BTUH. A 10-ton system will remove 120,000 BTUH.*
16. What is the absolute temperature of 50°F? *510°R*
17. What is the difference between boiling, evaporation, and sublimation? *Boiling is a temperature at which a fluid changes rapidly from a liquid to a vapor. Evaporation is the slow process of changing liquid to a vapor. Sublimation is a change of state from a solid to a vapor without going through the liquid phase.*
18. What is matter? What are the three states of matter? *Matter is any substance that can be found as a solid, liquid, or a gas.*
19. What are the five basic laws of nature?
    Law 1: *Heat exists in the air down to absolute zero, which is −460°F.*

Law 2: *Heat flows from a higher temperature to a lower temperature regardless of how small the temperature difference might be.*
Law 3: *Due to friction between molecules, all gases become warmer when compressed.*
Law 4: *Matter can be in a solid (ice), liquid (water), or gas (vapor) state.*
Law 5: *The temperature at which a material changes from a liquid to a gas or from a gas to a liquid depends on the pressure at which it is contained.*

20. What are the two laws of thermodynamics?
    - *Energy cannot be created or destroyed, but it can be converted from one form of energy to another form of energy.*
    - *Energy degrades into low-level heat energy. (Heat always flows from a higher temperature to a lower temperature.)*

Chapter 5

1. Refer to Figure 5-39. On a separate piece of paper, redraw, connect, and label the refrigerant lines. *See figure Unf 5-1 as the answer.*
2. Draw the refrigeration cycle. Label the major components and refrigerant lines. *See figure Unf 5-1 as the answer.*
3. Which components create a pressure change in the refrigeration cycle? *Compressor and metering device*
4. Which component rejects heat? *Condenser*
5. Which component is known as the heart of the system? *Compressor*
6. Which component absorbs heat from the air? *Evaporator*
7. Which component uses cool refrigerant in order to condense moisture from the air? *Evaporator*
8. Which component condenses the refrigerant in order to reject heat? *Condenser*
9. Which components add heat to the refrigerant? *Evaporator and compressor*
10. Which component has hot, superheated vapor at its outlet? *Compressor*
11. Which component has warm, subcooled liquid at its outlet? *Condenser*
12. For R-22, what is the corresponding pressure for the boiling point at 35°F? *Approximately 61 psi*
13. For R-410A, what is the corresponding boiling point temperature for 130 psig? *Approximately 45°F*
14. For R-22, what is the corresponding boiling point temperature for 300 psig? *Approximately 131°F*
15. For R-410A, what is the corresponding boiling point temperature for 400 psig? *Approximately 117°F*
16. What is superheat? *Heat added to the substance or refrigerant above its boiling point or saturation temperature*
17. What is the superheat of an R-22 air conditioning system if the suction pressure is 70 psig and the suction line measures 50°F? *9°F*
18. What is subcooling? *Cooling the liquid refrigerant below its condensing temperature*
19. What is the subcooling of an R-410A air conditioning system if the liquid line pressure is 477 psig and the liquid line temperature is 110°F? *20°F*
20. Draw the refrigeration cycle. Label all the components and the direction of refrigerant flow. Label the conditions of the refrigerant as it enters and leaves each component in the cycle. *Use figure Unf 5-1 as a guide. The answer should also include the general pressures (high or low) and refrigerant conditions (superheat, subcooling, mixture) entering and leaving each component.*

Chapter 6

1. Why is it important to check the air filter? *To ensure clean air in the building, to keep the evaporator coil clean, and to maintain desired airflow.*
2. Why is it important to check the system pressures? *To aid in system diagnosis.*
3. Why is it important to check for air leaks? *Air leaks in supply ducts will lose cooled or heated air. Air leaks in return ducts can pull unconditioned air into the air conditioning system. Air leaks can alter the building pressure zones.*
4. What are two reasons for scheduled maintenance? *Efficient operation and full capacity rating; also, better customer comfort, longer equipment life, and preventing problems before they occur.*
5. What is meant by "exceeding the expectations of the customer"? *Doing more than what is expected by the customer*
6. What device is used to measure the dry bulb and wet bulb temperatures? *Psychrometer*
7. What are important considerations when using coil-cleaning chemicals? *Read and follow directions on the chemical cleaner container. Be sure to have an MSDS sheet for the product available. Ensure that harmful/noxious fumes are not introduced into the building.*
8. How are air filters rated? What is the size of the filtered particles in this rating? *Filters have a MERV rating. The filter is tested to trap particles between 0.3 and 10 microns.*
9. What is the acceptable voltage range (percentage) measured at a condensing unit? *±10% of the rated input voltage*
10. List ten important things to check in a scheduled maintenance program.
    - *Air filter*
    - *Charge—including pressures, superheat, and/or subcooling*
    - *Condenser coil*
    - *Evaporator coil*
    - *Duct leaks*
    - *Thermostat operation*
    - *Compressor amperage*
    - *Motor lubricant and/or blower bearing lubricant*
    - *Wire connections*
    - *Visual inspection*

Chapter 7

1. Why is it important to review the problem with the customer prior to starting troubleshooting? *The customer may help the technician narrow down the symptoms and possible causes of the problem.*
2. What is the voltage measured across the thermostat in Figure 7-1 when the light operates? What is the voltage across the thermostat when the light is burned out? *120 volts is measured when the light is operating. 120 volts is measured when the light is burned out.*
3. Describe the meter-voltage-reading troubleshooting technique used to troubleshoot an open compressor internal thermostat found in Figure 7-2. *Turn the system to the on position and lower the thermostat several degrees below the building temperature setting. Attach one voltmeter probe to the right side, or L2. With the other test probe applied, voltage will be measured at L1; voltage will also be measured on the left side of contactor C and on the right side of contactor C. The applied voltage will be measured on the left side of COMP. Zero volts will be measured on the right side of COMP. Measuring directly across COMP will indicate the supply voltage. This means that voltage is supplied to the compressor, which fails to start. With wires disconnected at the compressor terminals, a resistance check with meter reveals that there is open circuit. Allow the compressor motor to cool, which may take up to 12 hours, prior to condemning the compressor. Then find the cause of the thermostat tripping.*
4. Describe general steps you would take to locate a problem in a residential air conditioning system. *Discuss the problem with the customer. Turn the thermostat up or down so that it can operate in mode for troubleshooting. Determine what is operating or not operating. List the symptoms and all probable causes to troubleshoot the problem.*
5. Why is productivity important to your company? *Productivity is important so that the company can make a profit to pay its technicians.*
6. The technician is unable to figure out the problem after completing a few basic checks. The technician is stumped. What would you do if you were in this position? *Call the supervisor*
7. What does a positive work ethic mean? What does honesty mean? *A positive work ethic means to give a day's work for a day's pay. Honesty is a refusal to lie, steal, or deceive in any way.*
8. Why is it important to check motor winding resistance at the terminals rather than at the contactor? *The wire from the contactor to the motor may be open. Caution: Never take voltage readings at the compressor terminals as an explosion may occur.*
9. On your preinspection of a no-cooling call, the compressor is not operating due to a burned terminal wire on the contactor switch. What would you do to repair this problem? *Replace the wire from the contactor to the compressor.*
10. How do you determine if the compressor is operating prior to hooking up gauges? *Listen for the sound of the compressor running. Place a clamp-on ammeter around the compressor lead to check for amp draw.*
11. Why is it important to use the right manifold gauge set on a condensing unit? *Using lower-pressure gauges on a high-pressure system may damage the gauges and cause an unsafe situation.*
12. Why is it important to use the correct refrigerant gauges on a system? *Some oil is trapped in the hoses, and the oil for use in one type of compressor may not be the same for a compressor that uses a different refrigerant. Using the wrong gauges can create oil cross-contamination. Also, the saturated temperatures on the gauge will not be correct if the gauge is designed for a different refrigerant.*

Chapter 8

1. What are four reasons that clearance is required around installed equipment? *Code compliance, manufacturer's installation instructions, airflow, and service/safety requirements*
2. What is the advantage of suspending an air handler? *To gain usable square footage within the building.*
3. What size platform is required on the service side of a furnace? *30 inch*
4. Why is it not a good idea to install a condensing unit under a roof overhang that does not have a gutter? *The rain water hitting the condenser fan will make noise and could damage the fan blade.*
5. What are the main reasons to insulate the suction line? *To prevent moisture damage, to prevent reduced system capacity due to increased heat load, and to avoid possible compressor damage due to insufficient motor cooling due to excessive superheat*
6. What is the difference between soldering and brazing? *Soldering is done at about 800°F or less. Brazing occurs at temperatures above 800°F, up to the melting point of the metal being joined.*
7. What is swaging? *Swaging is expanding one end of a tube, thus allowing another tube of the same size to slip inside it for brazing.*
8. Why is it important to wear safety glasses and gloves when grinding? *Glasses prevent sparks and metal debris from entering the eye. Gloves prevent burns when handling hot piping.*
9. What is the advantage of using a larger air-acetylene tip? *More heat is produced, allowing heating and soldering of larger piping.*
10. When is flux not required when brazing copper-to-copper piping? *When the brazing rod has phosphorous as one its elements*
11. What are five pieces of information that you might find in the manufacturer's condensing unit installation instructions? *Clearance around unit; installation recommendations; where and how to install; ground or*

roof clearance; correct charging requirements; and wiring recommendations
12. What are two reasons to use a valve core removal tool? *To change a leaking valve core or to speed up evacuation and charging*
13. What are three uses of temperature testers? *Measure air temperature, measure suction tube temperature (to calculate superheat), and measure liquid tube temperature (to calculate subcooling)*
14. What maintenance can a technician do to increase the life of the vacuum pump? *Change the pump oil after each use*
15. Airflow is stated in ___. *Cubic feet per minute (CFM)*
16. What is the approximate airflow requirement for a 5-ton system? *2,000 CFM*
17. What are two purposes of a digital scale? *Measure charge into a system and measure a cylinder to make sure it is not overcharged.*
18. What two tools could you use to bend soft copper tubing? *Spring bender and lever-type bending tools*
19. Why is it important not to uncoil excess soft copper tubing? *The tubing cannot be rolled back into its original shape.*
20. What are five reasons to correctly size refrigerant lines?
    - *Return oil to the compressor*
    - *Only liquid refrigerant enters the metering device*
    - *Minimize system capacity loss*
    - *Minimize refrigerant charge*
    - *Avoid excessive noise and pipe erosion*
21. List the three types of copper tubing in order from thinnest to thickest wall thickness. *Type M; Type L; Type K*
22. What is the purpose of a refrigerant trap? *Allow refrigerant oil to collect and to be swept into the refrigerant stream*
23. What are the purposes of duct insulation? *Prevent duct work from sweating (condensing) and reduce heat transfer*

Chapter 9

1. What is the difference between first cost, maintenance cost, and operating cost? *First cost is the money required to purchase and install the equipment. Maintenance cost includes repair and scheduled maintenance. Operating cost is associated with utility costs.*
2. How many watts are in 10 KW? *10,000 watts*
3. How many BTUH are in 5 KW? *17,000 BTUH*
4. How many BTUs are in 1 cubic foot of natural gas? How many BTUs are in 1 CCF? *One cubic foot of natural gas has approximately 1,000 BTUs, and 1 CCF has 100,000 BTUs.*
5. What is EER? How is it calculated? *EER is Energy Efficiency Ratio. It is equal to the ratio of the cooling capacity divided by the watts at full load design conditions.*
6. What is SEER? How is it calculated? *SEER is Seasonal Energy Efficiency Ratio. It is the total amount of heat removed in a cooling season divided by the total watts used during that cooling period.*
7. What is COP? How is it calculated? *COP is Coefficiency of Performance. It is the ratio of total heat output divided by watts input converted to heat.*
8. What is HSPF? How is it calculated? *HSPF is Heating Season Performance Factor. It is equal to the total heat output of a heat pump during the heating season divided by the total electric power in watts.*
9. What rating is used to describe the efficiency of an air conditioner? *SEER*
10. What rating is used to describe the efficiency of a heat pump? *COP and HSPF*
11. List the ranges of the human comfort zone. *68°F–80°F; 25%–60% RH; and air velocity of less than 15 FPM.*
12. What are two ways that solar energy is used? *To generate DC voltage and heat water and air*
13. What is wind energy? *Wind turns a generator that creates voltage.*
14. What is the purpose of the energy code? *To promote energy conservation*
15. What is the purpose of the Energy Star program? *To promote energy conservation in appliances and installations of systems in structures*
16. What type of material is used to seal metal ductwork seams? *UL 181 approved mastic or approved duct tape*
17. What type of material is used to seal fiberglass duct board? *UL 181 duct tape*
18. As the outdoor temperature rises above design ambient, the cooling capacity of an air conditioner ___. *Decreases*
19. The total on a residential electrical bill is $300. The customer used 2,000 KWH in this billing period. What is the cost per KWH? *$0.15 per KWH*
20. The technician wants to determine the BTUH output of electric strip heat. The technician measures 40 amps at 240 volts. What is the BTUH output? *Approximately 32,640 BTUH*
21. According to the human comfort chart, as the relative humidity in the heating season decreases, the human comfort level ___. *Decreases*
22. Describe the effects of relative humidity on viruses and bacteria. *Low or high RH conditions create breeding opportunities for viruses and bacteria.*
23. Why is a heating and cooling load calculation important? *To determine the actual heating and cooling requirements of a building so that the proper equipment can be selected and installed.*
24. A 4-ton system is going to be installed in a small commercial business. Approximately how many CFM will the air handler need to move? *1,600 CFM*

Chapter 10

1. What is customer service? *Customer service is keeping the user of HVACR systems happy.*
2. List five features of good listening skills.
   - *Allow the customer to talk without interruption.*
   - *Repeat the complaint.*
   - *Empathize by placing yourself in the customer's shoes.*
   - *Apologize by stating that you are sorry to hear that the customer is having a problem.*
   - *Assure by telling the customer that you understand the problem and you will investigate to determine the cause.*
3. How is value added when dealing with a customer? *By exceeding the customer's expectations.*
4. What is value selling? *Value selling means explaining features and benefits that will meet the customer's needs and stay within his budget.*
5. When should a system be repaired? When should a system be replaced? *A system should be repaired if the problem is minor. A system should be replaced when it is experiencing numerous breakdowns or when it has served beyond its useful life. Working but inefficient systems should be replaced.*
6. What is the role of a technician as a salesperson? *To help the customer make the right decision.*
7. Describe flat-rate pricing. Include charges for the service call and parts/equipment. *Flat-rate pricing establishes a set price for a service call and troubleshooting. It also includes a price for time and material for common service procedures.*
8. What is the purpose of a company uniform? *Uniforms standardize the appearance of the company's employees and make the technician part of a team. Uniforms are part of a professional look.*
9. Why is it important to watch your language on a job and at the office? *Inappropriate language may offend the customer or fellow employees.*
10. Why is customer interaction important? *The customer wants to know what is going on in the job. Interaction makes the customer feel comfortable. The customer wants to feel that they are the one in control.*
11. Who is the target audience of ACCA? *HVACR contractors*
12. PHCC supports what professions? *Plumbing, heating, and cooling professionals*
13. What are two main goals of RSES? *Training and certification of technicians*
14. How does NATE help the HVAC industry? *Offers industry-developed certification*
15. How does HVAC Excellence help the HVAC industry? *Offers training materials and certification*

Glossary

Absolute pressure. Gauge pressure plus the pressure of the atmosphere, normally 14.696 psi at sea level at 70°F.

Presión absoluta. La presión del calibrador más la presión de la atmósfera, que generalmente es 14,696 psi al nivel del mar a 70°F (21.11°C).

Absolute zero temperature. The lowest obtainable temperature where molecular motion stops, –460°F and –273°C.

Temperatura del cero absoluto. La temperatura más baja obtenible donde se detiene el movimiento molecular, –460°F y –273°C.

Absorption. The process by which one substance is absorbed by another.

Absorción. Proceso mediante el cual una sustancia es absorbida por otra.

Accumulator. A storage tank located in the suction line to a compressor. It allows small amounts of liquid refrigerant to boil away before entering the compressor. Sometimes used to store excess refrigerant in heat pump systems during the winter cycle.

Acumulador. Tanque de almacenaje ubicado en el conducto de aspiración a un compresor. Permite que pequeñas cantidades de refrigerante líquido se evaporen antes de entrar al compresor. Algunas veces se utiliza para almacenar exceso de refrigerante en sistemas de bombas de calor durante el ciclo de invierno.

Acetylene. A gas often used with air or oxygen for welding, brazing, or soldering applications.

Acetileno. Gas usado con frecuencia, junto con aire u oxígeno, en trabajos de soldadura y soldaduras de cobre.

Acid-contaminated system. A refrigeration system that contains acid due to contamination.

Sistema contaminado de ácido. Sistema de refrigeración que, debido a la contaminación, contiene ácido.

"A" coil. An evaporator coil that can be used for upflow, downflow, and horizontal-flow applications. It actually consists of two coils shaped like a letter "A."

Serpentín en forma de "A". Serpentín de evaporación que puede utilizarse para aplicaciones de flujo ascendente, descendente y horizontal. En realidad, consiste en dos serpentines en forma de "A".

ACR tubing. Air-conditioning and refrigeration tubing that is very clean, dry, and normally charged with dry nitrogen. The tubing is sealed at the ends to contain the nitrogen.

Tubería ACR. Tubería para el acondicionamiento de aire y la refrigeración que es muy limpia y seca, y que por lo general está cargada de nitrógeno seco. La tubería se sella en ambos extremos para contener el nitrógeno.

Activated alumina. A chemical desiccant used in refrigerant driers.

Alúmina activada. Disecante químico utilizado en secadores de refrigerantes.

Activated charcoal. A substance manufactured from coal or coconut shells into pellets. It is often used to adsorb solvents, other organic materials, and odors.

Carbón activado. Sustancia fabricada utilizando carbón o cáscaras de coco, en forma de gránulos. Se emplea para adsorber disolventes, así como otros materiales orgánicos y olores.

Active recovery. Recovering refrigerant with the use of a recovery machine that has its own built-in compressor.

Recuperación activa. El hecho de recuperar refrigerante utilizando un recuperador que dispone de su propio compresor.

Adsorption. The process by which a thin film of a liquid or gas adheres to the surface of a solid substance.

Adsorción. Proceso mediante el cual una fina capa de líquido o gas se adhiere a la superficie de una sustancia sólida.

Air-acetylene. A mixture of air and acetylene gas that when ignited is used for soldering, brazing, and other applications.

Aire-acetilénico. Mezcla de aire y de gas acetileno que se utiliza en la soldadura, la broncosoldadura y otras aplicaciones al ser encendida.

Air conditioner. Equipment that conditions air by cleaning, cooling, heating, humidifying, or dehumidifying it. A term often applied to comfort cooling equipment.

Acondicionador de aire. Equipo que acondiciona el aire limpiándolo, enfriándolo, calentándolo, humidificándolo o deshumidificándolo. Término comúnmente aplicado al equipo de enfriamiento para comodidad.

Air conditioning. A process that maintains comfort conditions in a defined area.

Acondicionamiento de aire. Proceso que mantiene condiciones agradables en un área definida.

Air-cooled condenser. One of the four main components of an air-cooled refrigeration system. It receives hot gas from the compressor and rejects heat to a place where it makes no difference.

Condensador enfriado por aire. Uno de los cuatro componentes principales de un sistema de refrigeración enfriado por aire. Recibe el gas caliente del compresor y dirige el calor a un lugar donde no afecte la temperatura.

Air gap. The clearance between the rotating rotor and the stationary winding on an open motor. Known as a vapor gap in a hermetically sealed compressor motor.

Espacio de aire. Espacio libre entre el rotor giratorio y el devanado fijo en un motor abierto. Conocido como espacio de vapor en un motor de compresor sellado herméticamente.

Air handler. The device that moves the air across the heat exchanger in a forced air system—normally considered to be the fan and its housing.

Tratante de aire. Dispositivo que dirige el aire a través del intercambiador de calor en un sistema de aire forzado—considerado generalmente como el ventilador y su alojamiento.

Air heat exchanger. A device used to exchange heat between air and another medium at different temperature levels, such as air-to-air, air-to-water, or air-to-refrigerant.
Intercambiador de aire y calor. Dispositivo utilizado para intercambiar el calor entre el aire y otro medio, como por ejemplo aire y aire, aire y agua or aire y refrigerante, a diferentes niveles de temperatura.
Air loop. The heat pump's heating and cooling ducted air system, which exchanges heat with the refrigerant loop.
Circuito de aire. Sistema de tubería del aire de calentamiento y de refrigeración de la bomba de calor, que sirve para intercambiar el calor con el circuito de refrigeración.
Air, standard. Dry air at 70°F and 14.696 psi, at which it has a mass density of 0.075 lb/ft^3 and a specific volume of 13.33 ft^3/lb, ASHRAE.
Aire, estándar. Aire seco a 70°F (21.11°C) y 14,696 psi (libra por pulgada cuadrada); a dicha temperatura tiene una densidad de masa de 0,075 libra/pies^3 y un volumen específico de 13,33 pies^3/libra, ASHRAE.
Air vent. A fitting used to vent air manually or automatically from a system.
Válvula de aire. Accesorio utilizado para darle al aire salida manual o automática de un sistema.
Algae. A form of green or black, slimy plant life that grows in water systems.
Alga. Tipo de planta legamosa de color verde o negro que crece en sistemas acuáticos.
Allen head. A recessed hex head in a fastener.
Cabeza allen. Cabeza de concavidad hexagonal en un asegurador.
All-weather system. System providing year-round conditioning of the air.
Sistema para todo el año. Sistema que proporciona una aclimatación ambiental todo el año.
Alternating current. An electric current that reverses its direction at regular intervals.
Corriente alterna. Corriente eléctrica que invierte su dirección a intervalos regulares.
Alternative refrigerant. One of the newer refrigerants that are replacing the traditional CFC or HCFC refrigerants that have been used for many years. Many of these refrigerants have very low ozone depletion and global warming indices. Some are completely chlorine-free.
Refrigerante alternativo. Cualquiera de los nuevos productos que sirve para sustituir a los refrigerantes basados en CFC o HCFC que han sido utilizados durante muchos años. Muchos de estos nuevos refrigerantes tienen un índice muy bajo de desgaste de la capa de ozono y de calentamiento de la superficie terrestre. Algunos no utilizan cloro.
Ambient temperature. The surrounding air temperature.
Temperatura ambiente. Temperatura del aire circundante.
American standard pipe thread. Standard thread used on pipe to prevent leaks.
Rosca estándar estadounidense para tubos. Rosca estándar utilizada en tubos para evitar fugas.
Ammeter. A meter used to measure current flow in an electrical circuit.

Amperímetro. Instrumento utilizado para medir el flujo de corriente en un circuito eléctrico.
Amperage. Amount (quantity) of electron or current flow (the number of electrons passing a point in a given time) in an electrical circuit.
Amperaje. Cantidad de flujo de electrones o de corriente (el número de electrones que sobrepasa un punto específico en un tiempo fijo) en un circuito eléctrico.
Ampere. Unit of current flow.
Amperio. Unidad de flujo de corriente.
Analog electronic devices. Devices that generate continuous or modulating signals within a certain control range.
Aparatos electrónicos analógicos. Aparatos que generan señales continuas o modulares adentro de cierto registro de control.
Analog VOM. A volt-ohm-milliampere meter constructed so that the meter indicator is a needle over a printed surface.
VOM analógico. Medidor de voltios-ohmios-miliamperios, construido de forma que el indicador consiste en una aguja que se mueve encima de una superficie impresa.
Angle valve. Valve with one opening at a 90° angle from the other opening.
Válvula en ángulo. Válvula con una abertura a un ángulo de 90° con respecto a la otra abertura.
Annual Fuel Utilization Efficiency (AFUE). The U.S. Federal Trade Commission requires furnace manufacturers to provide this rating so consumers may compare furnace performances before purchasing.
Eficacia de uso de combustible anual (AFUE en inglés). La Comisión Federal de Comercio (Federal Trade Commission) de los EE.UU. exige que los fabricantes de hornos indiquen este valor con el fin de que los consumidores puedan comparar los rendimientos de los hornos antes de adquirirlos.
ANSI. Abbreviation for the American National Standards Institute.
ANSI. Acrónimo de American National Standards Institute (Instituto Nacional Americano de Normas).
Atmospheric pressure. The weight of the atmosphere's gases pressing down on the earth. Equal to 14.696 psi at sea level and 70°F.
Presión atmosférica. El peso de la presión ejercida por los gases de la atmósfera sobre la tierra, equivalente a 14,696 psi al nivel del mar a 70°F.
Atom. The smallest particle of an element.
átomo. Partícula más pequeña de un elemento.
Atomize. Using pressure to change liquid to small particles of vapor.
Atomizar. Utilizar la presión para cambiar un líquido a partículas pequeñas de vapor.
Automatic changeover thermostat. A thermostat that changes from cool to heat automatically by room temperature.
Termostato de cambio automático. Un termostato que cambia de enfriar a calentar automáticamente con la temperatura del cuarto.
Automatic combination gas valve. A gas valve for gas furnaces that incorporates a manual control, gas supply for the

pilot, adjustment and safety features for the pilot, pressure regulator, and the controls for and the main gas valve.

Válvula de gas de combinación automática. Válvula de gas para hornos de gas que incorpora un regulador manual, suministro de gas para la llama piloto, ajuste y dispositivos de seguridad, regulador de presión, la válvula de gas principal y los reguladores de la válvula.

Automatic control. Controls that react to a change in conditions to cause the condition to stabilize.

Regulador automático. Reguladores que reaccionan a un cambio en las condiciones para provocar la estabilidad de dicha condición.

Automatic expansion valve. A refrigerant control valve that maintains a constant pressure in an evaporator.

Válvula de expansión automática. Válvula de regulación del refrigerante que mantiene una presión constante en un evaporador.

Automatic pump-down system. A control scheme in refrigeration consisting of a thermostat and liquid-line solenoid valve that clears refrigerant from the compressor's crankcase, evaporator, and suction line just before the compressor off cycle.

Sistema de bombeo hacia abajo automático. Un sistema de control en refrigeración que consiste de un termostato y una válvula de línea de líquido solenoide que vacía el refrigerante del cárter del cigüeñal del compresor, evaporador y línea de succión justo antes del ciclo de apagar del compresor.

Auxiliary drain pan. A separate drain pan that is placed under an air-conditioner evaporator to catch condensate in the event that the primary drain pan runs over.

Plato de drenaje auxiliar. Un plato separado de drenaje debajo del evaporador de un aire acondicionador para recoger la condensación en caso de que el plato principal se derrame.

Back electromotive force (BEMF). This is the voltage-generating effect of an electric motor's rotor turning within the motor.

Fuerza contraelectromotriz (BEMF en inglés). El efecto de generar tensión provocado por el rotor de un motor eléctrico al girar dentro del motor.

Back pressure. The pressure on the low-pressure side of a refrigeration system (also known as suction pressure).

Contrapresión. La presión en el lado de baja presión de un sistema de refrigeración (conocido también como presión de aspiración).

Back seat. The position of a refrigeration service valve when the stem is turned away from the valve body and seated, shutting off the service port.

Asiento trasero. Posición de una válvula de servicio de refrigeración cuando el vástago está orientado fuera del cuerpo de la válvula y aplicado sobre su asiento, cerrando así la apertura de servicio.

Baffle. A plate used to keep fluids from moving back and forth at will in a container.

Deflector. Placa utilizada para evitar el libre movimiento de líquidos en un recipiente.

Balanced-port TXV. A valve that will meter refrigerant at the same rate when the condenser head pressure is low.

Válvula electrónica de expansión con conducto equilibrado. Válvula que medirá el refrigerante a la misma proporción cuando la presión en la cabeza del condensador sea baja.

Ball check valve. A valve with a ball-shaped internal assembly that only allows fluid flow in one direction.

Válvula de retención de bolas. Válvula con un conjunto interior en forma de bola que permite el flujo de fluido en una sola dirección.

Ball valve. A valve with an internal part that is shaped like a sphere with a hole through the center. When turned 90°, the hole is crossways of the flow and stops the flow.

Válvula de bola. Una válvula con una parte interna en forma de esfera con un hueco a través del centro. Cuando vira 90°, el hueco es transversal al flujo y el flujo para.

Baseboard heating. Convection heaters providing wholehouse, spot, or individual room heating. The heat is normally provided by electrical resistance or hot water.

Calefacción de zócalo. Calentadores por convección que proporcionan calefacción a toda la casa, en un punto específico o en una sola habitación. Normalmente, el calor se obtiene mediante una resistencia eléctrica o agua caliente.

Battery. A device that produces electricity from the interaction of metals and acid.

Pila. Dispositivo que genera electricidad de la interacción entre metales y el ácido.

Bearing. A device that surrounds a rotating shaft and provides a low-friction contact surface to reduce wear from the rotating shaft.

Cojinete. Dispositivo que rodea un árbol giratorio y provee una superficie de contacto de baja fricción para disminuir el desgaste de dicho árbol.

Bearing washout. A cleaning of the compressor's bearing surfaces, which causes lack of lubrication. It is usually caused by liquid refrigerant mixing with the compressor's crankcase oil due to liquid floodback or migration.

Derrubio del cojinete. Una limpieza de las superficies del cojinete del compresor, lo que lleva a falta de lubricación. Generalmente causado por refrigerante líquido mezclado con aceite del cárter del cigüeñal del compresor debido a inundación o migración del líquido.

Bellows. An accordion-like device that expands and contracts when internal pressure changes.

Fuelles. Dispositivo en forma de acordeón con pliegues que se expanden y contraen cuando la presión interna sufre cambios.

Bellows seal. A method of sealing a rotating shaft or valve stem that allows rotary movement of the shaft or stem without leaking.

Cierre hermético de fuelles. Método de sellar un árbol giratorio o el vástago de una válvula que permite el movimiento giratorio del árbol o del vástago sin producir fugas.

Belly-band mount motor. An electric motor mounted with a strap around the motor secured with brackets on the strap.

Motor con barriguera. Motor eléctrico que se monta colocando una cincha a su alrededor y fijándola con abrazaderas.

Bending spring. A coil spring that can be fitted inside or outside a piece of tubing to prevent its walls from collapsing when being formed.

Muelle de flexión. Muelle helicoidal que puede acomodarse dentro o fuera de una pieza de tubería para evitar que sus paredes se doblen al ser formadas.

Bimetal. Two dissimilar metals fastened together to create a distortion of the assembly with temperature changes.

Bimetal. Dos metales distintos fijados entre sí para producir una distorción del conjunto al ocurrir cambios de temperatura.

Bimetal strip. Two dissimilar metal strips fastened back-to-back.

Banda bimetálica. Dos bandas de metales distintos fijadas entre sí en su parte posterior.

Binary. Consisting of 1s and 0s. The 1s and 0s represent numbers, words, or signals that can be stored in the computer's memory for future use. Calculators and computers are digital systems.

Binario. Consiste de 1s y 0s. Los 1s y 0s representan números, palabras o señales que pueden almacenarse en la memoria de la computadora para uso futuro. Las calculadoras y computadoras son sistemas digitales.

Bleeding. Allowing pressure to move from one pressure level to another very slowly.

Sangradura. Proceso a través del cual se permite el movimiento de presión de un nivel a otro de manera muy lenta.

Bleed valve. A valve with a small port usually used to bleed pressure from a vessel to the atmosphere.

Válvula de descarga. Válvula con un conducto pequeño utilizado normalmente para purgar la presión de un depósito a la atmósfera.

Blocked suction. A method of cylinder unloading. The suction line passage to a cylinder in a reciprocating compressor is blocked, thus causing that cylinder to stop pumping.

Aspiración obturada. Método de descarga de un cilindro. El paso del conducto de aspiración a un cilindro en un compresor alternativo se obtura, provocando así que el cilindro deje de bombear.

Blowdown. A system in a cooling tower whereby some of the circulating water is bled off and replaced with fresh water to dilute the sediment in the sump.

Vaciado. Sistema en una torre de refrigeración por medio del cual se purga parte del agua circulante y se reemplaza con agua fresca para diluir el sedimento en el sumidero.

Boiler. A container in which a liquid may be heated using any heat source. When the liquid is heated to the point that vapor forms and is used as the circulating medium, it is called a steam boiler.

Cardera. Recipiente en el que se puede calentar un líquido utilizando cualquier fuente de calor. Cuando se calienta el líquido al punto en que se produce vapor y se utiliza éste como el medio para la circulación, se llama caldera de vapor.

Boiling point. The temperature level of a liquid at which it begins to change to a vapor. The boiling temperature is controlled by the vapor pressure above the liquid.

Punto de ebullición. El nivel de temperatura de un líquido al que el líquido empieza a convertirse en vapor. La temperatura de ebullición se regula por medio de la presión del vapor sobre el líquido.

Boiling temperature. The boiling temperature of the liquid can be controlled by controlling the pressure. The standard boiling pressure for water is an atmospheric pressure of 29.92 in. Hg (mercury) where water boils at 212°F.

Temperatura de ebullición. La temperatura de ebullición de un líquido puede controlarse controlando la presión. La presión de ebullición estándar para agua es una presión atmosférica de 29.92 pulgadas de mercurio donde el agua hierve a 212°F.

Booster pump. An additional pump that is used to build the pressure above what the primary pump can accomplish.

Bomba promotora. Una bomba adicional que se usa para aumentar la presión por encima de lo que la bomba primaria puede.

Boot. The connection between the branch line duct and the floor register. It transitions the branch duct to the register size. It may be from rectangular to another size of rectangle or from round to rectangular.

Manguito. La conexión entre el conducto de la línea ramal y el contador en el piso. Él hace la transición de la línea ramal al tamaño del contador. Puede conectar un rectángulo a otro rectángulo de tamaño diferente o círculo a un rectángulo.

Bore. The inside diameter of a cylinder.

Calibre. Diámetro interior de un cilindro.

Bourdon tube. C-shaped tube manufactured of thin metal and closed on one end. When pressure is increased inside, it tends to straighten. It is used in a gauge to indicate pressure.

Tubo Bourdon. Tubo en forma de C fabricado de metal delgado y cerrado en uno de los extremos. Al aumentarse la presión en su interior, el tubo tiende a enderezarse. Se utiliza dentro de un calibrador para indicar la presión.

Brazing. High-temperature (above 800°F) melting of a filler metal for the joining of two metals.

Broncesoldadura. El derretir de un metal de relleno para fusionar dos metales a temperaturas altas (sobre los 800°F o 430°C).

Breaker. A heat-activated electrical device used to open an electrical circuit to protect it from excessive current flow.

Interruptor. Dispositivo eléctrico activado por el calor que se utiliza para abrir en circuito eléctrico a fin de protegerlo de un flujo excesivo de corriente.

British thermal unit. The amount (quantity) of heat required to raise the temperature of 1 lb of water 1°F.

Unidad térmica británica. Cantidad de calor necesario para elevar en 1°F (−17.56°C) la temperatura de una libra inglesa de agua.

BTU. Abbreviation for British thermal unit.

BTU. Abreviatura de unidad térmica británica.

Bubble point. The refrigerant temperature at which bubbles begin to appear in a saturated liquid.

Punto de burbujear. La temperatura refrigerada en que burbujas empiezan a aparecer en un liquído saturado.

Bulb sensor. The part of a sealed automatic control used to sense temperature.

Bombilla sensora. Pieza de un regulador automático sellado que se utiliza para advertir la temperatura.

Burner. A device used to prepare and burn fuel.

Quemador. Dispositivo utilizado para la preparación y la quema de combustible.

Burr. Excess material squeezed into the end of tubing or pipe after a cut has been made. This burr must be removed.

Rebaba. Exceso de material introducido por fuerza en el extremo de una tubería después de hacerse un corte. Esta rebaba debe removerse.

Calibrate. To adjust instruments or gauges to the correct setting for known conditions.

Calibrar. Ajustar instrumentos o calibradores en posición correcta para su operación en condiciones conocidas.

Calorimeter. An instrument of laboratory-grade quality used to measure heat absorbed into a substance.

Calorímetro. Instrumento de calidad laboratorio que sirve para medir el calor absorbido por una sustancia.

Capacitance. The term used to describe the electrical storage ability of a capacitor.

Capacitancia. Término utilizado para describir la capacidad de almacenamiento eléctrico de un capacitador.

Capacitive circuit. When the current in a circuit leads the voltage by 90°.

Circuito capacitivo. Un circuito en que el corriente mueve el tensión por un 90°.

Capacitor. An electrical storage device used to start motors (start capacitor) and to improve the efficiency of motors (run capacitor).

Capacitador. Dispositivo de almacenamiento eléctrico utilizado para arrancar motores (capacitador de arranque) y para mejorar el rendimiento de motores (capacitador de funcionamiento).

Capacitor-start–capacitor-run motor. A single-phase motor that has a start capacitor in series with the start winding that is disconnected after start-up and a run capacitor that is also in parallel with the start windings that stays in the circuit while running. This capacitor is built for full-time duty and uses the potential voltage generated by the start winding to give the run winding more efficiency.

Motor de arranque capacitivo–marcha capacitiva. Motor de fase sencilla que tiene un condensador de arranque en serie con la bobina de arranque que se desconecta después del encendido, y un condensador de marcha que está en paralelo con las bobinas de arranque y que permanece en el circuito mientras está en marcha. Este condensador está hecho para servicio a tiempo completo y usa el voltaje que genera la bobina de arranque para darle mayor eficiencia a la bobina de marcha.

Capacitor-start motor. A single-phase motor with a start and run winding that has a capacitor in series with the start winding, which remains in the circuit until the motor gets up to about 75% the run speed.

Motor de arranque capacitivo. Motor de fase sencilla con una bobina de arranque y marcha que tiene un condensador en serie con la bobina de arranque, el cual permanece en el circuito hasta que el motor alcanza alrededor de 75% de la velocidad de marcha.

Capacity. The rating system of equipment used to heat or cool substances.

Capacidad. Sistema de clasificación de equipo utilizado para calentar o enfriar sustancias.

Capillary attraction. The attraction of a liquid material between two pieces of material such as two pieces of copper or copper and brass. For instance, in a joint made up of copper tubing and a brass fitting, the solder filler material has a greater attraction to the copper and brass than to itself and is drawn into the space between them.

Atracción capilar. Atracción de un material líquido entre dos piezas de material, como por ejemplo dos piezas de cobre o cobre y latón. Por ejemplo, en una junta fabricada de tubería de cobre y un accesorio de latón, el material de relleno de la soldadura tiene mayor atracción al cobre y al latón que a sí mismo y es arrastrado hacia el espacio entre éstos.

Capillary tube. A fixed-bore metering device. This is a small-diameter tube that can vary in length from a few inches to several feet. The amount of refrigerant flow needed is predetermined and the length and diameter of the capillary tube is sized accordingly.

Tubo capilar. Dispositivo de medición de calibre fijo. Éste es un tubo de diámetro pequeño cuyo largo puede oscilar entre unas cuantas pulgadas y varios pies. La cantidad de flujo de refrigerante requerida es predeterminada y, de acuerdo a esto, se fijan el largo y el diámetro del tubo capilar.

Carbon dioxide. A by-product of natural gas combustion that is not harmful.

Dióxido de carbono. Subproducto de la combustión del gas natural que no es nocivo.

Carbon monoxide. A poisonous, colorless, odorless, tasteless gas generated by incomplete combustion.

Monóxido de carbono. Gas mortífero, inodoro, incoloro e insípido que se desprende en la combustión incompleta del carbono.

Cavitation. A vapor formed due to a drop in pressure in a pumping system. Vapor at a pump inlet may be caused at a cooling tower if the pressure is low and water is turned to vapor.

Cavitación. Vapor producido como consecuencia de una caída de presión en un sistema de bombeo. El vapor a la entrada de una bomba puede ser producido en una torre de refrigeración si la presión es baja y el agua se convierte en vapor.

Cellulose. A substance formed in wood plants from glucose or sugar.

Celulosa. Sustancia presente en plantas de madera y que se forma a partir de glucosa y azúcar.

Celsius scale. A temperature scale with 100-degree graduations between water freezing (0°C) and water boiling (100°C).

Escala Celsio. Escala dividida en cien grados, con el cero marcado a la temperatura de fusión del hielo (0°C) y el cien a la de ebullición del agua (100°C).

Centigrade scale. See Celsius scale.

Centígrado. Véase escala Celsio.

Centrifugal compressor. A compressor used for large refrigeration systems that uses centrifugal force to accomplish compression. It is not positive displacement, but it is similar to a blower.

Compresor centrífugo. Compresor utilizado en sistemas grandes de refrigeración que usa fuerza centrífuga para lograr compresión. No es desplazamiento positivo, pero es similar a un soplador.

Centrifugal pump. A pump that uses centrifugal force to move a fluid. An impeller is rotated rapidly within the pump, causing the fluid to fly away from the center, which forces the fluid through a piping system.

Bomba centrífuga. Bomba que utiliza la fuerza centrífuga para desplazar un fluido. Un propulsor dentro de la bomba gira a alta velocidad alejando el líquido del centro e impulsándolo a través de una tubería.

Centrifugal switch. A switch that uses a centrifugal action to disconnect the start windings from the circuit.

Conmutador centrífugo. Conmutador que utiliza una acción centrífuga para desconectar los devanados de arranque del circuito.

Change of state. The condition that occurs when a substance changes from one physical state to another, such as ice to water and water to steam.

Cambio de estado. Condición que ocurre cuando una sustancia cambia de un estado físico a otro, como por ejemplo el hielo a agua y el agua a vapor.

Charge of refrigerant. The quantity of refrigerant in a system.

Carga de refrigerante. Cantidad de refrigerante en un sistema.

Charging curve. A graphical method of assisting a service technician with charging an air-conditioning or heat pump system.

Curva de recarga. Un método gráfico para asistir al técnico de servicio con el recargar de un sistema de aire acondicionado o de bomba de calor.

Charging cylinder. A device that allows the technician to accurately charge a refrigeration system with refrigerant.

Cilindro cargador. Dispositivo que le permite al mecánico cargar correctamente un sistema de refrigeración con refrigerante.

Charging scale. A scale used to weigh refrigerant when charging a refrigeration or air-conditioning system.

Báscula de plancha. Báscula que se utiliza para pesar el refrigerante durante la carga de un sistema de refrigeración o de aire acondicionado.

Check valve. A device that permits fluid flow in one direction only.

Válvula de retención. Dispositivo que permite el flujo de fluido en una sola dirección.

Chilled water system. An air-conditioning system that circulates refrigerated water to the area to be cooled. The refrigerated water picks up heat from the area, thus cooling the area.

Sistema de agua enfriada. Sistema de acondicionamiento de aire que hace circular agua refrigerada al área que será enfriada. El agua refrigerada atrapa el calor del área y la enfria.

Chiller purge unit. A system that removes air or noncondensables from a low-pressure chiller.

Unidad enfriadora de purga. Sistema que remueve el aire o sustancias no condensables de un enfriador de baja presión.

Chill factor. A factor or number that is a combination of temperature, humidity, and wind velocity that is used to compare a relative condition to a known condition.

Factor de frío. Factor o número que es una combinación de la temperatura, la humedad y la velocidad del viento utilizado para comparar una condición relativa a una condición conocida.

Chimney. A vertical shaft used to convey flue gases above the rooftop.

Chimenea. Cañón vertical utilizado para conducir los gases de combustión por encima del techo.

Chimney effect. A term used to describe air or gas when it expands and rises when heated.

Efecto de chimenea. Término utilizado para describir el aire o el gas cuando se expande y sube al calentarse.

Chlorofluorocarbons (CFCs). Those refrigerants thought to contribute to the depletion of the ozone layer.

Clorofluorocarburos (CFCs en inglés). Líquidos refrigerantes que, según algunos, han contribuido a la reducción de la capa de ozono.

Circuit. An electron or fluid-flow path that makes a complete loop.

Circuito. Electrón o trayectoria del flujo de fluido que hace un ciclo completo.

Circuit breaker. A device that opens an electric circuit when an overload occurs.

Interruptor para circuitos. Dispositivo que abre un circuito eléctrico cuando ocurre una sobrecarga.

Clamp-on ammeter. An instrument that can be clamped around one conductor in an electrical circuit and measure the current.

Amperímetro fijado con abrazadera. Instrumento que puede fijarse con una abrazadera a un conductor en un circuito eléctrico y medir la corriente.

Clearance volume. The volume at the top of the stroke in a reciprocating compressor cylinder between the top of the piston and the valve plate.

Volumen de holgura. Volumen en la parte superior de una carrera en el cilindro de un compresor recíproco entre la parte superior del pistón y la placa de una válvula.

Closed circuit. A complete path for electrons to flow on.

Circuito cerrado. Circuito de trayectoria ininterrumpida que permite un flujo continuo de electrones.

Closed-circuit cooling tower. May have a wet/dry mode, an adiabatic mode, and a dry mode.

Torre de enfriamiento de circuito cerrado. Puede tener un modo seco/mojado, un modo adiabático y un modo seco.

Closed loop. Piping circuit that is complete and not open to the atmosphere.

Ciclo cerrado. Circuito de tubería completo y no abierto a la atmósfera.

Closed-loop heat pump. Heat pump system that reuses the same heat transfer fluid, which is buried in plastic pipes within the earth or within a lake or pond for the heat source.

Bomba de calor de circuito cerrado. Sistema de bomba de calor que reutiliza el mismo líquido de transferencia térmica que se encuentra en tubos de plástico enterrados o sumergidos en un lago o estanque de los que obtiene el calor.

CO_2 indicator. An instrument used to detect the quantity of carbon dioxide in flue gas for efficiency purposes.

Indicador del CO_2. Instrumento utilizado para detectar la cantidad de dióxido de carbono en el gas de combustión a fin de lograr un mejor rendimiento.

Coaxial heat exchanger. A tube-within-a-tube liquid heat exchanger. Typically it is used for water-source heat pumps and small water-cooled air conditioners.

Intercambiador de calor coaxial. Intercambiador de calor líquido con un tubo dentro de un tubo. Típicamente se usa para bombas de calor en fuentes de agua y acondicionadores de aire pequeños enfriados por agua.

Code. The local, state, or national rules that govern safe installation and service of systems and equipment for the purpose of safety of the public and trade personnel.

Código. Reglamentos locales, estatales o federales que rigen la instalación segura y el servicio de sistemas y equipo con el propósito de garantizar la seguridad del personal público y profesional.

Coefficient of performance (COP). The ratio of usable output energy divided by input energy.

Coeficiente de rendimiento (COP en inglés). Relación de la de energía de salida utilizable dividida por la energía de entrada.

Cold. The word used to describe heat at lower levels of intensity.

Frío. Término utilizado para describir el calor a niveles de intensidad más bajos.

Cold anticipator. A fixed resistor in a thermostat that is wired in parallel with the cooling contacts. This starts the cooling system before the thermostat calls for cooling, which allows the system to get up to capacity before the cooling is actually needed.

Anticipador de frío. Resistor fijo en un termostato, conectado en paralelo con los contactos de enfriamiento. Dicho resistor pone en marcha el sistema de enfriamiento antes de que lo haga el termostato, permitiendo así que el sistema alcance su plena capacidad antes de que realmente se necesite el enfriamiento.

Cold wall. The term used in comfort heating to describe a cold outside wall and its effect on human comfort.

Pared fría. Término utilizado en la calefacción para comodidad que describe una pared exterior fría y sus efectos en la comodidad de una persona.

Combustion. A reaction called rapid oxidation or burning produced with the right combination of a fuel, oxygen, and heat.

Combustión. Reacción conocida como oxidación rápida o quema producida con la combinación correcta de combustible, oxígeno y calor.

Combustion analyzer. An instrument used to measure oxygen concentrations within flue gases. This analyzer can test smoke and test for carbon monoxide and other gases.

Analizador de combustión. Instrumento que se utiliza para medir la concentración del oxígeno en los gases de escape. Este tipo de analizador permite medir el contenido de monóxido de carbono y otros gases en los humos.

Comfort. People are said to be comfortable when they are not aware of the ambient air surrounding them. They do not feel cool or warm or sweaty.

Comodidad. Se dice que las personas están cómodas cuando no están conscientes del aire ambiental que les rodea. No sienten frío ni calor, ni están sudadas.

Comfort chart. A chart used to compare the relative comfort of one temperature and humidity condition to another condition.

Esquema de comodidad. Esquema utilizado para comparar la comodidad relativa de una condición de temperatura y humedad a otra condición.

Compound gauge. A gauge used to measure the pressure above and below the atmosphere's standard pressure. It is a Bourdon tube sensing device and can be found on all gauge manifolds used for air conditioning and refrigeration service work.

Calibrador compuesto. Calibrador utilizado para medir la presión mayor y menor que la presión estándar de la atmósfera. Es un dispositivo sensor de tubo Bourdon que puede encontrarse en todos los distribuidores de calibrador utilizados para el servicio de sistemas de acondicionamiento de aire y de refrigeración.

Compression. A term used to describe a vapor when pressure is applied and the molecules are compacted closer together.

Compresión. Término utilizado para describir un vapor cuando se aplica presión y se compactan las moléculas.

Compression ratio. A term used with compressors to describe the actual difference in the low- and high-pressure sides of the compression cycle. It is absolute discharge pressure divided by absolute suction pressure.

Relación de compresión. Término utilizado con compresores para describir la diferencia real en los lados de baja y alta presión del ciclo de compresión. Es la presión absoluta de descarga dividida por la presión absoluta de aspiración.

Compressor. A vapor pump that pumps vapor (refrigerant or air) from one pressure level to a higher pressure level.

Compresor. Bomba de vapor que bombea el vapor (refrigerante o aire) de un nivel de presión a un nivel de presión más alto.

Compressor crankcase. The internal part of the compressor that houses the crankshaft and lubricating oil.

Cárter del compresor. Parte interna del compresor donde se aloja el cigüeñal y el aceite de lubricación.

Compressor displacement. The internal volume of a compressor's cylinders, used to calculate the pumping capacity of the compressor.

Desplazamiento del compresor. Volumen interno de los cilindros de un compresor, utilizado para calcular la capacidad de bombeo del mismo.

Compressor head. The component that sits on top of the compressor cylinder and holds the components together.

Cabeza del compresor. El componente que está en la parte superior del cilindro y que mantiene a los componentes unidos.

Compressor oil cooler. One or more piping systems used for cooling the crankcase oil.

Enfriador del aceite del compresor. Uno o varios sistemas de tubería que sirven para enfriar el aceite del cárter.

Compressor shaft seal. The seal that prevents refrigerant inside the compressor from leaking around the rotating shaft.

Junta de estanqueidad del árbol del compresor. La junta de estanqueidad que evita la fuga, alrededor del árbol giratorio, del refrigerante en el interior del compresor.

Condensate. The moisture collected on an evaporator coil.

Condensado. Humedad acumulada en la bobina de un evaporador.

Condensate pump. A small pump used to pump condensate to a higher level.

Bomba para condensado. Bomba pequeña utilizada para bombear el condensado a un nivel más alto.

Condensation. Liquid formed when a vapor condenses.

Condensación. El líquido formado cuando se condensa un vapor.

Condense. Changing a vapor to a liquid.

Condensar. Convertir un vapor en líquido.

Condenser. The component in a refrigeration system that transfers heat from the system by condensing refrigerant.

Condensador. Componente en un sistema de refrigeración que transmite el calor del sistema al condensar el refrigerante.

Condenser flooding. An automatic method of maintaining the correct head pressure in mild weather by using refrigerant from an auxiliary receiver.
Inundación del condensador. Método automático de mantener una presión correcta en la cabeza en un tiempo suave utilizando refrigerante de un receptor auxiliar.
Condensing pressure. The pressure that corresponds to the condensing temperature in a refrigeration system.
Presión para condensación. La presión que corresponde a la temperatura de condensación en un sistema de refrigeración.
Condensing temperature. The temperature at which a vapor changes to a liquid.
Temperatura de condensación. Temperatura a la que un vapor se convierte en líquido.
Condensing unit. A complete unit that includes the compressor and the condensing coil.
Conjunto del condensador. Unidad completa que incluye el compresor y la bobina condensadora.
Conduction. Heat transfer from one molecule to another within a substance or from one substance to another.
Conducción. Transmisión de calor de una molécula a otra dentro de una sustancia o de una sustancia a otra.
Conductivity. The ability of a substance to conduct electricity or heat.
Conductividad. Capacidad de una sustancia de conducir electricidad o calor.
Conductor. A path for electrical energy to flow on.
Conductor. Trayectoria que permite un flujo continuo de energía eléctrica.
Connecting rod. A rod that connects the piston to the crankshaft.
Barra conectiva. Barra que conecta al pistón con el cigüeñal.
Contactor. A larger version of the relay. It can be repaired or rebuilt and has movable and stationary contacts.
Contactador. Versión más grande del relé. Puede ser reparado o reconstruido. Tiene contactos móviles y fijos.
Contaminant. Any substance in a refrigeration system that is foreign to the system, particularly if it causes damage.
Contaminante. Cualquier sustancia en un sistema de refrigeración extraña a éste, principalmente si causa averías.
Control. A device to stop, start, or modulate flow of electricity or fluid to maintain a preset condition.
Regulador. Dispositivo para detener, poner en marcha o modular el flujo de electricidad o de fluido a fin de mantener una condición establecida con anticipación.
Controlled device. May be any control that stops, starts, or modulates fuel, fluid flow, or air to provide expected conditions in the conditioned space.
Aparato controlado. Un aparato que puede ser cualquier control que detiene, enciende o modula el combustible, flujo de fluido o aire para proveer las condiciones esperadas en un espacio acondicionado.
Controller. A device that provides the output to the controlled device.
Controlador. Aparato que provee la salida para el aparato controlado.

Control system. A network of controls to maintain desired conditions in a system or space.
Sistema de regulación. Red de reguladores que mantienen las condiciones deseadas en un sistema o un espacio.
Convection. Heat transfer from one place to another using a fluid.
Convección. Transmisión de calor de un lugar a otro por medio de un fluido.
Conversion factor. A number used to convert from one equivalent value to another.
Factor de conversión. Número utilizado en la conversión de un valor equivalente a otro.
Cooler. A walk-in or reach-in refrigerated box.
Nevera. Caja refrigerada donde se puede entrar o introducir la mano.
Cooling tower. The final device in many water-cooled systems, which rejects heat from the system into the atmosphere by evaporation of water.
Torre de refrigeración. Dispositivo final en muchos sistemas enfriados por agua, que dirige el calor del sistema a la atmósfera por medio de la evaporación de agua.
Copper plating. Small amounts of copper are removed by electrolysis and deposited on the ferrous metal parts in a compressor.
Encobrado. Remoción de pequeñas cantidades de cobre por medio de electrólisis que luego se colocan en las piezas de metal férreo en un compresor.
Corrosion. A chemical action that eats into or wears away material from a substance.
Corrosión. Acción química que carcoma o desgasta el material de una sustancia.
Cotter pin. Used to secure a pin. The cotter pin is inserted through a hole in the pin, and the ends spread to retain it.
Pasador de chaveta. Se usa para asegurar una clavija. El pasador se inserta a través de un roto en la clavija y sus extremos se abren para asegurarla.
Counter EMF. Voltage generated or induced above the applied voltage in a single-phase motor.
Contra EMF. Tensión generada o inducida sobre la tensión aplicada en un motor unifásico.
Counterflow. Two fluids flowing in opposite directions.
Contraflujo. Dos fluidos que fluyen en direcciones opuestas.
Coupling. A device for joining two fluid-flow lines. Also the device connecting a motor driveshaft to the driven shaft in a direct-drive system.
Acoplamiento. Dispositivo utilizado para la conexión de dos conductos de flujo de fluido. Es también el dispositivo que conecta un árbol de mando del motor al árbol accionado en un sistema de mando directo.
CPVC (chlorinated polyvinyl chloride). Plastic pipe similar to PVC except that it can be used with temperatures up to 180°F at 100 psig.
CPVC (cloruro de polivinilo clorado). Tubo plástico similar al PVC, pero que puede utilizarse a temperaturas de hasta 180°F (82°C) a 100 psig [indicador de libras por pulgada cuadrada].
Crackage. Small spaces in a structure that allow air to infiltrate the structure.

Formación de grietas. Espacios pequeños en una estructura que permiten la infiltración del aire dentro de la misma.

Cradle-mount motor. A motor with a mounting cradle that fits the motor end housing on each end and is held down with a bracket.

Motor montado con cuña. Motor equipado de una cuña adaptada a la caja en los dos extremos y sujetado por fijaciones.

Crankcase heat. Heat provided to the compressor crankcase.

Calor para el cárter del cigüeñal. Calor suministrado al cárter del cigüeñal del compresor.

Crankcase pressure regulator (CPR). A valve installed in the suction line, usually close to the compressor. It is used to keep a low-temperature compressor from overloading on a hot pull down by limiting the pressure to the compressor.

Regulador de la presión del cárter del cigüeñal (CPR en inglés). Válvula instalada en el conducto de aspiración, normalmente cerca del compresor. Se utiliza para evitar la sobrecarga en un compresor de temperatura baja durante un arrastre caliente hacia abajo limitando la presión al compresor.

Crankshaft. In a reciprocating compressor, the crankshaft changes the round-and-round motion into the reciprocating back-and-forth motion of the pistons using off-center devices called throws.

Cigüeñal. En un compresor alternativo, el cigüeñal cambia el movimiento circular en un movimiento alternativo hacia delante y hacia atrás de los pistones usando aparatos descentrados llamados cigüeñas.

Crankshaft seal. Same as the compressor shaft seal.

Junta de estanqueidad del árbol del cigüeñal. Exactamente igual que la junta de estanqueidad del árbol del compresor.

Crankshaft throw. The off-center portion of a crankshaft that changes rotating motion to reciprocating motion.

Excentricidad del cigüeñal. Porción descentrada de un cigüeñal que cambia el movimiento giratorio a un movimiento alternativo.

Cross charge. A control with a sealed bulb that contains two different fluids that work together for a common specific condition.

Carga transversal. Regulador con una bombilla sellada compuesta de dos fluidos diferentes que pueden funcionar juntos para una condición común específica.

Cross liquid charge bulb. A type of charge in the sensing bulb of the TXV that has different characteristics from the system refrigerant. This is designed to help prevent liquid refrigerant from flooding to the compressor at start-up.

Bombilla de carga del líquido transversal. Tipo de carga en la bombilla sensora de la válvula electrónica de expansión que tiene características diferentes a las del refrigerante del sistema. La carga está diseñada para ayudar a evitar que el refrigerante líquido se derrame dentro del compresor durante la puesta en marcha.

Cross vapor charge bulb. Similar to the vapor charge bulb but contains a fluid different from the system refrigerant. This is a special-type charge and produces a different pressure/temperature relationship under different conditions.

Bombilla de carga del vapor transversal. Similar a la bombilla de carga del vapor pero contiene un fluido diferente al del refrigerante del sistema. Ésta es una carga de tipo especial y produce una relación diferente entre la presión y la temperatura bajo condiciones diferentes.

Crystallization. When a salt solution becomes too concentrated and part of the solution turns to salt.

Cristalización. Condición que ocurre cuando una solución salina se concentra demasiado y una parte de la solución se convierte en sal.

Current, electrical. Electrons flowing along a conductor.

Corriente eléctrica. Electrones que fluyen a través de un conductor.

Current relay. An electrical device activated by a change in current flow.

Relé para corriente. Dispositivo eléctrico accionado por un cambio en el flujo de corriente.

Current sensing relay. An inductive relay coil usually located around a wire used to sense current flowing through the wire. Its action usually opens or closes a set of contacts in the starting of single phase induction motors.

Relé detector de corriente. Una bobina de relé inductiva que generalmente está ubicada cerca de un cable y se usa para detectar el flujo de corriente a través del cable. Su acción generalmente abre o cierra una serie de contactos al encender los motores de fase sencilla.

Cut-in and cut-out. The two points at which a control opens or closes its contacts based on the condition it is supposed to maintain.

Puntos de conexión y desconexión. Los dos puntos en los que un regulador abre o cierra sus contactos según las condiciones que debe mantener.

Cycle. A complete sequence of events (from start to finish) in a system.

Ciclo. Secuencia completa de eventos, de comienzo a fin, que ocurre en un sistema.

Cylinder. A circular container with straight sides used to contain fluids or to contain the compression process (the piston movement) in a compressor.

Cilindro. Recipiente circular con lados rectos, utilizado para contener fluidos o el proceso de compresión (movimiento del pistón) en un compresor.

Cylinder, compressor. The part of the compressor that contains the piston and its travel.

Cilindro del compresor. Pieza del compresor que contiene el pistón y su movimiento.

Cylinder head, compressor. The top to the cylinder on the high-pressure side of the compressor.

Culata del cilindro del compresor. Tapa del cilindro en el lado de alta presión del compresor.

Cylinder, refrigerant. The container that holds refrigerant.

Cilindro del refrigerante. El recipiente que contiene el refrigerante.

Cylinder unloading. A method of providing capacity control by causing a cylinder in a reciprocating compressor to stop pumping.

Descarga del cilindro. Método de suministrar regulación de capacidad provocando que el cilindro en un compresor alternativo deje de bombear.

Damper. A component in an air distribution system that restricts airflow for the purpose of air balance.
Desviador. Componente en un sistema de distribución de aire que limita el flujo de aire para mantener un equilibrio de aire.
DC converter. A type of rectifier that changes alternating current (AC) to direct current (DC).
Convertidor CD. Tipo de rectificador que cambia la corriente alterna (CA) a corriente directa (CD).
DC motor. A motor that operates on direct current (DC).
Motor CD. Un motor que opera con corriente directa (CD).
Deep vacuum. An attained vacuum that is below 250 microns.
Vacío profundo. Un vacío que se obtiene lo cual es menor de 250 micrones.
Defrost. Melting of ice.
Descongelar. Convertir hielo en líquido.
Defrost condensate. The condensate or water from a defrost application of a refrigeration system.
Condensado de descongelación. Condensado o agua causada por el dispositivo de descongelación en un sistema de refrigeración.
Dehumidify. To remove moisture from air.
Deshumidificar. Remover la humedad del aire.
Dehydrate. To remove moisture from a sealed system or a product.
Deshidratar. Remover la humedad de un sistema sellado o un producto.
Delta-T. The temperature difference at two different points, such as the inlet and outlet temperature difference across a water chiller.
Delta-T. La diferencia en temperatura en dos puntos diferentes, tales como la diferencia de temperatura entre la entrada y la salida de un enfriador de agua.
Demand metering. In this system, the power company charges the customer based on the highest usage for a prescribed period of time during the billing period. The prescribed time for demand metering may be any 15- or 30-minute period within the billing period.
Medición por demanda. Un sistema utilizado por la compañía de electricidad en el cual cobran al consumidor por el período de facturación basado en el uso más alto durante un período de tiempo prescrito durante el período de facturación. El tiempo prescrito para la medición por demanda puede ser cualquier período de 15 ó 30 minutos durante el período de facturación.
Density. The weight per unit of volume of a substance.
Densidad. Relación entre el peso de una sustancia y su volumen.
Department of Transportation (DOT). The governing body of the U.S. government that makes the rules for transporting items, such as volatile liquids.
Departamento de Transportación (DOT en inglés). El cuerpo regente del gobierno de los Estados Unidos que crea las reglas para transportar artículos tales como líquidos volátiles.
Desiccant. Substance in a refrigeration system drier that collects and holds moisture.
Disecante. Sustancia en el secador de un sistema de refrigeración que acumula y guarda la humedad.

Desiccant drier. A device that dehumidifies compressed air for use in controls or processing.
Secador desecante. Un aparato que se usa para deshumedecer el aire comprimido que se usa en los controles o procesos.
Design pressure. The pressure at which the system is designed to operate under normal conditions.
Presión de diseño. Presión a la que el sistema ha sido diseñado para funcionar bajo condiciones normales.
De-superheating. Removing heat from the superheated hot refrigerant gas down to the condensing temperature.
Des sobrecalentamiento. Reducir el calor del gas caliente del refrigerante sobrecalentado hasta alcanzar la temperatura de condensación.
Detector. A device used to search and find.
Detector. Dispositivo de búsqueda y detección.
Detent or snap action. The quick opening and closing of an electrical switch.
Acción de detén o de encaje. El abrir y cerrar rápido de un interruptor eléctrico.
Dew. Moisture droplets that form on a cool surface.
Rocío. Gotitas de humedad que se forman en una superficie fría.
Dew point. The exact temperature at which moisture begins to form.
Punto de rocío. Temperatura exacta a la que la humedad comienza a formarse.
Diaphragm. A thin flexible material (metal, rubber, or plastic) that separates two pressure differences.
Diafragma. Material delgado y flexible, como por ejemplo el metal, el caucho o el plástico, que separa dos presiones diferentes.
Die. A tool used to make an external thread such as on the end of a piece of pipe.
Troquel. Herramienta utilizada para formar una rosca externa, como por ejemplo en el extremo de un tubo.
Differential. The difference in the cut-in and cut-out points of a control, pressure, time, temperature, or level.
Diferencial. Diferencia entre los puntos de conexión y desconexión de un regulador, una presión, un intervalo de tiempo, una temperatura o un nivel.
Diffuser. The terminal or end device in an air distribution system that directs air in a specific direction using louvers.
Placa difusora. Punto o dispositivo terminal en un sistema de distribución de aire que dirige el aire a una dirección específica, utilizando aberturas tipo celosía.
Diffuse radiation. Radiation from the sun that reaches the earth after it is reflected from other substances, such as moisture or other particles in space.
Radiación difusa. Radiación solar que alcanza la tierra después de haber sido reflejada por otras sustancias como gotas de agua u otras partículas presentes en el aire.
Digital electronic devices. Devices that generate strings of data or groups of logic consisting of 1s and 0s.
Aparatos electrónicos digitales. Aparatos que generan cadenas de data o grupos de lógica que consisten de unos y ceros.
Digital electronic signal. An electrical signal, usually 0 to 10 V DC or 0 to 20 milliamps DC, that is used to control system conditions.

Señal electrónica digital. Una señal eléctrica, generalmente de 0 a 10 voltios CD o de 0 a 20 miliamperes CD, que se usa para controlar las condiciones del sistema.

Digital VOM. A volt-ohm-milliampere meter that displays the reading in digits or numbers.

VOM digital. Medidor de voltios-ohmios-miliamperios que indica la lectura en dígitos o números.

Diode. A solid-state device composed of both P-type and N-type material. When connected in a circuit one way, current will flow. When the diode is reversed, current will not flow.

Diodo. Dispositivo de estado sólido compuesto de material P y de material N. Cuando se conecta a un circuito de una manera, la corriente fluye. Cuando la dirección del diodo cambia, la corriente deja de fluir.

DIP (dual inline pair) switch. A very small low-amperage, single-pole, double-throw switch used in electronic circuits to set up the program in the circuit.

Interruptor de doble paquete en línea (DIP en inglés). Un interruptor muy pequeño, de bajo amperaje, unipolar de doble tiro que se usa en los circuitos electrónicos para preparar el programa en un circuito.

Direct current. Electricity in which all electron flow is continuously in one direction.

Corriente continua. Electricidad en la que todos los electrones fluyen continuamente en una sola dirección.

Direct digital control (DDC). Very low-voltage control signal, usually 0 to 10 V DC or 0 to 20 milliamps DC.

Control digital directo (DDC en inglés). Señal control de voltaje bien bajo generalmente 0 a 20 voltios CD o de 0 a 20 miliamperes CD.

Direct-drive compressor. A compressor that is connected directly to the end of the motor shaft. No pulleys are involved.

Compresor de conducción directa. Un compresor que está conectado directamente a un extremo del eje de un motor, sin usar poleas.

Direct-drive motor. A motor that is connected directly to the load, such as an oil burner motor or a furnace fan motor.

Motor de conducción directa. Un motor que está conectado directamente a la carga, tal como un motor de un quemador de aceite o el motor de un ventilador en un calefactor.

Direct expansion. The term used to describe an evaporator with an expansion device other than a low-side float type.

Expansión directa. Término utilizado para describir un evaporador con un dispositivo de expansión diferente al tipo de dispositivo flotador de lado bajo.

Direct radiation. The energy from the sun that reaches the earth directly.

Radiación directa. Energía solar que alcanza la tierra directamente.

Discharge pressure. The pressure on the high-pressure side of a compressor.

Presión de descarga. La presión en el extremo de alta presión de un compresor.

Discharge valve. The valve at the top of a compressor cylinder that shuts on the downstroke to prevent high-pressure gas from reentering the refrigerant cylinder, allowing low-pressure gas to enter.

Válvula de descarga. La válvula que está en la parte superior del cilindro de un compresor que se cierra en el recorrido hacia abajo del pistón para evitar que el gas a alta presión regrese al cilindro del refrigerante y permitir que el gas a baja presión entre.

Distributor. A component installed at the outlet of the expansion valve that distributes the refrigerant to each evaporator circuit.

Distribuidor. Componente instalado a la salida de la válvula de expansión que distribuye el refrigerante a cada circuito del evaporador.

Double flare. A connection used on copper, aluminum, or steel tubing that folds tubing wall to a double thickness.

Abocinado doble. Conexión utilizada en tuberías de cobre, aluminio o acera que pliega la pared de la tubería y crea un espesor doble.

Dowel pin. A pin, which may or may not be tapered, used to align and fasten two parts.

Pasador de espiga. Pasador, que puede o no ser cónico, utilizado para alinear y fijar dos piezas.

Downflow furnace. This furnace sometimes is called a counterflow furnace. The air intake is at the top, and the discharge air is at the bottom.

Horno de corriente descendente. También conocido como horno de contracorriente. La entrada del aire está en la parte superior y la salida en la parte inferior.

Drier. A device used in a refrigerant line to remove moisture.

Secador. Dispositivo utilizado en un conducto de refrigerante para remover la humedad.

Drip pan. A pan shaped to collect moisture condensing on an evaporator coil in an air-conditioning or refrigeration system.

Colector de goteo. Un colector formado para acumular la humedad que se condensa en la bobina de un evaporador en un sistema de acondicionamiento de aire o de refrigeración.

Dry-bulb temperature. The temperature measured using a plain thermometer.

Temperatura de bombilla seca. Temperatura que se mide con un termómetro sencillo.

Dry well. A well used for the discharged water in an open-loop geothermal heat pump.

Pozo seco. Pozo que se utiliza para depositar agua de descarga en una bomba de calor geotérmica de circuito abierto.

Duct. A sealed channel used to convey air from the system to and from the point of utilization.

Conducto. Canal sellado que se emplea para dirigir el aire del sistema hacia y desde el punto de utilización.

Dust mites. Microscopic spiderlike insects. Dust mites and their remains are thought to be a primary irritant to some people.

Ácaros del polvo. Insectos microscópicos parecidos a arañas. Los ácaros del polvo y sus restos se consideran irritantes principales para algunas personas.

Eccentric. An off-center device that rotates in a circle around a shaft.

Excéntrico. Dispositivo descentrado que gira en un círculo alrededor de un árbol.

ECM. An electronically commutated motor. This DC motor uses electronics to commutate the rotor instead of brushes. It is typically built for under 1 hp.

CEM (ECM en inglés). Un motor conmutado electrónicamente. Este motor CD usa electrónica para conmutar el rotor en lugar de escobillas. Típicamente están hechos para menos de 1 caballo de fuerza.

Eddy current test. A test with an instrument to find potential failures in evaporator or condenser tubes.

Prueba para la corriente de Foucault. Prueba que se realiza con un instrumento para detectar posibles fallas en los tubos del evaporador o del condensador.

Effective temperature. Different combinations of temperature and humidity that provide the same comfort level.

Temperatura efectiva. Diferentes combinaciones de temperatura y humedad que proveen el mismo nivel de comodidad.

Electrical power. Electrical power is measured in watts. One watt is equal to one ampere flowing with a potential of one volt. Watts = Volts × Amperes ($P = E \times I$).

Potencia eléctrica. La potencia eléctrica se mide en watios. Un watio equivale a un amperio que fluye con una potencia de un voltio. Watios = voltios × amperios ($P = E \times I$).

Electrical shock. When an electrical current travels through a human body.

Sacudida eléctrica. Paso brusco de una corriente eléctrica a través del cuerpo humano.

Electric forced-air furnace. An electrical resistance type of heating furnace used with a duct system to provide heat to more than one room.

Horno eléctrico de aire soplado. Horno con resistencia eléctrica que se utiliza con un sistema de conductos para proporcionar calefacción a varias habitaciones.

Electric heat. The process of converting electrical energy, using resistance, into heat.

Calor eléctrico. Proceso de convertir energía eléctrica en calor a través de la resistencia.

Electric hydronic boiler. A boiler using electrical resistance heat, which often has a closed-loop piping system to distribute heated water for space heating.

Caldera hidrónica eléctrica. Caldera que utiliza calor proporcionado por una resistencia eléctrica. A menudo cuenta con un sistema cerrado para distribuir agua caliente para usos de calefacción.

Electrodes. Electrodes carry high voltage to the tips, where an arc is created for the purpose of ignition for oil or gas furnaces.

Electrodos. Los electrodos llevan alto voltaje hasta las puntas, donde se crea un arco con el propósito de encender los calefactores de aceite o de gas.

Electromagnet. A coil of wire wrapped around a soft iron core that creates a magnet.

Electroimán. Bobina de alambre devanado alrededor de un núcleo de hierro blando que crea un imán.

Electromechanical controls. Electromechanical controls convert some form of mechanical energy to operate an electrical function, such as a pressure-operated switch.

Controles electromecánicos. Los controles electromecánicos convierten algún tipo de energía mecánica para operar una función eléctrica, tal como un interruptor operado por presión.

Electromotive force. A term often used for voltage indicating the difference of potential in two charges.

Fuerza electromotriz. Término empleado a menudo para el voltaje, indicando la diferencia de potencia entre dos cargas.

Electron. The smallest portion of an atom that carries a negative charge and orbits around the nucleus of an atom.

Electrón. La parte más pequeña de un átomo, con carga negativa y que sigue una órbita alrededor del núcleo de un átomo.

Electronic air filter. A filter that charges dust particles using a high-voltage direct current and then collects these particles on a plate of an opposite charge.

Filtro de aire electrónico. Filtro que carga partículas de polvo utilizando una corriente continua de alta tensión y luego las acumula en una placa de carga opuesta.

Electronic charging scale. An electronically operated scale used to accurately charge refrigeration systems by weight.

Escala electrónica para carga. Escala accionada electrónicamente que se utiliza para cargar correctamente sistemas de refrigeración por peso.

Electronic circuit board. A phenolic type of plastic board that electronic components are mounted on. Typically, the circuits are routed on the back side of the board and the components are mounted on the front with prongs of wire that are soldered to the circuits on the back. These can be mass-produced and coated with a material that keeps the circuits separated if moisture and dust accumulate.

Tarjeta de circuitos electrónicos. Una tarjeta plástica tipo fenólico en la cual se montan los componentes electrónicos. La tarjeta generalmente tiene los circuitos trazados en la parte de atrás de la tarjeta y los componentes se colocan en el frente con cables que se sueldan al circuito por detrás. Éstas pueden producirse en masa y pueden recubrirse con un material que mantiene separado a los circuitos si se acumula humedad y polvo.

Electronic controls. Controls that use solid-state semiconductors for electrical and electronic functions.

Controles electrónicos. Controles que usan semiconductores de estado sólido para las funciones eléctricas y electrónicas.

Electronic expansion valve (EXV). A metering valve that uses a thermistor as a temperature-sensing element that varies the voltage to a heat motor-operated valve.

Válvula electrónica de expansión (EXV en inglés). Válvula de medición que utiliza un termistor como elemento sensor de temperatura para variar la tensión a una válvula de calor accionada por motor.

Electronic leak detector. An instrument used to detect gases in very small portions by using electronic sensors and circuits.

Detector electrónico de fugas. Instrumento que se emplea para detectar cantidades de gases sumamente pequeñas utilizando sensores y circuitos electrónicos.

Electronic or programmable thermostat. A space thermostat that is electronic in nature with semiconductors that provide different timing programs for cycling the equipment.

Termostato electrónico o programable. Un termostato de espacio que es de naturaleza electrónica con semiconductores que proveen diferentes programas de cronometraje para ciclar el equipo.

Electronic relay. A solid-state relay with semiconductors used to stop, start, or modulate power in a circuit.

Relé electrónico. Un relé de estado sólido con semiconductores para detener, iniciar o modular la electricidad en un circuito.
Electronics. The use of electron flow in conductors, semiconductors, and other devices.
Electrónica. La utilización del flujo de electrones en conductores, semiconductores y otros dispositivos.
Electrostatic precipitator. Another term for an electronic air cleaner.
Precipitador electrostático. Otro término para un limpiador eléctrico del aire.
Emitter. A terminal on a semiconductor.
Emisor. Punto terminal en un semiconductor.
End bell. The end structure of an electric motor that normally contains the bearings and lubrication system.
Extremo acampanado. Estructura terminal de un motor eléctrico que generalmente contiene los cojinetes y el sistema de lubricación.
End-mount motor. An electric motor mounted with tabs or studs fastened to the motor housing end.
Motor con montaje en los extremos. Motor eléctrico con lengüetas o espigas de montaje en el extremo de su caja.
End play. The amount of lateral travel in a motor or pump shaft.
Holgadura. Amplitud de movimiento lateral en un motor o en el árbol de una bomba.
Energy. The capacity for doing work.
Energía. Capacidad para realizar un trabajo.
Energy efficiency ratio (EER). An equipment efficiency rating that is determined by dividing the output in BTU/h by the input in watts. This does not take into account the start-up and shutdown for each cycle.
Relación del rendimiento de energía (EER en inglés). Clasificación del rendimiento de un equipo que se determina al dividir la salida en BTU/h por la entrada en watios. Esto no toma en cuenta la puesta en marcha y la parada de cada ciclo.
Energy management. The use of computerized or other methods to manage the power to a facility. This may include cycling off nonessential equipment, such as water fountain pumps or lighting, when it may not be needed. The air-conditioning and heating system is also operated at optimum times when needed instead of around the clock.
Manejo de energía. El uso de métodos computarizados o de otro tipo para manejar el consumo de energía en la facilidad. Esto puede incluir apagar los equipos no esenciales en ciclos, tales como las bombas de las fuentes de agua o las luces cuando no son necesarias. Los sistemas de aire acondicionado y de calefacción también se operan en tiempos óptimos cuando son necesarios en vez de todo el tiempo.
Enthalpy. The amount of heat a substance contains from a predetermined base or point.
Entalpía. Cantidad de calor que contiene una sustancia, establecida desde una base o un punto predeterminado.
Environment. Our surroundings, including the atmosphere.
Medio ambiente. Nuestros alrededores, incluyendo la atmósfera.
Environmental Protection Agency (EPA). A branch of the federal government dealing with the control of ozone-depleting refrigerants and other chemicals and the overall welfare of the environment.
Agencia de Protección Ambiental (EPA en inglés). Una rama del gobierno federal que trata con el control de los refrigerantes y otros químicos que repletan el ozono, y el bienestar completo del ambiente.
EPA. Abbreviation for the Environmental Protection Agency.
EPA. Acrónimo en inglés de Environmental Protection Agency (Agencia de protección ambiental).
Ester. A popular synthetic lubricant that performs best with HFCs and HFC-based blends.
Ester. Lubricante sintético de uso común que da óptimos resultados con HFC y mezclas basadas en HFC.
Ethane gas. The fossil fuel, natural gas, used for heat.
Gas etano. Combustible fósil, gas natural, utilizado para generar calor.
Evacuation. The removal of any gases not characteristic to a system or vessel.
Evacuación. Remoción de los gases no característicos de un sistema o depósito.
Evaporation. The condition that occurs when heat is absorbed by liquid and it changes to vapor.
Evaporación. Condición que ocurre cuando un líquido absorbe calor y se convierte en vapor.
Evaporative condenser (cooling tower). A combination water cooling tower and condenser. The refrigerant from the compressor is routed to the cooling tower where the tower evaporates water to cool the refrigerant. In the evaporative condenser the refrigerant is routed to the tower and the water circulates only in the tower. In a cooling tower, the water is routed to the condenser at the compressor location.
Condensador evaporatorio (torre de enfriamiento). Una combinación de una torre de enfriamiento de agua y un condensador. El refrigerante del compresor se desvía a la torre de enfriamiento donde la torre evapora agua para enfriar al refrigerante. En el condensador evaporatorio el refrigerante se lleva a la torre y el agua circula sólo en la torre. En una torre de enfriamiento, el agua se pasa por el condensador donde está ubicado el compresor.
Evaporative cooling. Devices that provide this type of cooling use fiber mounted in a frame with water slowly running down the fiber. Fresh air is drawn in and through the water-soaked fiber and cooled by evaporation to a point close to the wet-bulb temperature of the air.
Enfriamiento por formación de vapor. Los dispositivos que proporcionan este tipo of enfriamiento utilizan agua que corre sobre fibra colocada en un marco. Se pasa aire nuevo a través de la fibra húmeda, el aire se enfría por evaporación a una temperatura próxima a la de una bombilla húmeda.
Evaporator. The component in a refrigeration system that absorbs heat into the system and evaporates the liquid refrigerant.
Evaporador. El componente en un sistema de refrigeración que absorbe el calor hacia el sistema y evapora el refrigerante líquido.
Evaporator fan. A forced convector used to improve the efficiency of an evaporator by air movement over the coil.
Abanico del evaporador. Convector forzado que se utiliza para mejorar el rendimiento de un evaporador por medio del movimiento de aire a través de la bobina.

Evaporator pressure regulator (EPR). A mechanical control installed in the suction line at the evaporator outlet that keeps the evaporator pressure from dropping below a certain point.
Regulador de presión del evaporador (EPR en inglés). Regulador mecánico instalado en el conducto de aspiración de la salida del evaporador; evita que la presión del evaporador caiga hasta alcanzar un nivel por debajo del nivel específico.
Evaporator types. Flooded—an evaporator where the liquid refrigerant level is maintained to the top of the heat exchange coil. Dry type—an evaporator coil that achieves the heat exchange process with a minimum of refrigerant charge.
Clases de evaporadores. Inundado—un evaporador en el que se mantiene el nivel del refrigerante líquido en la parte superior de la bobina de intercambio de calor. Seco—una bobina de evaporador que logra el proceso de intercambio de calor con una mínima cantidad de carga de refrigerante.
Even parallel system. Parallel compressors of equal sizes mounted on a steel rack and controlled by a microprocessor.
Sistema paralelo homogéneo. Compresores de capacidades iguales, montados en paralelo en un bastidor de acero y controlados mediante un microprocesador.
Exhaust valve. The movable component in a refrigeration compressor that allows hot gas to flow to the condenser and prevents it from refilling the cylinder on the downstroke.
Válvula de escape. Componente móvil en un compresor de refrigeración que permite el flujo de gas caliente al condensador y evita que este gas rellene el cilindro durante la carrera descendente.
Expansion joint. A flexible portion of a piping system or building structure that allows for expansion of the materials due to temperature changes.
Junta de expansión. Parte flexible de un sistema de tubería o de la estructura de un edificio que permite la expansión de los materiales debido a cambios de temperatura.
Expansion (metering) device. The component between the high-pressure liquid line and the evaporator that feeds the liquid refrigerant into the evaporator.
Dispositivo de (medición) de expansión. Componente entre el conducto de líquido de alta presión y el evaporador que alimenta el refrigerante líquido hacia el evaporador.
Explosion-proof motor. A totally sealed motor and its connections that can be operated in an explosive atmosphere, such as in a natural gas plant.
Motor a prueba de explosión. Un motor totalmente sellado y con conexiones que puede operarse en una atmósfera explosiva, tal como dentro de una planta de gas natural.
External drive. An external type of compressor motor drive, as opposed to a hermetic compressor.
Motor externo. Motor tipo externo de un compresor, en comparación con un compresor hermético.
External equalizer. The connection from the evaporator outlet to the bottom of the diaphragm on a thermostatic expansion valve.
Equilibrador externo. Conexión de la salida del evaporador a la parte inferior del diafragma en una válvula de expansión termostática.

External heat defrost. A defrost system for a refrigeration system where the heat comes from some external source. It might be an electric strip heater in an air coil, or water in the case of an ice maker. *External* means other than hot gas defrost.
Descongelador de calor externo. Un sistema de descongelación para un sistema de refrigeración en el cual el calor viene de una fuente externa. La misma puede ser un calentador de tira eléctrico en un serpentín o agua en caso de una hielera. *Externo* se refiere a otro tipo de descongelación aparte del de gas caliente.
External motor protection. Motor overload protection that is mounted on the outside of the motor.
Protección externa para motor. Protección para un motor contra la sobrecarga y que está montada en el exterior del motor.

Fahrenheit scale. The temperature scale that places the boiling point of water at 212°F and the freezing point at 32°F.
Escala Fahrenheit. Escala de temperatura en la que el punto de ebullición del agua se encuentra a 212°F y el punto de fusión del hielo a 32°F.
Fan. A device that produces a pressure difference in air to move it.
Abanico. Dispositivo que produce una diferencia de presión en el aire para moverlo.
Fan cycling. The use of a pressure control to turn a condenser fan on and off to maintain a correct pressure within the system.
Funcionamiento cíclico. La utilización de un regulador de presión para poner en marcha y detener el abanico de un condensador a fin de mantener una presión correcta dentro del sistema.
Fan relay coil. A magnetic coil that controls the starting and stopping of a fan.
Bobina de relé del ventilador. Bobina magnética que regula la puesta en marcha y la parada de un ventilador.
Farad. The unit of capacity of a capacitor. Capacitors in our industry are rated in microfarads.
Faradio. Unidad de capacidad de un capacitador. En nuestro medio, los capacitadores se clasifican en microfaradios.
Feedback loop. The circular data route in a control loop that usually travels from the control medium's sensor to the controller, then to the controlled device, and back into the controlled process to the sensor again as a change in the control point.
Circuito de retroalimentación. Ruta circular de la data en un circuito de control que generalmente va desde el sensor del medio de control al controlador y luego al aparato controlado, y de regreso al proceso controlado y al sensor nuevamente como un cambio en el punto de control.
Female thread. The internal thread in a fitting.
Rosca hembra. Rosca interna en un accesorio.
Fill or wetted-surface method. Water in a cooling tower is spread out over a wetted surface while air is passed over it to enhance evaporation.
Método de relleno o de superficie mojada. El agua en una torre de refrigeración se extiende sobre una superficie mojada mientras el aire se dirige por encima de la misma para facilitar la evaporación.
Film factor. The relationship between the medium giving up heat and the heat exchange surface (evaporator). This relates to the velocity of the medium passing over the evaporator. When

the velocity is too slow, the film between the air and the evaporator becomes greater and becomes an insulator, which slows the heat exchange.

Factor de capa. Relación entre el medio que emite calor y la superficie del intercambiador de calor (evaporador). Esto se refiere a la velocidad del medio que pasa sobre el evaporador. Cuando la velocidad es demasiado lenta, la capa entre el aire y el evaporador se expande y se convierte en un aislador, disminuyendo así la velocidad del intercambio del calor.

Filter. A fine mesh or porous material that removes particles from passing fluids.

Filtro. Malla fina o material poroso que remueve partículas de los fluidos que pasan por él.

Filter drier. A type of refrigerant filter that includes a desiccant material that has an attraction for moisture. The filter drier will remove particles and moisture from refrigerant and oil.

Secador de filtro. Un tipo de filtro de refrigerante que incluye un material desecante el cual tiene una atracción a la humedad. El secador de filtro removerá las partículas y la humedad del aceite y del refrigerante.

Fin comb. A hand tool used to straighten the fins on an air-cooled condenser.

Herramienta para aletas. Herramienta manual utilizada para enderezar las aletas en un condensador enfriado por aire.

Finned-tube evaporator. A copper or aluminum tube that has fins, usually made of aluminum, pressed onto the copper lines to extend the surface area of the tubes.

Evaporador de tubo con aletas. Un tubo de cobre o aluminio que tiene aletas, generalmente de aluminio, colocadas a presión contra las líneas de cobre para extender el área de superficie de los tubos.

Fixed-bore device. An expansion device with a fixed diameter that does not adjust to varying load conditions.

Dispositivo de calibre fijo. Dispositivo de expansión con un diámetro fijo que no se ajusta a las condiciones de carga variables.

Fixed resistor. A nonadjustable resistor. The resistance cannot be changed.

Resistor fijo. Resistor no ajustable. La resistencia no se puede cambiar.

Flapper valve. See reed valve.

Chapaleta. Véase válvula de lámina.

Flare. The angle that may be fashioned at the end of a piece of tubing to match a fitting and create a leak-free connection.

Abocinado. Ángulo que puede formarse en el extremo de una pieza de tubería para emparejar un accesorio y crear una conexión libre de fugas.

Flare nut. A threaded connector used in a flare assembly for tubing.

Tuerca abocinada. Conector de rosca utilizado en un conjunto abocinado para tuberías.

Flash gas. A term used to describe the pressure drop in an expansion device when some of the liquid passing through the valve is changed quickly to a gas and cools the remaining liquid to the corresponding temperature.

Gas instantáneo. Término utilizado para describir la caída de la presión en un dispositivo de expansión cuando una parte del líquido que pasa a través de la válvula se convierte rápidamente en gas y enfría el líquido restante a la temperatura correspondiente.

Floating head pressure. Letting the head pressure (condensing pressure) fluctuate with the ambient temperature from season to season for lower compression ratios and better efficiencies.

Fluctuar la presión de la carga. Hecho de dejar que la presión de la carga (presión de condensación) fluctúe con la temperatura del ambiente en cada temporada del año, para obtener relaciones de compresión más bajas y mejores rendimientos.

Float valve or switch. An assembly used to maintain or monitor a liquid level.

Válvula o conmutador de flotador. Conjunto utilizado para mantener o controlar el nivel de un líquido.

Flooded evaporator. A refrigeration system operated with the liquid refrigerant level very close to the outlet of the evaporator coil for improved heat exchange.

Sistema inundado. Sistema de refrigeración que funciona con el nivel del refrigerante líquido bastante próximo a la salida de la bobina del evaporador para mejorar el intercambio de calor.

Flooding. The term applied to a refrigeration system when the liquid refrigerant reaches the compressor.

Inundación. Término aplicado a un sistema de refrigeración cuando el nivel del refrigerante líquido llega al compresor.

Flue. The duct that carries the products of combustion out of a structure for a fossil- or a solid-fuel system.

Conducto de humo. Conducto que extrae los productos de combustión de una estructura en sistemas de combustible fósil o sólido.

Flue-gas analysis instruments. Instruments used to analyze the operation of fossil fuel–burning equipment such as oil and gas furnaces by analyzing the flue gases.

Instrumentos para el análisis del gas de combustión. Instrumentos utilizados para llevar a cabo un análisis del funcionamiento de los quemadores de combustible fósil, como por ejemplo hornos de aceite pesado o gas, a través del estudio de los gases de combustión.

Fluid. The state of matter of liquids and gases.

Fluido. Estado de la materia de líquidos y gases.

Fluid expansion device. Using a bulb or sensor, tube, and diaphragm filled with fluid, this device will produce movement at the diaphragm when the fluid is heated or cooled. A bellows may be added to produce more movement. These devices may contain vapor and liquid.

Dispositivo para la expansión del fluido. Utilizando una bombilla o sensor, un tubo y un diafragma lleno de fluido, este dispositivo generará movimiento en el diafragma cuando se caliente o enfríe el fluido. Se le puede agregar un fuelle para generar aún más movimiento. Dichos dispositivos pueden contener vapor y líquido.

Flush. The process of using a fluid to push contaminants from a system.

Descarga. Proceso de utilizar un fluido para remover los contaminantes de un sistema.

Flux. A substance applied to soldered and brazed connections to prevent oxidation during the heating process.

Fundente. Sustancia aplicada a conexiones soldadas y bronce-soldadas para evitar la oxidación durante el proceso de calentamiento.

Foaming. A term used to describe oil when it has liquid refrigerant boiling out of it.

Espumación. Término utilizado para describir el aceite cuando el refrigerante líquido se derrama del mismo.

Foot-pound. The amount of work accomplished by lifting 1 lb of weight 1 ft; a unit of energy.

Libra-pie. Medida de la cantidad de energía o fuerza que se requiere para levantar una libra a una distancia de un pie; unidad de energía.

Force. Energy exerted.

Fuerza. Energía ejercida sobre un objeto.

Forced convection. The movement of fluid by mechanical means.

Convección forzada. Movimiento de fluido por medios mecánicos.

Forced-draft cooling tower. A water cooling tower that has a fan on the side of the tower that pushes air through the tower, as opposed to an induced-draft tower, which has the fan on the side and draws air through the tower.

Torre de enfriamiento de ventilación forzada. Una torre de enfriamiento que tiene un ventilador en el lado de la torre y empuja aire a través de la torre, contrario a una torre de ventilación inducida que tiene un ventilador en el lado de la torre y jala el aire a través de la torre.

Forced-draft evaporator. An evaporator over which air is forced to spread the cooling more efficiently. This term usually refers to a domestic refrigerator or freezer.

Evaporador de tiro forzado. Evaporador encima del cual se envía aire soplado para obtener una distribución más eficaz del frío. Este término suele aplicarse a refrigeradores o congeladores domésticos.

Fossil fuels. Natural gas, oil, and coal formed millions of years ago from dead plants and animals.

Combustibles fósiles. El gas natural, el petróleo y el carbón que se formaron hace millones de años de plantas y animales muertos.

Four-way valve. The valve in a heat pump system that changes the direction of the refrigerant flow between the heating and cooling cycles.

Válvula con cuatro vías. Válvula en un sistema de bomba de calor que cambia la dirección del flujo de refrigerante entre los ciclos de calentamiento y enfriamiento.

Fractionation. When a zeotropic refrigerant blend phase changes, the different components in the blend all have different vapor pressures. This causes different vaporization and condensation rates and temperatures as they phase-change.

Fraccionación. Cuando se produce un cambio en la fase de una mezcla zeotrópica de refrigerantes, los diferentes componentes que forman la mezcla tienen presiones de vapor diferentes. Esto provoca diferentes temperaturas y tasas de vaporización y de condensación según cambien de fase.

Freezer burn. The term applied to frozen food when it becomes dry and hard from dehydration due to poor packaging.

Quemadura del congelador. Término aplicado a la comida congelada cuando se seca y endurece debido a la deshidratación ocacionada por el empaque de calidad inferior.

Freeze-up. Excess ice or frost accumulation on an evaporator to the point that airflow may be affected.

Congelación. Acumulación excesiva de hielo o congelación en un evaporador a tal extremo que el flujo de aire puede ser afectado.

Freezing. The change of state of water from a liquid to a solid.

Congelamiento. Cambio de estado del agua de líquido a sólido.

Freon. The previous trade name for refrigerants manufactured by E. I. du Pont de Nemours & Co., Inc.

Freón. Marca registrada previa para refrigerantes fabricados por la compañía E. I. du Pont de Nemours, S.A.

Frequency. The cycles per second (cps) of the electrical current supplied by the power company. This is normally 60 cps in the United States.

Frecuencia. Ciclos por segundo (cps), generalmente 60 cps en los Estados Unidos, de la corriente eléctrica suministrada por la empresa de fuerza motriz.

Friction loss. The loss of pressure in a fluid flow system (air or water) due to the friction of the fluid rubbing on the sides. It is typically measured in feet of equivalent loss.

Pérdida por fricción. La pérdida de presión en un sistema de flujo de fluido (aire o agua) debido a la fricción del fluido al frotar contra los lados. Típicamente se mide en pies de pérdida equivalente.

Front seated. A position on a service valve that will not allow refrigerant flow in one direction.

Sentado delante. Posición en una válvula de servicio que no permite el flujo de refrigerante en una dirección.

Frost back. A condition of frost on the suction line and even on the compressor body.

Obturación por congelación. Condición de congelación que ocurre en el conducto de aspiración e inclusive en el cuerpo del compresor.

Frostbite. When skin freezes.

Quemadura por frío. Congelación de la piel.

Frozen. The term used to describe water in the solid state; also used to describe a rotating shaft that will not turn.

Congelado. Término utilizado para describir el agua en un estado sólido; utilizado también para describir un árbol giratorio que no gira.

Fuel oil. The fossil fuel used for heating; a petroleum distillate.

Aceite pesado. Combustible fósil utilizado para calentar; un destilado de petróleo.

Full-load amperage (FLA). The current an electric motor draws while operating under a full-load condition. Also called the run-load amperage and rated-load amperage.

Amperaje de carga total (FLA en inglés). Corriente que un motor eléctrico consume mientras funciona en una condición de carga completa. Conocido también como amperaje de carga de funcionamiento y amperaja de carga estándar.

Furnace. Equipment used to convert heating energy, such as fuel oil, gas, or electricity, to usable heat. It usually contains a heat exchanger, a blower, and the controls to operate the system.

Horno. Equipo utilizado para la conversión de energía calórica, como por ejemplo el aceite pesado, el gas o la electricidad, en calor utilizabe. Normalmente contiene un intercambiador de calor, un soplador y los reguladores para accionar el sistema.

Fuse. A safety device used in electrical circuits for the protection of the circuit conductor and components.

Fusible. Dispositivo de seguridad utilizado en circuitos eléctricos para la protección del conductor y los componentes del circuito.

Fusible link. An electrical safety device normally located in a furnace that burns and opens the circuit during an overheat situation.

Cartucho de fusible. Dispositivo eléctrico de seguridad ubicado por lo general en un horno, que quema y abre el circuito en caso de sobrecalentamiento.

Fusible plug. A device (made of low-melting-temperature metal) used in pressure vessels that is sensitive to high temperatures and relieves the vessel contents in an overheating situation.

Tapón de fusible. Dispositivo utilizado en depósitos en presión, hecho de un metal que tiene una temperatura de fusión baja. Este dispositivo es sensible a temperaturas altas y alivia el contenido del depósito en caso de sobrecalentamiento.

Gas. The vapor state of matter.

Gas. Estado de vapor de una materia.

Gasket. A thin piece of flexible material used between two metal plates to prevent leakage.

Guarnición. Pieza delgada de material flexible utilizada entre dos piezas de metal para evitar fugas.

Gas-pressure switch. Used to detect gas pressure before gas burners are allowed to ignite.

Conmutador de presión del gas. Utilizado para detectar la presión del gas antes de que los quemadores de gas puedan encenderse.

Gas valve. A valve used to stop, start, or modulate the flow of natural gas.

Válvula de gas. Válvula utilizada para detener, poner en marcha o modular el flujo de gas natural.

Gauge. An instrument used to indicate pressure.

Calibrador. Instrumento utilizado para indicar presión.

Gauge manifold. A tool that may have more than one gauge with a valve arrangement to control fluid flow.

Distribuidor de calibrador. Herramienta que puede tener más de un calibrador con las válvulas arregladas a fin de regular el flujo de fluido.

Gauge port. The service port used to attach a gauge for service procedures.

Orificio de calibrador. Orificio de servicio utilizado con el propósito de fijar un calibrador para procedimientos de servicio.

Geothermal well. A well dedicated to a geothermal heat pump that draws water from the top of the water column and returns the same water to the bottom of the water column.

Pozo geotermal. Pozo utilizado por una bomba de calor geotérmica que extrae agua de la parte superior de la columna de agua y la devuelve en la parte inferior de la misma columna.

Global warming. An earth-warming process caused by the atmosphere's absorption of the heat energy radiated from the earth's surface.

Calentamiento de la tierra. Proceso de calentamiento de la tierra provocado por la absorción de la energía radiada de la superficie de la tierra, por la atmósfera, en forma de calor.

Global warming potential (GWP). An index that measures the direct effect of chemicals emitted into the atmosphere.

Potencial de calentamiento de la tierra (GWP en inglés). Índice que mide el efecto directo de los productos químicos que se emiten a la atmósfera.

Glow coil. A device that automatically reignites a pilot light if it goes out.

Bobina encendedora. Dispositivo que automáticamente vuelve a encender la llama piloto si ésta se apaga.

Glycol. Antifreeze solution used in the water loop of geothermal heat pumps.

Glicol. Líquido anticongelante que se emplea en el circuito de agua de las bombas de calor geotérmicas.

Graduated cylinder. A cylinder with a visible column of liquid refrigerant used to measure the refrigerant charged into a system. Refrigerant temperatures can be dialed on the graduated cylinder.

Cilindro graduado. Cilindro con una columna visible de refrigerante líquido utilizado para medir el refrigerante inyectado al sistema. Las temperaturas del refrigerante pueden marcarse en el cilindro graduado.

Grain. Unit of measure. One pound = 7,000 grains.

Grano. Unidad de medida. Una libra equivale a 7.000 granos.

Gram. Metric measurement term used to express weight.

Gramo. Término utilizado para referirse a la unidad básica de peso en el sistema métrico.

Grille. A louvered, often decorative, component in an air system at the inlet or the outlet of the airflow.

Rejilla. Componente con celosías, comúnmente decorativo, en un sistema de aire que se encuentra a la entrada o a la salida del flujo de aire.

Grommet. A rubber, plastic, or metal protector usually used where wire or pipe goes through a metal panel.

Guardaojal. Protector de caucho, plástico o metal normalmente utilizado donde un alambre o un tubo pasa a través de una base de metal.

Ground, electrical. A circuit or path for electron flow to earth ground.

Tierra eléctrica. Circuito o trayectoria para el flujo de electrones a la puesta a tierra.

Ground fault circuit interrupter (GFCI). A circuit breaker that can detect very small leaks to ground, which, under certain circumstances, could cause an electrical shock. This small leak, which may not be detected by a conventional circuit breaker, will cause the GFCI circuit breaker to open the circuit.

Disyuntor por pérdidas a tierra (GFCI en inglés). Disyuntor capaz de detectar fugas muy pequeñas hacia tierra, que en determinadas circunstancias pueden provocar sacudidas eléctricas. Existe la posibilidad de que un disyuntor convencional no sea capaz de detectar las fugas pequeñas, en cuyo caso el disyuntor GFCI abre el circuito.

Ground loop. These loops of plastic pipe are buried in the ground in a closed-loop geothermal heat pump system and contain a heat transfer fluid.

Circuito de tierra. Circuitos de tubos de plástico enterrados y que forman parte de un sistema de bomba geotérmica de circuito cerrado. Dichos tubos contienen un fluido que permite la transferencia de calor.

Ground wire. A wire from the frame of an electrical device to be wired to the earth ground.

Alambre a tierra. Alambre que va desde el armazón de un dispositivo eléctrico para ser conectado a la puesta a tierra.

Guide vanes. Vanes used to produce capacity control in a centrifugal compressor. Also called prerotation guide vanes.

Paletas directrices. Paletas utilizadas para producir la regulación de capacidad en un compresor centrífugo. Conocidas también como paletas directrices para prerotación.

Halide refrigerants. Refrigerants that contain halogen chemicals: R-12, R-22, R-500, and R-502 are among them.

Refrigerantes de hálido. Refrigerantes que contienen productos químicos de halógeno: entre ellos se encuentran el R-12, R-22, R-500 y R-502.

Halide torch. A torch-type leak detector used to detect the halogen refrigerants.

Soplete de hálido. Detector de fugas de tipo soplete utilizado para detectar los refrigerantes de halógeno.

Halogens. Chemical substances found in many refrigerants containing chlorine, bromine, iodine, and fluorine.

Halógenos. Sustancias químicas presentes en muchos refrigerantes que contienen cloro, bromo, yodo y flúor.

Hand truck. A two-wheeled piece of equipment that can be used for moving heavy objects.

Vagoneta para mano. Equipo con dos ruedas que puede utilizarse para transportar objetos pesados.

Hanger. A device used to support tubing, pipe, duct, or other components of a system.

Soporte. Dispositivo utilizado para apoyar tuberías, tubos, conductos u otros componentes de un sistema.

Head. Another term for pressure, usually referring to gas or liquid.

Carga. Otro término para presión, refiriéndose normalmente a gas o líquido.

Header. A pipe or containment to which other pipe lines are connected.

Conductor principal. Tubo o conducto al que se conectan otras conexiones.

Head pressure control. A control that regulates the head pressure in a refrigeration or air-conditioning system.

Regulador de la presión de la carga. Regulador que controla la presión de la carga en un sistema de refrigeración o de acondicionamiento de aire.

Heat. Energy that causes molecules to be in motion and to raise the temperature of a substance.

Calor. Energía que ocasiona el movimiento de las moléculas provocando un aumento de temperatura en una sustancia.

Heat anticipator. A device that anticipates the need for cutting off the heating system prematurely so the system does not overshoot the set point temperature.

Anticipador de calor. Dispositivo que anticipa la necesidad de detener la marcha del sistema de calentamiento para que el sistema no exceda la temperatura programada.

Heat coil. A device made of tubing or pipe designed to transfer heat to a cooler substance by using fluids.

Bobina de calor. Dispositivo hecho de tubos, diseñado para transmitir calor a una sustancia más fría por medio de fluidos.

Heat exchanger. A device that transfers heat from one substance to another.

Intercambiador de calor. Dispositivo que transmite calor de una sustancia a otra.

Heat fusion. A process that will permanently join sections of plastic pipe together.

Fusión térmica. Proceso mediante el cual se unen dos piezas de tubo de plástico permanentemente.

Heat of compression. That part of the energy from the pressurization of a gas or a liquid converted to heat.

Calor de compresión. La parte de la energía generada de la presurización de un gas o un líquido que se ha convertido en calor.

Heat of fusion. The heat released when a substance is changing from a liquid to a solid.

Calor de fusión. Calor liberado cuando una sustancia se convierte de líquido a sólido.

Heat of respiration. When oxygen and carbon hydrates are taken in by a substance or when carbon dioxide and water are given off. Associated with fresh fruits and vegetables during their aging process while stored.

Calor de respiración. Cuando se admiten oxígeno e hidratos de carbono en una sustancia o cuando se emiten dióxido de carbono y agua. Se asocia con el proceso de maduración de frutas y legumbres frescas durante su almacenamiento.

Heat pump. A refrigeration system used to supply heat or cooling using valves to reverse the refrigerant gas flow.

Bomba de calor. Sistema de refrigeración utilizado para suministrar calor o frío mediante válvulas que cambian la dirección del flujo de gas del refrigerante.

Heat reclaim. Using heat from a condenser for purposes such as space and domestic water heating.

Reclamación de calor. La utilización del calor de un condensador para propósitos tales como la calefacción de espacio y el calentamiento doméstico de agua.

Heat recovery ventilator. Units that recover heat only, used primarily in winter.

Ventilador de recuperación de calor. Unidades que recuperan calor solamente, usados principalmente en el invierno.

Heat sink. A low-temperature surface to which heat can transfer.

Fuente fría. Superficie de temperatura baja a la que puede transmitírsele calor.

Heat tape. Electric resistance wires embedded into a flexible housing usually wrapped around a pipe to keep it from freezing.

Cinta calefactora. Resistencia eléctrica incrustada en una cubierta flexible normalmente instalada alrededor de un tubo para impedir su congelación.

Heat transfer. The transfer of heat from a warmer to a colder substance.

Transmisión de calor. Cuando se transmite calor de una sustancia más caliente a una más fría.

Helix coil. A bimetal formed into a helix-shaped coil that provides longer travel when heated.

Bobina en forma de hélice. Bimetal encofrado en una bobina en forma de hélice que provee mayor movimiento al ser calentado.

HEPA filter. An abbreviation for high-efficiency particulate arrestor. These filters are used when a high degree of filtration is desired or required.

Filtro HEPA. Abreviatura para filtro de partículas de alto rendimiento. Este tipo de filtro se utiliza cuando se requiere un elevado grado de filtrado.

Hermetic compressor. A motor and compressor that are totally sealed by being welded in a container.

Compresor hermético. Un motor y un compresor que están totalmente sellados al ser soldados al contenedor.

Hermetic system. An enclosed refrigeration system where the motor and compressor are sealed within the same system with the refrigerant.

Sistema hermético. Sistema de refrigeracíon cerrado donde el motor y el compresor se obturan dentro del mismo sistema con el refrigerante.

Hertz. Cycles per second.

Hertz. Ciclos por segundo.

Hg. Abbreviation for the element mercury.

Hg. Abreviatura del elemento mercurio.

High-pressure control. A control that stops a boiler heating device or a compressor when the pressure becomes too high.

Regulador de alta presión. Regulador que detiene la marcha del dispositivo de calentamiento de una caldera o de un compresor cuando la presión alcanza un nivel demasiado alto.

High side. A term used to indicate the high-pressure or condensing side of the refrigeration system.

Lado de alta presión. Término utilizado para indicar el lado de alta presión o de condensación del sistema de refrigeración.

High-temperature refrigeration. A refrigeration temperature range starting with evaporator temperatures no lower than 35°F, a range usually used in air conditioning (cooling).

Refrigeración a temperatura alta. Margen de la temperatura de refrigeración que comienza con temperaturas de evaporadores no menores de 35°F (2°C). Este margen se utiliza normalmente en el acondicionamiento de aire (enfriamiento).

High-vacuum pump. A pump that can produce a vacuum in the low micron range.

Bomba de vacío alto. Bomba que puede generar un vacío dentro del margen de micrón bajo.

Horsepower. A unit equal to 33,000 ft-lb of work per minute.

Potencia en caballos. Unidad equivalente a 33.000 libras-pies de trabajo por minuto.

Hot gas. The refrigerant vapor as it leaves the compressor. This is often used to defrost evaporators.

Gas caliente. El vapor del refrigerante al salir del compresor. Esto se utiliza con frecuencia para descongelar evaporadores.

Hot gas bypass. Piping that allows hot refrigerant gas into the cooler low-pressure side of a refrigeration system, usually for system capacity control.

Desviación de gas caliente. Tubería que permite la entrada de gas caliente del refrigerante en el lado más frío de baja presión de un sistema de refrigeración, normalmente para la regulación de la capacidad del sistema.

Hot gas defrost. A system where the hot refrigerant gases are passed through the evaporator to defrost it.

Descongelación con gas caliente. Sistema en el que los gases calientes del refrigerante se pasan a través del evaporador para descongelarlo.

Hot gas line. The tubing between the compressor and the condenser.

Conducto de gas caliente. Tubería entre el compresor y el condensador.

Hot water heat. A heating system using hot water to distribute the heat.

Calor de agua caliente. Sistema de calefacción que utiliza agua caliente para la distribución del calor.

Hot wire. The wire in an electrical circuit that has a voltage potential between it and another electrical source or between it and ground.

Conductor electrizado. Conductor en un circuito eléctrico a través del cual fluye la tensión entre éste y otra fuente de electricidad o entre éste y la tierra.

Humidifier. A device used to add moisture to the air.

Humedecedor. Dispositivo utilizado para agregarle humedad al aire.

Humidistat. A control operated by a change in humidity.

Humidistato. Regulador activado por un cambio en la humedad.

Humidity. Moisture in the air.

Humedad. Vapor de agua existente en el ambiente.

Hunting. The open and close throttling of a valve that is searching for its set point.

Caza. El abrir y cerrar de una válvula que está buscando su punto de ajuste.

Hydraulics. Producing mechanical motion by using liquids under pressure.

Hidráulico. Generación de movimiento mecánico por medio de líquidos bajo presión.

Hydrocarbons. Organic compounds containing hydrogen and carbon found in many heating fuels.

Hidrocarburos. Compuestos orgánicos que contienen el hidrógeno y el carbón presentes en muchos combustibles de calentamiento.

Hydrochlorofluorocarbons (HCFCs). Refrigerants containing hydrogen, chlorine, fluorine, and carbon, thought to contribute to the depletion of the ozone layer, although not to the extent of chlorofluorocarbons.

Hidroclorofluorocarburos (HCFCs en inglés). Líquidos refrigerantes que contienen hidrógeno, cloro, flúor y carbono, y que, según algunos, han contribuido a la reducción de la capa de ozono aunque no en tal grado como los clorflurocarburos.

Hydrofluorocarbon (HFC). A chlorine-free refrigerant containing hydrogen, fluorine, and carbon with zero ozone depletion potential.

Hidroflurocarbono (HFC en inglés). Refrigerante libre de cloro, compuesto de hidrógeno, fluoro y carbono que no tiene efectos perjudiciales sobre la capa de ozono.

Hydrometer. An instrument used to measure the specific gravity of a liquid.

Hidrómetro. Instrumento utilizado para medir la gravedad específica de un líquido.

Hydronic. Usually refers to a hot water heating system.

Hidrónico. Normalmente se refiere a un sistema de calefacción de agua caliente.

Hygrometer. An instrument used to measure the amount of moisture in the air.

Higrómetro. Instrumento utilizado para medir la cantidad de humedad en el aire.

Impedance. A form of resistance in an alternating current circuit.
Impedancia. Forma de resistencia en un circuito de corriente alterna.
Impeller. The rotating part of a pump that causes the centrifugal force to develop fluid flow and pressure difference.
Impulsor. Pieza giratoria de una bomba que hace que la fuerza centrífuga desarrolle flujo de fluido y una differencia en presión.
Inclined water manometer. Indicates air pressures in very low-pressure systems.
Manómetro de agua inclinada. Señala las presiones de aire en sistemas de muy baja presión.
Indoor air quality (IAQ). This term generally refers to the study or research of air quality within buildings and the procedures used to improve air quality.
Calidad del aire en el interior (IAQ en inglés). Generalmente, este término hace referencia al estudio o investigación de la calidad del aire en el interior de los edificios, así como a los procesos empleados para su mejora.
Induced magnetism. Magnetism produced, usually in a metal, from another magnetic field.
Magnetismo inducido. Magnetismo generado, normalmente en un metal, desde otro campo magnético.
Inductance. An induced voltage producing a resistance in an alternating current circuit.
Inductancia. Tensión inducida que genera una resistencia en un circuito de corriente alterna.
Induction motor. An alternating current motor where the rotor turns from induced magnetism from the field windings.
Motor inductor. Motor de corriente alterna donde el rotor gira debido al magnetismo inducido desde los devanados inductores.
Inductive circuit. When the current in a circuit lags the voltage by 90°.
Circuito inductivo. Cuando la corriente en un circuito está atrasada al voltaje por 90°.
Inductive reactance. A resistance to the flow of an alternating current produced by an electromagnetic induction.
Reactancia inductiva. Resistencia al flujo de una corriente alterna generada por una inducción electromagnética.
Inefficient equipment. Equipment that is not operating at its design level of capacity because of some fault in the equipment, such as a cylinder not pumping in a multicylinder compressor.
Equipo ineficiente. Equipo que no está operando al nivel de capacidad al que fue diseñado debido a alguna falla en el equipo, tal como que un cilindro no esté bombeando en un compresor de múltiples cilindros.
Inert gas. A gas that will not support most chemical reactions, particularly oxidation.
Gas inerte. Gas incapaz de resistir la mayoría de las reacciones químicas, especialmente la oxidación.
Infiltration. Air that leaks into a structure through cracks, windows, doors, or other openings due to less pressure inside the structure than outside the structure.
Infiltración. Penetración de aire en una estructura a través de grietas, ventanas, puertas u otras aberturas debido a que la presión en el interior de la estructura es menor que en el exterior.
Infrared humidifier. A humidifier that has infrared lamps with reflectors to reflect the infrared energy onto the water. The water evaporates rapidly into the duct airstream and is carried throughout the conditioned space.
Humidificador por infrarrojos. Humidificador equipado con lámparas de infrarrojos cuyos reflectores reflejan la energía infrarroja sobre el agua haciendo que ésta se evapore rápidamente hacia el conducto de aire para ser transportada en el espacio acondicionado.
Infrared rays. The rays that transfer heat by radiation.
Rayos infrarrojos. Rayos que transmiten calor por medio de la radiación.
Inherent motor protection. This is provided by internal protection such as a snap-disc or a thermistor.
Protección de motor inherente. Ésta es provista por una protección interna tal como un disco de encaje o un termisor.
In. Hg vacuum. The atmosphere will support a column of mercury 29.92 in. high. To pull a complete vacuum in a refrigeration system, the pressure inside the system must be reduced to 29.92 in. Hg vacuum.
Vacío en mm Hg. La atmósfera soporta una columna de mercurio de 760 mm. Para poder crear un vacío completo en un sistema de refrigeración, la presión interna debe descender a 760 mm Hg.
In-phase. When two or more alternating current circuits have the same polarity at all times.
En fase. Cuando dos o más circuitos de corriente alterna tienen siempre la misma polaridad.
Insulation, electric. A substance that is a poor conductor of electricity.
Aislamiento eléctrico. Sustancia que es un conductor pobre de electricidad.
Insulation, thermal. A substance that is a poor conductor of the flow of heat.
Aislamiento térmico. Sustancia que es un conductor pobre de flujo de calor.
Insulator. A material with several electrons in the outer orbit of the atom making them poor conductors of electricity or good insulators. Examples are glass, rubber, and plastic.
Aislante. Material con varios electrones en la órbita exterior del átomo, que los convierte en malos conductores de electricidad o en buenos aislantes, por ejemplo vidrio, caucho y plástico.
Interlocking components. Mechanical and electrical interlocks that are used to prevent a piece of equipment from starting before it is safe to start. For example, the chilled water pump and the condenser water pump must both be started before the compressor in a water-cooled chilled water system.
Componentes con enclavamiento. Enclavamientos mecánicos y eléctricos se usan para evitar que una pieza de equipo se encienda antes de ser seguro que encienda. Por ejemplo, la bomba de agua enfriada y la bomba de agua del condensador deben encenderse antes que el compresor en un sistema de enfriamiento de agua enfriado por agua.
Internal motor overload. An overload that is mounted inside the motor housing, such as a snap-disc or thermistor.

Sobrecarga interna del motor. Una sobrecarga que se monta dentro del cárter del motor tal como disco de encaje o un termisor.

Inverter. A device that alters the frequency of an electronically altered sine wave, which will affect the speed of an alternating current motor.

Inversor. Dispositivo que hace alternar la frecuencia de una onda de signos alternados electrónicamente. Esta inversión tiene un efecto sobre la velocidad de un motor de corriente alterna.

Joule. Metric measurement term used to express the quantity of heat.

Joule. Término utilizado para referirse a la unidad básica de cantidad de calor en el sistema métrico.

Junction box. A metal or plastic box within which electrical connections are made.

Caja de empalme. Caja metálica o plástica dentro de la cual se nacen conexiones eléctricas.

Kelvin. A temperature scale where absolute 0 equals 0 or where molecular motion stops at 0. It has the same graduations per degree of change as the Celsius scale.

Escala absoluta. Escala de temperaturas donde el cero absoluto equivale a 0 ó donde el movimiento molecular se detiene en 0. Tiene las mismas graduaciones por grado de cambio que la escala Celsio.

Kilopascal. A metric unit of measurement for pressure used in the air-conditioning, heating, and refrigeration field. There are 6.89 kilopascals in 1 psi.

Kilopascal. Unided métrica de medida de presión utilizada en el ramo del acondicionamiento de aire, calefacción y refrigeración. 6.89 kilopascales equivalen a 1 psi.

Kilowatt. A unit of electrical power equal to 1,000 watts.

Kilowatio. Unidad eléctrica de potencia equivalente a 1.000 watios.

Kilowatt-hour. 1 kilowatt (1,000 watts) of energy used for 1 hour.

Kilowatio hora. Unidad de energía equivalente a la que produce un kilowatio durante una hora.

King valve. A service valve at the liquid receiver's outlet in a refrigeration system.

Válvula maestra. Válvula de servicio ubicada en el receptor del líquido.

Lag shield anchors. Used with lag screws to secure screws in masonry materials.

Anclajes de tornillos barraqueros. Se usan con tornillos barraqueros para asegurar tornillos en materiales de albañilería.

Latent heat. Heat energy absorbed or rejected when a substance is changing state and there is no change in temperature.

Calor latente. Energía calórica absorbida o rechazada cuando una sustancia cambia de estado y no se experimentan cambios de temperatura.

Latent heat of condensation. The latent heat given off when refrigerant condenses.

Calor latente de la condensación. Calor latente producido por la condensación del refrigerante.

Latent heat of vaporization. The latent heat absorbed when refrigerant evaporates.

Calor latente de vaporización. Calor latente absorbido por la evaporación del refrigerante.

Leads. Extended surfaces inside a heat exchanger used to enhance the heat transfer qualities of the heat exchanger.

Extensiones. Superficies extendidas dentro de un intercambiador de calor que se usan para mejorar las cualidades de transferencia de calor del intercambiador de calor.

Leak detector. Any device used to detect leaks in a pressurized system.

Detector de fugas. Cualquier dispositivo utilizado para detectar fugas en un sistema presurizado.

Limit control. A control used to make a change in a system, usually to stop it when predetermined limits of pressure or temperature are reached.

Regulador de límite. Regulador utilizado para realizar un cambio en un sistema, normalmente para detener su marcha cuando se alcanzan niveles predeterminados de presión o de temperatura.

Limit switch. A switch that is designed to stop a piece of equipment before it does damage to itself or the surroundings, for example, a high limit on a furnace or an amperage limit on a motor.

Interruptor de límite. Interruptor que está diseñado para detener una pieza de equipo antes que ésta se haga daño o dañe sus alrededores, por ejemplo, un límite alto en un calefactor o un límite de amperaje en un motor.

Line set. A term used for tubing sets furnished by the manufacturer.

Juego de conductos. Término utilizado para referise a los juegos de tubería suministrados por el fabricante.

Line tap valve. A device that may be used for access to a refrigerant line.

Válvula de acceso en línea. Dispositivo que puede utilizarse para acceder al tubo del refrigerante.

Line-voltage thermostat. A thermostat that switches line voltage. For example, it is used for electric baseboard heat.

Termostato de voltaje de línea. Un termostato que interrumpe el voltaje de línea. Por ejemplo, para los calentadores eléctricos de rodapié.

Line wiring diagram. Sometimes called a ladder diagram, this type of diagram shows the power-consuming devices between the lines. Usually, the right side of the diagram consists of a common line.

Diagrama del cableado de línea. También conocido como diagrama de escalera, este tipo de diagrama muestra los dispositivos de consumo de corriente que hay entre las líneas. Generalmente, la línea común está en el lado derecho del diagrama.

Liquefied petroleum. Liquefied propane, butane, or a combination of these gases. The gas is kept as a liquid under pressure until ready to use.

Petróleo licuado. Propano o butano licuados, o una combinación de estos gases. El gas se mantiene en estado líquido bajo presión hasta que se encuentre listo para usar.

Liquid. A substance where molecules push outward and downward and seek a uniform level.

Líquido. Sustancia donde las moléculas empujan hacia afuera y hacia abajo y buscan un nivel uniforme.

Liquid charge bulb. A type of charge in the sensing bulb of the thermostatic expansion valve. This charge is characteristic of the refrigerant in the system and contains enough liquid so that it will not totally boil away.

Bombilla de carga líquida. Tipo de carga en la bombilla sensora de la válvula de expansión termostática. Esta carga es característica del refrigerante en el sistema y contiene suficiente líquido para que el mismo no se evapore completamente.

Liquid-filled remote bulb. A remote bulb thermostat that is completely liquid filled, such as the mercury bulb on some gas furnace pilot safety devices.

Bombillo remoto lleno de líquido. Un termostato de bombillo remoto que está completamente lleno con líquido, tal como el bulbo de mercurio en algunos dispositivos de seguridad de los calefactores de gas.

Liquid floodback. Liquid refrigerant returning to the compressor's crankcase during the running cycle.

Regreso de líquido. El regresar del refrigerante líquido al cárter del cigüeñal del compresor durante el ciclo de marcha.

Liquid hammer. The momentum force of liquid causing a noise or a disturbance when hitting against an object.

Martillo líquido. La fuerza mecánica de un líquido que causa un ruido o un disturbio cuando choca con un objeto.

Liquid line. A term applied in the industry to refer to the tubing or piping from the condenser to the expansion device.

Conducto de líquido. Término aplicado en nuestro medio para referirse a la tubería que va del condensador al dispositivo de expansión.

Liquid nitrogen. Nitrogen in liquid form.

Nitrógeno líquido. Nitrógeno en forma líquida.

Liquid receiver. A container in the refrigeration system where liquid refrigerant is stored.

Receptor del líquido. Recipiente en el sistema de refrigeración donde se almacena el refrigerante líquido.

Liquid refrigerant charging. The process of allowing liquid refrigerant to enter the refrigeration system through the liquid line to the condenser and evaporator.

Carga para refrigerante líquido. Proceso de permitir la entrada del refrigerante líquido al condensador y al evaporador en el sistema de refrigeración a través del conducto de líquido.

Liquid refrigerant distributor. This device is used between the expansion valve and the evaporator on multiple circuit evaporators to evenly distribute the refrigerant to all circuits.

Distribuidor de refrigerante líquido. Este aparato se usa entre la válvula de expansión y el evaporador en los evaporadores de múltiples circuitos para distribuir el refrigerante a todos los circuitos.

Liquid slugging. A large amount of liquid refrigerant in the compressor cylinder, usually causing immediate damage.

Relleno de líquido. Acumulación de una gran cantidad de refrigerante líquido en el cilindro del compresor, que normalmente provoca una avería inmediata.

Lithium-bromide. A type of salt solution used in an absorption chiller.

Bromuro de litio. Tipo de solución salina utilizada en un enfriador por absorción.

Load matching. Trying to always match the capacity of the refrigeration or air-conditioning system with that of the heat load put on the evaporators.

Adaptación de carga. Hecho de intentar adaptar la capacidad del sistema de refrigeración o aire acondicionado a la carga térmica que deben soportar los evaporadores.

Load shed. Part of an energy management system where various systems in a structure may be cycled off to conserve energy.

Despojo de carga. Parte de un sistema de manejo de energía en el cual varios sistemas en una estructura pueden apagarse para conservar energía.

Locked-rotor amperage (LRA). The current an electric motor draws when it is first turned on. This is normally five times the full-load amperage.

Amperaje de rotor bloqueado (LRA en inglés). Corriente que un motor eléctrico consume al ser encendido, la cual generalmente es cinco veces mayor que el amperaje de carga completa.

Low ambient control. Various types of controls that are used to control head pressure in air-cooled air-conditioning and refrigeration systems that must operate year-round or in cold weather.

Control de ambiente bajo. Varios tipos de controles que se usan para controlar la presión en los sistemas de aire acondicionado y refrigeración, enfriados por aire, y que tienen que operar todo el año o en climas fríos.

Low-boy furnace. This furnace is approximately 4 ft high, and the air intake and discharge are both at the top.

Horno bajo. Este tipo de horno tiene una altura de aproximadamente 1,2 metros, con la toma y evacuación del aire situadas en la parte de arriba.

Low-loss fitting. A fitting that is fastened to the end of a gauge manifold that allows the technician to connect and disconnect gauge lines with a minimum of refrigerant loss.

Acoplamiento de poca pérdida. Un tipo de acoplamiento que se conecta en un extremo de un colector de calibración y que le permite al técnico conectar y desconectar las líneas de calibración con una pérdida mínima de refrigerante.

Low-pressure control. A pressure switch that can provide low charge protection by shutting down the system on low pressure. It can also be used to control space temperature.

Regulador de baja presión. Conmutador de presión que puede proveer protección contra una carga baja al detener el sistema si éste alcanza una presión demasiado baja. Puede utilizarse también para regular la temperatura de un espacio.

Low side. A term used to refer to that part of the refrigeration system that operates at the lowest pressure, between the expansion device and the compressor.

Lado bajo. Término utilizado para referirse a la parte del sistema de refrigeración que funciona a niveles de presión más baja, entre el dispositivo de expansión y el compresor.

Low-temperature refrigeration. A refrigeration temperature range starting with evaporator temperatures no higher than 0°F for storing frozen food.

Refrigeración a temperatura baja. Margen de la temperatura de refrigeración que comienza con temperaturas de evaporadores no mayores de 0°F (−18°C) para almacenar comida congelada.

Low-voltage thermostat. The typical thermostat used for residential and commercial air-conditioning and heating equipment to control space temperature. The supplied voltage is 24 V.
Termostato de bajo voltaje. El termostato típico que se usa en el equipo de aire acondicionado y de calefacción comercial y residencial para controlar la temperatura de un espacio. El voltaje suplido es de 24 voltios.

LP fuel. Liquefied petroleum, propane, or butane. A substance used as a gas for fuel. It is transported and stored in the liquid state.
Combustible PL. Petróleo licuado, propano o butano. Sustancia utilizada como gas para combustible. El petróleo licuado se transporta y almacena en estado líquido.

Magnetic field. A field or space where magnetic lines of force exist.
Campó magnético. Campo o espacio donde existen líneas de fuerza magnética.

Magnetic overload protection. This protection reads the actual current draw of the motor and is able to shut it off based on actual current, versus the heat-operated thermal overloads, which are sensitive to the ambient heat of a hot cabinet.
Protección de sobrecarga magnética. Esta protección lee la toma de corriente actual del motor y es capaz de apagarlo basado en la corriente actual; esto es contrario a las sobrecargas termales operadas por calor, las cuales son sensibles al calor ambiental de un gabinete caliente.

Magnetism. A force causing a magnetic field to attract ferrous metals, or where like poles of a magnet repel and unlike poles attract each other.
Magnetismo. Fuerza que hace que un campo magnético atraiga metales férreos, o cuando los polos iguales de un imán se rechazan y los opuestos se atraen.

Makeup air. Air, usually from outdoors, provided to make up for the air used in combustion.
Aire de compensación. Aire, normalmente procedente del exterior, que se utiliza para compensar aquél utilizado en la combustión.

Makeup water. Water that is added back into any circulating water system due to loss of water. Makeup water in a cooling tower may be quite a large volume.
Agua de compensación. Agua que se añade a cualquier sistema de circulación de agua debido a la pérdida de agua. El agua de compensación en una torre de enfriamiento puede ser un volumen bastante grande.

Male thread. A thread on the outside of a pipe, fitting, or cylinder; an external thread.
Rosca macho. Rosca en la parte exterior de un tubo, accesorio o cilindro; rosca externa.

Manifold. A device where multiple outlets or inlets can be controlled with valves or other devices. Our industry typically uses a gas manifold with orifices for gas-burning appliances and gauge manifolds used by technicians.
Colector. Aparato desde el cual se pueden controlar varias entradas y salidas con válvulas u otros aparatos. Nuestra industria generalmente usa un colector de gas con orificios para los enseres que queman gas y colectores de calibración que los usan los técnicos.

Manometer. An instrument used to check low vapor pressures. The pressures may be checked against a column of mercury or water.
Manómetro. Instrumento utilizado para revisar las presiones bajas de vapor. Las presiones pueden revisarse comparándolas con una columna de mercurio o de agua.

Manual reset. A safety control that must be reset by a person, as opposed to automatically reset, to call attention to the problem. An electrical breaker is a manual reset device.
Reinicio manual. Un control de seguridad que una persona tiene que reiniciar, contrario a un reinicio automático, para llamar atención al problema. Un cortacircuito eléctrico es un aparato de reinicio manual.

Mapp gas. A composite gas similar to propane that may be used with air.
Gas Mapp. Gas compuesto similar al propano que puede utilizarse con aire.

Mass. Matter held together to the extent that it is considered one body.
Masa. Materia compacta que se considera un solo cuerpo.

Mass spectrum analysis. An absorption machine factory leak test performed using helium.
Análisis del límite de masa. Prueba para fugas y absorción llevada a cabo en la fábrica utilizando helio.

Matter. A substance that takes up space and has weight.
Materia. Sustancia que ocupa espacio y tiene peso.

Mechanical controls. A control that has no connection to power, such as a water-regulating valve or a pressure relief valve.
Controles mecánicos. Un control que no tiene conexión a corriente, tales como una válvula reguladora de agua o una válvula de alivio de presión.

Medium-temperature refrigeration. Refrigeration where evaporator temperatures are 32°F or below, normally used for preserving fresh food.
Refrigeración a temperatura media. Refrigeración, donde las temperaturas del evaporador son 32°F (0°C) o menos, utilizada generalmente para preservar comida fresca.

Megohmmeter. An instrument that can detect very high resistances, in millions of ohms. A megohm is equal to 1,000,000 ohms.
Megaohmnímetro. Un instrumento que puede detectar resistencias muy altas, de millones de ohmios. Un megaohmio es equivalente a 1.000.000 de ohmios.

Melting point. The temperature at which a substance will change from a solid to a liquid.
Punto de fusión. Temperatura a la que una sustancia se convierte de sólido a líquido.

Mercury bulb. A glass bulb containing a small amount of mercury and electrical contacts used to make and break the electrical circuit in a low-voltage thermostat.
Bombilla de mercurio. Bombilla de cristal que contiene una pequeña cantidad de mercurio y que funciona como contacto eléctrico, utilizada para conectar y desconectar el circuito eléctrico en un termostato de baja tensión.

Metering device. A valve or small fixed-size tubing or orifice that meters liquid refrigerant into the evaporator.

Dispositivo de medida. Válvula o tubería pequeña u orificio que mide la cantidad de refrigerante líquido que entra en el evaporador.

Methane. Natural gas composed of 90% to 95% methane, a combustible hydrocarbon.

Metano. El gas natural se compone de un 90% a un 95% de metano, un hidrocarburo combustible.

Metric system. System International (SI); system of measurement used by most countries in the world.

Sistema métrico. Sistema internacional; el sistema de medida utilizado por la mayoría de los países del mundo.

Micro. A prefix meaning $\frac{1}{1,000,000}$.

Micro. Prefijo que significa una parte de un millón.

Microfarad. Capacitor capacity equal to $\frac{1}{1,000,000}$ of a farad.

Microfaradio. Capacidad de un capacitador equivalente a $\frac{1}{1.000.000}$ de un faradio.

Micrometer. A precision measuring instrument.

Micrómetro. Instrumento de precisión utilizado para medir.

Micron. A unit of length equal to $\frac{1}{1,000}$ of a millimeter or $\frac{1}{1,000,000}$ of a meter.

Micrón. Unidad de largo equivalente a $\frac{1}{1.000}$ de un milímetro, o $\frac{1}{1.000.000}$ de un metro.

Micron gauge. A gauge used when it is necessary to measure pressure close to a perfect vacuum.

Calibrador de micrón. Calibrador utilizado cuando es necesario medir la presión de un vacío casi perfecto.

Microprocessor. A small, preprogrammed, solid-state microcomputer that acts as a main controller.

Microprocesador. Un microordenador de estado sólido preprogramado que actúa de controlador principal.

Midseated (cracked). A position on a service valve that allows refrigerant flow in all directions.

Sentado en el medio (agrietado). Posición en una válvula de servicio que permite el flujo de refrigerante en cualquier dirección.

Migration of oil or refrigerant. When the refrigerant moves to some place in the system where it is not supposed to be, such as when oil migrates to an evaporator or when refrigerant migrates to a compressor crankcase.

Migración de aceite o refrigerante. Cuando el refrigerante se mueve a cualquier lugar en el sistema donde no debe estar, como cuando aceite migra a un evaporador o cuando refrigerante se transplanta al cárter del cigüeñal del compresor.

Milli. A prefix meaning $\frac{1}{1,000}$.

Mili. Prefijo que significa una parte de mil.

Mineral oil. A traditional refrigeration lubricant used in CFC and HCFC systems.

Aceite mineral. Lubricante de refrigeración utilizado tradicionalmente en los sistemas de CFC y HCFC.

Minimum efficiency reporting value (MERV). Air filter rating system ranging from 1 to 20, with upper levels providing the most filtering.

Valor mínimo de reporte de eficacia (MERV en inglés). Sistema de clasificación de filtros de aire que va desde el 1 al 20 con los niveles mayores indicando más filtración.

Modulating flow. Controlling the flow between maximum or no flow. For example, the accelerator on a car provides modulating flow.

Flujo modulante. Controlado el flujo entre flujo que no es flujo máximo o ningún flujo. Por ejemplo, el acelerador de un carro provee un flujo modulante.

Modulator. A device that adjusts by small increments or changes.

Modulador. Dispositivo que se ajusta por medio de incrementos o cambios pequeños.

Moisture indicator. A device for determining moisture.

Indicador de humedad. Dispositivo utilizado para determinar la humedad.

Mold. A fungus found where there is moisture that develops and releases spores. Can be harmful to humans.

Moho. Un hongo encontrado donde hay humedad que se desarrolla y libera esporas. Puede ser nocivo a los seres humanos.

Molecular motion. The movement of molecules within a substance.

Movimiento molecular. Movimiento de moléculas dentro de una sustancia.

Molecule. The smallest particle that a substance can be broken into and still retain its chemical identity.

Molécula. La particula más pequeña en la que una sustancia puede dividirse y aún conservar sus propias características.

Monochlorodifluoromethane. The refrigerant R-22.

Monoclorodiflorometano. El refrigerante R-22.

Montreal Protocol. An agreement signed in 1987 by the United States and other countries to control the release of ozone-depleting gases.

Protocolo de Montreal. Un acuerdo firmado en 1987 por los Estados Unidos y otros países para controlar la liberación de gases que destruyen el ozono.

Motor service factor. A factor above an electric motor's normal operating design parameters, indicated on the nameplate, under which it can operate.

Factor de servicio del motor. Factor superior a los parametros de diseño normales de funcionamiento de un motor eléctrico, indicados en el marbete; este factor indica su nivel de funcionamiento.

Motor starter. Electromagnetic contactors that contain motor protection and are used for switching electric motors on and off.

Arrancador de motor. Contactadores electromagnéticos que contienen protección para el motor y se utilizan para arrancar y detener motores eléctricos.

Motor temperature-sensing thermostat. A thermostat that monitors the motor temperature and shuts it off for the motor's protection.

Termostato que detecta la temperatura del motor. Un termostato que vigila la temperatura del motor y lo apaga para la protección del motor.

Muffler, compressor. Sound absorber at the compressor.

Silenciador del compresor. Absorbedor de sonido ubicado en el compresor.

Mullion. Stationary frame between two doors.

Parteluz. Armazón fijo entre dos puertas.

Mullion heater. Heating element mounted in the mullion of a refrigerator to keep moisture from forming on it.

Calentador del parteluz. Elemento de calentamiento montado en el parteluz de un refrigerador para evitar la formación de humedad en el mismo.

Multimeter. An instrument that will measure voltage, resistance, and milliamperes.

Multímetro. Instrumento que mide la tensión, la resistencia y los miliamperios.

Multiple circuit coil. An evaporator or condenser coil that has more than one circuit because of the coil length. When the coil is too long, there will be an unacceptable pressure drop and loss of efficiency.

Serpentín de circuito múltiple. Un serpentín de evaporador o condensador que tiene más de un circuito por causa de la longitud del serpentín. Cuando el serpentín es demasiado largo, habrá una pérdida de presión y eficacia inaceptable.

Multiple evacuation. A procedure for evacuating a system. A vacuum is pulled, a small amount of refrigerant allowed into the system, and the procedure duplicated. This is often done three times.

Evacuación múltiple. Procedimiento para evacuar o vaciar un sistema. Se crea un vacío, se permite la entrada de una pequeña cantidad de refrigerante al sistema, y se repite el procedimiento. Con frecuencia esto se lleva a cabo tres veces.

National Electrical Code® **(NEC®).** A publication that sets the standards for all electrical installations, including motor overload protection.

Código estadounidense de electridad. Publicación que establece las normas para todas las instalaciones eléctricas, incluyendo la protección contra la sobrecarga de un motor.

National Fire Protection Association (NFPA). An association organized to prevent fires through establishing standards, providing research, and providing public education.

Asociación nacional para la protección contra incendios (NFPA en inglés). Asociación cuyo objetivo es prevenir incendios estableciendo normativas y facilitando la investigación y concienciación del público.

National pipe taper (NPT). The standard designation for a standard tapered pipe thread.

Cono estadounidense para tubos (NPT en inglés). Designación estándar para una rosca cónica para tubos estándar.

Natural convection. The natural movement of a gas or fluid caused by differences in temperature.

Convección natural. Movimiento natural de un gas o fluido ocasionado por diferencias en temperatura.

Natural-draft tower. A water cooling tower that does not have a fan to force air over the water. It relies on the natural breeze or airflow.

Torre de corriente de aire natural. Una torre de enfriamiento de agua que no tiene un ventilador para forzar el aire sobre el agua; depende de la brisa o flujo natural del aire.

Natural gas. A fossil fuel formed over millions of years from dead vegetation and animals that were deposited or washed deep into the earth.

Gas natural. Combustible fósil formado a través de millones de años de la vegetación y los animales muertos que fueron depositados o arrastrados a una gran profundidad dentro la tierra.

Near-azeotropic blend. Two or more refrigerants mixed together that will have a small range of boiling and/or condensing points for each system pressure. Small fractionation and temperature glides will occur but are often negligible.

Mezcla casi-azeotrópica. Mezcla de dos o más refrigerantes, que tiene un bajo rango de punto de ebullición y/o condensación para cada presión del sistema. Puede producirse una cierta fraccionación y variación de temperatura, aunque suelen ser insignificantes.

Needlepoint valve. A device having a needle and a very small orifice for controlling the flow of a fluid.

Válvula de aguja. Dispositivo que tiene una aguja y un orificio bastante pequeño para regular el flujo de un fluido.

Negative electrical charge. An atom or component that has an excess of electrons.

Carga eléctrica negativa. Átomo o componente que tiene un exceso de electrones.

Neoprene. Synthetic flexible material used for gaskets and seals.

Neopreno. Material sintético flexible utilizado en guarniciones y juntas de estanqueidad.

Net oil pressure. Difference in the suction pressure and the compressor oil pump outlet pressure.

Presión neta del aceite. Diferencia en la presión de aspiración y la presión a la salida de la bomba de aceite del compresor.

Net refrigeration effect (NRE). The quantity of heat in BTU/lb that the refrigerant absorbs from the refrigerated space to produce useful cooling.

Efecto neto de refrigeración (NRE en inglés). La cantidad de calor expresado en BTU/lb que el refrigerante absorbe del espacio refrigerado para producir refrigeración útil.

Net stack temperature. The temperature difference between the ambient temperature and the flue gas temperature, typically for oil- and gas-burning equipment.

Temperatura neta de chimenea. La diferencia en temperatura entre la temperatura ambiental y la del conducto de gas, normalmente para equipo que quema aceite y gas.

Neutralizer. A substance used to counteract acids.

Neutralizador. Sustancia utilizada para contrarrestar ácidos.

Neutron. Neutrons and protons are located at the center of the nucleus of an atom. Neutrons have no charge.

Neutrón. Los neutrones y protones están situados en le centro del núcleo del átomo. Los neutrones carecen de carga.

Nitrogen. An inert gas often used to "sweep" a refrigeration system to help ensure that all refrigerant and contaminants have been removed.

Nitrógeno. Gas inerte utilizado con frecuencia para purgar un sistema de refrigeración. Esta gas ayuda a asegurar la remoción de todo el refrigerante y los contaminantes del sistema.

Nominal. A rounded-off stated size. The nominal size is the closest rounded-off size.

Nominal. Tamaño redondeado establecido. El tamaño nominal es el tamaño redondeado más cercano.

Noncondensable gas. A gas that does not change into a liquid under normal operating conditions.

Gas no condensable. Gas que no se convierte en líquido bajo condiciones de funcionamiento normales.

Nonferrous. Metals containing no iron.

No férreos. Metales que no contienen hierro.

North Pole, magnetic. One end of a magnet or the magnetic north pole of the earth.
Polo norte magnético. El extremo de un imán o el polo norte magnético del mundo.
Nozzle. A drilled opening that measures liquid flow, such as an oil burner nozzle.
Tobera. Una apertura taladrada que mide el flujo de líquido, tal como la tobera de un quemador de aceite.
Nut driver. These tools have a socket head used primarily to turn hex head screws on air-conditioning, heating, and refrigeration cabinets.
Extractor de tuercas. Estas herramientas tienen una cabeza hueca hexagonal usadas principalmente para darle vuelta a tuercas de cabeza hexagonal en gabinetes de acondicionamiento de aire, de calefacción y de refrigeración.

Off cycle. A period when a system is not operating.
Ciclo de apagado. Período de tiempo cuando un sistema no está en funcionamiento.
Off-cycle defrost. Used for medium-temperature refrigeration where the evaporator coil operates below freezing but the air in the cooler is above freezing. The coil is defrosted by the air inside the cooler while the compressor is off cycle.
Descongelación de período de reposo. Se usa para refrigeración de temperatura media en la cual el serpentín del evaporador funciona por debajo del punto de congelación, pero el aire en el enfriador está por encima del punto de congelación. El serpentín se descongela por el aire dentro del enfriador mientras el compresor está en reposo.
Offset. The absolute (not signed + or −) difference between the set point and the control point of a control process.
Compensación. La diferencia absoluta (sin signo + o −) entre el punto de ajuste y el punto de control de un proceso de control.
Offset. The position of ductwork that must be rerouted around an obstacle.
Desviación. El posición de un conducto que tiene que desviarse alrededor de un obstáculo.
Ohm. A unit of measurement of electrical resistance.
Ohmio. Unidad de medida de la resistencia eléctrica.
Ohmmeter. A meter that measures electrical resistance.
Ohmiómetro. Instrumento que mide la resistencia eléctrica.
Ohm's Law. A law involving electrical relationships discovered by Georg Ohm: $E = I \times R$.
Ley de Ohm. Ley que define las relaciones eléctricas, descubierta por Georg Ohm: $E = I \times R$.
Oil level regulator. A needle valve and float system located on each compressor of a parallel compressor system. It senses the oil level in the compressor's crankcase and adds oil if necessary. It receives its oil from the oil reservoir.
Regulador del nivel de aceite. Sistema de válvula de aguja y flotador que se encuentra en cada compresor de un sistema de compresores en paralelo. Detecta el nivel del aceite en el cárter del compresor y permite la entrada de más aceite procedente de un depósito, en caso necesario.
Oil-pressure safety control (switch). A control used to ensure that a compressor has adequate oil lubricating pressure.
Regulador de seguridad para la presión de aceite (conmutador). Regulador utilizado para asegurar que un compresor tenga la presión de lubrificación de aceita adecuada.
Oil, refrigeration. Oil used in refrigeration systems.
Aceite de refrigeración. Aceite utilizado en sistemas de refrigeración.
Oil reservoir. A storage cylinder for oil usually used on parallel compressor systems. It is located between the oil separator and the oil level regulators. It receives its oil from the oil separator.
Depósito de aceite. Cilindro en el que se almacena el aceite utilizado en los sistemas de compresores en paralelo. Está ubicado entre el separador de aceite y el regulador del nivel. Recibe el aceite del separador.
Oil separator. Apparatus that removes oil from a gaseous refrigerant.
Separador de aceite. Aparato que remueve el aceite de un refrigerante gaseoso.
One-time relief valve. A pressure relief valve that has a diaphragm that blows out due to excess pressure. It is set at a higher pressure than the spring-loaded relief valve in case it fails.
Válvula de alivio de una vez. Una válvula de alivio de presión que tiene un diafragma que revienta debido a la presión excesiva. La misma se fija a una presión más alta que la válvula de alivio de resorte para en caso de que ésta falle.
Open compressor. A compressor with an external drive.
Compresor abierto. Compresor con un motor externo.
Open-loop heat pump. Heat pump system that uses the water in the earth as the heat transfer medium and then expels the water back to the earth in some manner.
Bomba de calor de circuito abierto. Sistema de bomba de calor que utiliza el agua de la tierra como medio de transferencia del calor y luego devuelve el agua a la tierra de cierta manera.
Open winding. The condition that exists when there is a break and no continuity in an electric motor winding.
Devanado abierto. Condición que se presenta cuando hay una interrupción en la continuidad del devanado de un motor.
Operating pressure. The actual pressure under operating conditions.
Presión de funcionamiento. La presión real bajo las condiciones de funcionamiento.
Organic. Materials formed from living organisms.
Orgánico. Materiales formados de organismos vivos.
Orifice. A small opening through which fluid flows.
Orificio. Pequeña abertura a través de la cual fluye un fluido.
Outward clinch tacker. A stapler or tacker that will anchor staples outward and can be used with soft materials.
Grapadora de agarre hacia fuera. Grapadora o tachueladora que ancla las grapas hacia fuera y que puede usarse con materiales suaves.
Overload protection. A system or device that will shut down a system if an overcurrent condition exists.
Protección contra sobrecarga. Sistema o dispositivo que detendrá la marcha de un sistema si existe una condición de sobreintensidad.
Oxidation. The combining of a material with oxygen to form a different substance. This results in the deterioration of the original substance. Rust is oxidation.

Oxidación. La combinación de un material con oxígeno para formar una sustancia diferente, lo que ocasiona el deterioro de la sustancia original. Herrumbre es oxidación.

Ozone. A form of oxygen (O_3). A layer of ozone is in the stratosphere that protects the earth from certain of the sun's ultraviolet wavelengths.

Ozono. Forma de oxígeno (O_3). Una capa de ozono en la estratosfera protege la tierra de ciertos rayos ultravioletas del sol.

Ozone depletion. The breaking up of the ozone molecule by the chlorine atom in the stratosphere. Stratosphere ozone protects us from ultraviolet radiation emitted by the sun.

Reducción del ozono. Descomposición de la molécula de ozono por el átomo de cloro en la estratosfera. El ozono presente en la estratosfera nos protege de las radiaciones ultravioletas del sol.

Ozone depletion potential (ODP). A scale used to measure how much a substance will deplete stratospheric ozone.

Potencial de depleción de ozono (ODP en inglés). Una escala que se usa para medir cuánta depleción del ozono de la estratosfera una sustancia va a causar.

Package unit. A refrigerating system where all major components are located in one cabinet.

Unidad completa. Sistema de refrigeración donde todos los componentes principales se encuentran en un solo gabinete.

Packing. A soft material that can be shaped and compressed to provide a seal. It is commonly applied around valve stems.

Empaquetadura. Material blando que puede formarse y comprimirse para proveer una junta de estanqueidad. Comúnmente se aplica alrededor de los vástagos de válvulas.

Paraffinic oil. A refrigeration mineral oil containing some paraffin wax, which is refined from eastern U.S. crude oil.

Aceite de parafina. Aceite mineral que contiene parafina y que se utiliza en sistemas de refrigeración. El aceite se obtiene mediante el refinado de petróleo extraído en los EE.UU. del este.

Parallel circuit. An electrical or fluid circuit where the current or fluid takes more than one path at a junction.

Circuito paralelo. Corriente eléctrica o fluida donde la corriente o el fluido siguen más de una trayectoria en un empalme.

Parallel compressor. Many compressors piped in parallel and mounted on a steel rack. The compressors are usually cycled by a microprocessor.

Compresor en paralelo. Varios compresores conectados en paralelo y montados en un bastidor de acero. Normalmente, se sirve de un microprocesador para activarlos y desactivarlos.

Parallel flow. A flow path in which many paths exist for the fluid to flow.

Flujo paralelo. Vía de flujo que consta de varias vías que permiten el flujo de un fluido.

Part-winding start. A large motor that is actually two motors in one housing. It starts on one and then the other is energized. This is to reduce inrush current at start-up. For example, a 100-hp motor may have two 50-hp motors built into the same winding. It will start using one motor followed by the start of the other one. They will both run under the load.

Arranque de bobina parcial. Un motor grande que es, en efecto, dos motores bajo un mismo cárter. El mismo arranca con uno y luego se activa el segundo. El propósito de esto es reducir la corriente interna al arrancar. Por ejemplo, un motor de 100 caballos de fuerza puede tener dos motores de 50 caballos de fuerza construidos con la misma bobina. El motor arrancará usando un motor, seguido por el arranque del otro motor. Los dos motores correrán bajo carga.

Pascal. A metric unit of measurement of pressure.

Pascal. Unidad métrica de medida de presión.

Passive recovery. Recovering refrigerant with the use of the refrigeration system's compressor or internal vapor pressure.

Recuperación pasiva. Recuperación de un refrigerante utilizando el compresor del sistema de refrigeración o la presión del vapor interno.

Passive solar design. The use of nonmoving parts of a building to provide heat or cooling, or to eliminate certain parts of a building that cause inefficient heating or cooling.

Diseño solar pasivo. La utilización de piezas fijas de un edificio para proveer calefacción o enfriamiento, o para eliminar ciertas piezas de un edificio que causan calefacción o enfriamiento ineficientes.

PE (polyethylene). Plastic pipe used for water, gas, and irrigation systems.

Polietileno. Tubo plástico utilizado en sistemas de agua, de gas y de irrigación.

Percent refrigerant quality. Percent vapor.

Calidad porcentual de refrigerante. Porcentaje de vapor.

Permanent magnet. An object that has its own permanent magnetic field.

Imán permanente. Objeto que tiene su propio campo magnético permanente.

Permanent split-capacitor motor (PSC). A split-phase motor with a run capacitor only. It has a very low starting torque.

Motor permanente de capacitador separado (PSC en inglés). Motor de fase separada que sólo tiene un capacitador de funcionamiento. Su par de arranque es sumamente bajo.

Phase. One distinct part of a cycle.

Fase. Una parte específica de un ciclo.

Phase-change loop. The loop of piping, usually in a geothermal heat pump system, where there is a change of phase of the heat transfer fluid from liquid to vapor or vapor to liquid.

Circuito de cambio de fase. Circuito de tubería, generalmente en un sistema de bombeo de calor geotermal, en el cual hay un cambio de fase del fluido de transferencia de calor de líquido a vapor o de vapor a líquido.

Phase failure protection. Used on three-phase equipment to interrupt the power source when one phase becomes de-energized. The motors cannot be allowed to run on the two remaining phases or damage will occur.

Protección de fallo de fase. Se usa en equipo trifásico para interrumpir la fuente de potencia cuando se energiza una fase. No se puede permitir que los motores corran con las dos fases restantes o podría ocurrir una avería.

Phase reversal. Phase reversal can occur if someone switches any two wires on a three-phase system. Any system with a three-phase motor will reverse, and this cannot be allowed on some equipment.

Inversión de fase. La inversión de fase puede ocurrir si por cualquier razón alguien intercambia cualquier par de cables en un sistema trifásico. Cualquier sistema con un motor trifásico irá en dirección contraria y esto no puede permitirse en algunos sistemas.

Pictorial wiring diagram. This type of diagram shows the location of each component as it appears to the person installing or servicing the equipment.

Diagrama representativo del cableado. Este tipo de diagrama indica la ubicación de cada componente tal y como lo verá el personal técnico.

Piercing valve. A device that is used to pierce a pipe or tube to obtain a pressure reading without interrupting the flow of fluid. Also called a line tap valve.

Válvula punzante. Aparato que se usa para punzar un tubo para obtener una lectura de la presión sin interrumpir el flujo del fluido. También se llama una válvula de toma.

Pilot duty relay. A small relay that is used in control circuits for switching purposes. It is small and cannot take a lot of current flow, such as to start a motor.

Relé de función piloto. Un pequeño relé que se usa en los circuitos de control con propósitos de interrupción. Es pequeño y no puede tolerar un flujo alto de corriente, como para arrancar un motor.

Pilot light. The flame that ignites the main burner on a gas furnace.

Llama piloto. Llama que enciende el quemador principal en un horno de gas.

Piston. The part that moves up and down in a cylinder.

Pistón. La pieza que asciende y desciende dentro de un cilindro.

Piston displacement. The volume within the cylinder that is displaced with the movement of the piston from top to bottom.

Desplazamiento del pistón. Volumen dentro del cilindro que se desplaza de arriba a abajo con el movimiento del pistón.

Plenum. A sealed chamber at the inlet or outlet of an air handler. The duct attaches to the plenum.

Plenum. Cámara sellada a la entrada o a la salida de un tratante de aire. El conducto se fija al plenum.

Pneumatic controls. Controls operated by low-pressure air, typically 20 psig.

Controles neumáticos. Controles que se operan por aire de baja presión, generalmente 20 libras por pulgada cuadrada de presión de manómetro (psig).

Polyalkylene glycol. A popular synthetic glycol-based lubricant used with HFC refrigerants, mainly in automotive systems. This was the first generation of oil used with HFC refrigerants.

Glicol polialkilénico. Lubricante sintético de uso común basado en glicol, usado con refrigerantes HFC, principalmente en sistemas de automóviles. Ésta fue la primera generación de aceites usados con refrigerantes HFC.

Polybutylene. A material used for the buried piping in geothermal heat pumps.

Polibutileno. Material utilizado para la fabricación de tubos enterrados, en sistemas de bombas de calor geotérmicas.

Polyethylene. A material used for the buried piping in geothermal heat pumps.

Polietileno. Material utilizado para la fabricación de tubos enterrados, en sistemas de bombas de calor geotérmicas.

Polyol ester. A very popular ester-based lubricant often used in HFC refrigerant systems.

Poliol éster. Lubricante muy popular basado en éster usado frecuentemente en sistemas con refrigerantes HFC.

Polyphase. Three or more phases.

Polifase. Tres o más fases.

Polyphosphate. A scale inhibitor with many phosphate molecules.

Polifosfato. Un inhibidor de escama que contiene muchas moléculas de fosfato.

Porcelain. A ceramic material.

Porcelana. Material cerámico.

Positive displacement. A term used with a pumping device such as a compressor that is designed to move all matter from a volume such as a cylinder or it will stall, possibly causing failure of a part.

Desplazamiento positivo. Término utilizado con un dispositivo de bombeo, como por ejemplo un compresor, diseñado para mover toda la materia de un volumen, como un cilindro, o se bloqueará, posiblemente causándole fallas a una pieza.

Positive electrical charge. An atom or component that has a shortage of electrons.

Carga eléctrica positiva. Átomo o componente que tiene una insuficiencia de electrones.

Positive temperature coefficient start device. A thermistor used to provide start assistance to a permanent split-capacitor motor.

Dispositivo de arranque de coeficiente de temperatura positiva. Termistor utilizado para ayudar a arrancar un motor permanente de capacitador separado.

Potential relay. A switching device used with hermetic motors that breaks the circuit to the start capacitor and/or start windings after the motor has reached approximately 75% of its running speed.

Relé de potencial. Dispositivo de conmutación utilizado con motores herméticos que interrumpe el circuito del capacitador y/o de los devanados de arranque antes de que el motor haya alcanzado aproximadamente un 75% de su velocidad de marcha.

Potential voltage. The voltage measured across the start winding in a single-phase motor while it is turning at full speed. This voltage is much greater than the applied voltage to the run winding. For example, the run winding may have 230 V applied to it, and a measured voltage across the start winding may be 300 V. This is created by the motor stator turning in the magnetic field of the run winding. *Voltage potential* is the difference in voltage between any two parts of a circuit.

Potencial de voltaje. El voltaje medido a través de la bobina de arranque en un motor de fase sencilla mientras está girando a velocidad completa. Este voltaje es mucho mayor que el voltaje aplicado a la bobina de marcha. Por ejemplo, a la bobina de marcha se le puede aplicar 230 V, y el voltaje medido a través de la bobina de arranque puede ser 300 V. Este voltaje lo crea el estator del motor al girar en el campo magnético de la bobina de marcha. *Potencial de voltaje* es la diferencia en voltaje entre cualesquiera dos partes del circuito.

Potentiometer. An instrument that controls electrical current.

Potenciómetro. Instrumento que regula corriente eléctrica.

Powder actuated tool (PAT). A tool with a powder load that forces a pin, threaded stud, or other fastener into masonry.
Herramienta activada por pólvora (PAT en inglés). Una herramienta con una carga de pólvora que inserta una clavija, una clavija con rosca u otro sujetador en la albañilería.
Power. The rate at which work is done.
Potencia. Velocidad a la que se realiza un trabajo.
Power-consuming devices. A power-consuming device is considered the electrical load. For example, in a lightbulb circuit, the switch is a power-passing device that passes power to the lightbulb that consumes the power and produces light.
Aparatos consumidores de potencia. Un aparato consumidor de potencia se considera la carga eléctrica. Por ejemplo, en un circuito de bombilla de luz, el interruptor es un dispositivo que pasa corriente que pasa la corriente a la bombilla, la cual consume electricidad y produce luz.
Pressure. Force per unit of area.
Presión. Fuerza por unidad de área.
Pressure access ports. Places in a system where pressure can be taken or registered.
Puerto de acceso a presión. Lugares en un sistema donde se puede tomar o registrar la presión.
Pressure differential valve. A valve that senses a pressure differential and opens when a specific pressure differential is reached.
Válvula de presión diferencial. Válvula que detecta diferencia de presiones y se abre cuando se alcanza una diferencia específica.
Pressure drop. The difference in pressure between two points.
Caída de presión. Diferencia en presión entre dos puntos.
Pressure/enthalpy diagram. A chart indicating the pressure and heat content of a refrigerant and the extent to which the refrigerant is a liquid and vapor.
Diagrama de presión y entalpía. Esquema que indica la presión y el contenido de calor de un refrigerante y el punto en que el refrigerante es líquido y vapor.
Pressure limiter. A device that opens when a certain pressure is reached.
Dispositivo limitador de presión. Dispositivo que se abre cuando se alcanza una presión específica.
Pressure-limiting TXV. A valve designed to allow the evaporator to build only to a predetermined pressure when the valve will shut off the flow of refrigerant.
Válvula electrónica de expansión limitadora de presión. Válvula diseñada para permitir que la temperatura del evaporador alcance una presión predeterminada cuando la válvula detenga el flujo de refrigerante.
Pressure regulator. A valve capable of maintaining a constant outlet pressure when a variable inlet pressure occurs. Used for regulating fluid flow such as natural gas, refrigerant, and water.
Regulador de presión. Válvula capaz de mantener una presión constante a la salida cuando ocurre una presión variable a la entrada. Utilizado para regular el flujo de fluidos, como por ejemplo el gas natural, el refrigerante y el agua.
Pressure switch. A switch operated by a change in pressure.
Conmutador accionado por presión. Conmutador accionado por un cambio en presión.

Pressure tank. A pressurized tank for water storage located in the water piping of an open-loop geothermal heat pump system. It prevents short cycling of the well pump.
Depósito de presión. Depósito presurizado que sirve para almacenar el agua contenida en la tubería de un sistema de bomba de calor geotérmica de circuito abierto. Impide el funcionamiento de la bomba del pozo en ciclos cortos.
Pressure/temperature relationship. This refers to the pressure/temperature relationship of a liquid and vapor in a closed container. If the temperature increases, the pressure will also increase. If the temperature is lowered, the pressure will decrease.
Relación entre presión y temperatura. Se refiere a la relación entre la presión y la temperatura de un líquido y un vapor en un recipiente cerrado. Si la temperatura aumenta, la presión también aumentará. Si la temperatura baja, habrá una caída de presión.
Pressure transducer. A pressure-sensitive device located in the piping of a refrigeration system that will transform a pressure signal to an electronic signal. The electronic signal will then feed a microprocessor.
Transductor de presión. Dispositivo sensible a la presión, situado en la tubería del sistema de refrigeración y que convierte una señal de presión en señal eléctrica. Seguidamente, la señal eléctrica se envía a un microprocesador.
Primary air. Air that is introduced to a furnace's burner before the combustion process has taken place.
Aire primario. Aire que se introduce en el quemador de un calefactor antes de que ocurra el proceso de combustión.
Primary control. Controlling device for an oil burner to ensure ignition within a specific time span, usually 90 seconds.
Regulador principal. Dispositivo de regulación para un quemador de aceite pesado. El regulador principal asegura el encendido dentro de un período de tiempo específico, normalmente 90 segundos.
Programmable thermostat. An electronic thermostat that can be set up to provide desired conditions at desired times.
Termostato programable. Un termostato electrónico que se puede programar para proveer las condiciones deseadas en tiempos deseados.
Propane. An LP (liquefied petroleum) gas used for heat.
Propano. Gas de petróleo licuado que se utiliza para producir calor.
Propeller fan. This fan is used in exhaust fan and condenser fan applications. It will handle large volumes of air at low-pressure differentials.
Ventilador helicoidal. Se utiliza en ventiladores de evacuación y de condensador. Es capaz de mover grandes volúmenes de aire a bajas diferenciales de presión.
Proportional controller. A modulating control mode where the controller changes or modifies its output signal in proportion to the size of the change in the error.
Controlador proporcional. Un modo de control modulante donde el controlador cambia o modifica su señal de salida en proporción al tamaño del cambio en el error.
Propylene glycol. An antifreeze fluid cooled by a primary (phase-change) refrigerant, which then is circulated by pumps throughout the refrigeration system to absorb heat.

Glicol propílico. Líquido anticongelante enfriado por un refrigerante primario (cambio de fase). Seguidamente, una bomba lo hace circular por el sistema de refrigeración para que absorba el calor.

Proton. That part of an atom having a positive charge.

Protón. Parte de un átomo que tiene carga positiva.

Protozoa. A microscopic organism with a complex life cycle.

Protozoario. Un organismo microscópico con un ciclo de vida complejo.

PSC motor. See Permanent split-capacitor motor.

Motor PSC. Véase motor permanente de capacitador separado.

psi. Abbreviation for pounds per square inch.

psi. Abreviatura de libras por pulgada cuadrada.

psia. Abbreviation for pounds per square inch absolute.

psia. Abreviatura de libras por pulgada cuadrada absoluta.

psig. Abbreviation for pounds per square inch gauge.

psig. Abreviatura de indicador de libras por pulgada cuadrada.

Psychrometer. An instrument for determining relative humidity.

Sicrómetro. Instrumento para medir la humedad relativa.

Psychrometric chart. A chart that shows the relationship of temperature, pressure, and humidity in the air.

Esquema sicrométrico. Esquema que indica la relación entre la temperatura, la presión y la humedad en el aire.

Psychrometrics. The study of air and its properties, particularly the moisture content.

Sicrometría. El estudio del aire y sus propiedades, particularmente el contenido de humedad.

P-type material. Semiconductor material with a positive charge.

Material tipo P. Material con carga positiva utilizado en semiconductores.

Pump. A device that forces fluids through a system.

Bomba. Dispositivo que introduce fluidos por fuerza a través de un sistema.

Pump down. To use a compressor to pump the refrigerant charge into the condenser and/or receiver.

Extraer con bomba. Utilizar un compresor para bombear la carga del refrigerante dentro del condensador y/o receptor.

Pure compound. A substance formed in definite proportions by weight with only one molecule present.

Componente puro. Una sustancia formada en proporciones por peso definidas con sólo una molécula presente.

Purge. To remove or release fluid from a system.

Purga. Remover o liberar el fluido de un sistema.

PVC (polyvinyl chloride). Plastic pipe used in pressure applications for water and gas as well as for sewage and certain industrial applications.

Cloruro de polivinilo (PVC en inglés). Tubo plástico utilizado tanto en aplicaciones de presión para agua y gas, como en ciertas aplicaciones industriales y de aguas negras.

Quench. To submerge a hot object in a fluid for cooling.

Entriamiento por inmersión. Sumersión de un objeto caliente en un fluido para enfriarlo.

Quick-connect coupling. A device designed for easy connecting or disconnecting of fluid lines.

Acoplamiento de conexión rápida. Dispositivo diseñado para facilitar la conexión o desconexión de conductos de fluido.

R-12. Dichlorodifluoromethane, once a popular refrigerant for refrigeration systems. It can no longer be manufactured in the United States and many other countries.

R-12. Diclorodiflorometano, que fue una vez un refrigerante muy utilizado en sistemas de refrigeración. Ya no se puede fabricar ni en los Estados Unidos ni en muchos otros países.

R-22. Monochlorodifluoromethane, a popular HCFC refrigerant for air-conditioning systems.

R-22. Monoclorodiflorometano, refrigerante HCFC muy utilizado en sistemas de acondicionamiento de aire.

R-123. Dichlorotrifluoroethane, an HCFC refrigerant developed for low-pressure application.

R-123. Diclorotrifloroetano, refrigerante HCFC elaborado para aplicaciones de baja presión.

R-134a. Tetrafluoroethane, an HFC refrigerant developed for refrigeration systems and as a replacement for R-12.

R-134a. Tetrafloroetano, refrigerante HFC elaborado para sistemas de refrigeración y como sustituto del R-12.

R-410A. A mixture of difluoromethane and pentafluoroethane, a refrigerant developed to replace R-22 for air-conditioning systems.

R-410A. Una mezcla de difluorometano y pentafluoroetano, refrigerante desarrollado para remplazar el R-22 para los sistemas de aire acondicionado.

Rack system. Many compressors piped in parallel and mounted on a steel rack. The compressors are usually cycled by a microprocessor.

Sistema de bastidor. Varios compresores conectados en paralelo y montados en un bastidor de acero. Normalmente, un microprocesador se encarga de activar y desactivar los compresores.

Radiant heat. Heat that passes through air, heating solid objects that in turn heat the surrounding area.

Calor radiante. Calor que pasa a través del aire y calienta objetos sólidos que a su vez calientan el ambiente.

Radiation. Heat transfer. See Radiant heat.

Radiación. Transferencia de calor. Véase Calor radiante.

Radon. A colorless, odorless, and radioactive gas. Radon can enter buildings through cracks in concrete floors and walls, floor drains, and sumps.

Radón. Gas incoloro, inodoro y radioactivo. El radón puede penetrar en los edificios a través de las grietas en el hormigón y suelos, desagües y sumideros.

Random or off-cycle defrost. Defrost provided by the space temperature during the normal off cycle.

Descongelación variable o de ciclo apagado. Descongelación llevada a cabo por la temperatura del espacio durante el ciclo normal de apagado.

Range. The pressure or temperature settings of a control defining certain boundaries of temperature or pressure.

Rango. Los valores de presión o temperatura para un control que definen los límites de la temperatura o presión.

Rankine. The absolute Fahrenheit scale with 0 at the point where all molecular motion stops.

Rankine. Escala absoluta de Fahrenheit con el 0 al punto donde se detiene todo movimiento molecular.

Rapid oxidation. A reaction between the fuel, oxygen, and heat that is known as rapid oxidation or the process of burning.
Oxidación rápida. Reacción producida entre el combustible y el oxígeno. El calor producido se conoce como oxidación rápida o proceso de quemado.
Reactance. A type of resistance in an alternating current circuit.
Reactancia. Tipo de resistencia en un circuito de corriente alterna.
Reamer. Tool to remove burrs from inside a pipe after it has been cut.
Escariador. Herramienta utilizada para remover las rebabas de un tubo después de haber sido cortado.
Receiver-drier. A component in a refrigeration system for storing and drying refrigerant.
Receptor-secador. Componente en un sistema de refrigeración que almacena y seca el refrigerante.
Reciprocating. Back-and-forth motion.
Movimiento alternativa. Movimiento de atrás para adelante.
Reciprocating compressor. A compressor that uses a piston in a cylinder and a back-and-forth motion to compress vapor.
Compresor alternativo. Compresor que utiliza un pistón en un cilindro y un movimiento de atrás para adelante a fin de comprir el vapor.
Recirculated water system. A system where water is used over and over, such as a chilled water or cooling tower system.
Sistema de agua recirculada. Sistema donde el agua se usa una y otra vez, tal como en un sistema de agua enfriada o de torre enfriamiento.
Recovery cylinder. A cylinder into which refrigerant is transferred; should be approved by the Department of Transportation as a recovery cylinder. The color code for these cylinders is a yellow top with a gray body.
Cilindro de recuperación. Cilindro al que se transfiere el refrigerante y que debe ser homologado por el departamento de transporte como cilindro de recuperación. Este tipo de cilindro se identifica pintando su parte superior en amarillo y el resto del cuerpo en gris.
Rectifier. A device for changing alternating current to direct current.
Rectificador. Dispositivo utilizado para convertir corriente alterna en corriente continua.
Reed valve. A thin steel plate used as a valve in a compressor.
Válvula de lámina. Placa delgada de acero utilizada como una válvula en un compresor.
Refrigerant. The fluid in a refrigeration system that changes from a liquid to a vapor and back to a liquid at practical pressures.
Refrigerante. Fluido en un sistema de refrigeración que se convierte de líquido en vapor y nuevamente en líquido a presiones prácticas.
Refrigerant blend. Two or more refrigerants blended or mixed together to make another refrigerant. Blends can combine as either azeotropic or zeotropic blends.
Mezcla de refrigerante. Dos o más refrigerantes mezclados para crear otro. Las mezclas pueden ser de tipo azeotrópico o zeotrópico.
Refrigerant loop. The heat pump's refrigeration system, which exchanges energy with the fluid in the ground loop and the air side of the system.
Circuito de refrigeración. Sistema de refrigeración de la bomba de calor que sirve para intercambiar energía entre el fluido en el circuito de tierra y la parte del sistema que contiene el aire.
Refrigerant receiver. A storage tank in a refrigeration system where the excess refrigerant is stored. Since many systems use different amounts of refrigerant during the season, the excess is stored in the receiver tank when not needed. The refrigerant can also be pumped to the receiver when repairs on the low-pressure side of the system are made.
Recibidor de refrigerante. Tanque de almacenamiento en un sistema de refrigeración donde se almacena el exceso de refrigerante. Como muchos sistemas usan diferentes cantidades de refrigerantes durante la temporada, el exceso se almacena en el tanque cuando no se necesita. El refrigerante también puede bombearse al recibidor cuando se hacen reparaciones al extremo de baja presión del sistema.
Refrigerant reclaim. "To process refrigerant to new product specifications by means which may include distillation. It will require chemical analysis of the refrigerant to determine that appropriate product specifications are met. This term usually implies the use of processes or procedures available only at a reprocessing or manufacturing facility."
Recuperación del refrigerante. "Procesar refrigerante según nuevas especificaciones para productos a través de métodos que pueden incluir la destilación. Se requiere un análisis químico del refrigerante para asegurar el cumplimiento de las especificaciones para productos a través de métodos que pueden incluir la destilación. Se requiere un análisis químico del refrigerante para asegurar el cumplimiento de las especificaciones para productos adecuadas. Por lo general este término supone la utilización de procesos o de procedimientos disponibles solamente en fábricas de reprocesamiento o manufactura".
Refrigerant recovery. "To remove refrigerant in any condition from a system and store it in an external container without necessarily testing or processing it in any way."
Recobrar refrigerante líquido. "Remover refrigerante en cualquier estado de un sistema y almacenarlo en un recipiente externo sin ponerlo a prueba o elaborarlo de ninguna manera".
Refrigerant recycling. "To clean the refrigerant by oil separation and single or multiple passes through devices, such as replaceable core filter driers, which reduce moisture, acidity, and particulate matter. This term usually applies to procedures implemented at the job site or at a local service shop."
Recirculación de refrigerante. "Limpieza del refrigerante por medio de la separación del aceite y pasadas sencillas o múltiples a traves de dispositivos, como por ejemplo secadores filtros con núcleos reemplazables que disminuyen la humedad, la acidez y las partículas. Por lo general este término se aplica a los procedimientos utilizados en el lugar del trabajo o en un taller de servicio local".
Refrigerated air driers. A device that removes the excess moisture from compressed air.
Secadores de aire refrigerados. Aparato que remueve el exceso de humedad del aire comprimido.
Refrigeration. The process of removing heat from a place where it is not wanted and transferring that heat to a place where it makes little or no difference.

Refrigeración. Proceso de remover el calor de un lugar donde no es deseado y transferirlo a un lugar donde no afecte la temperatura.

Register. A terminal device on an air distribution system that directs air but also has a damper to adjust airflow.

Registro. Dispositivo de terminal en un sistema de distribución de aire que dirige el aire y además tiene un desviador para ajustar su flujo.

Regulator. A valve used to control the pressure in liquid systems to some value. Many households have a water pressure regulator to reduce the pressure from the main to a more usable pressure in the house. Gas systems all have pressure regulators to stabilize the pressure to the burners.

Regulador. Una válvula que se usa para controlar y fijar la presión en los sistemas líquidos a algún valor. Muchas casas tienen un regulador de presión de agua para reducir la presión de la tubería principal a una presión más útil en la casa. Todos los sistemas de gas tienen reguladores de presión para estabilizar la presión en el quemador.

Relative humidity. The amount of moisture contained in the air as compared to the amount the air could hold at that temperature.

Humedad relativa. Cantidad de humedad presente en el aire, comparada con la cantidad de humedad que el aire pueda contener a dicha temperatura.

Relay. A small electromagnetic device to control a switch, motor, or valve.

Relé. Pequeño dispositivo electromagnético utilizado para regular un conmutador, un motor o una válvula.

Relief valve. A valve designed to open and release vapors at a certain pressure.

Válvula para alivio. Válvula diseñada para abrir y liberar vapores a una presión específica.

Remote system. Often called a split system where the condenser is located away from the evaporator and/or other parts of the system.

Sistema remoto. Llamado muchas veces sistema separado donde el condensador se coloca lejos del evaporador y/o otras piezas del sistema.

Resilient-mount motor. Electric motor that uses various materials to isolate the motor noise from metal framework. This type of motor requires a ground strap.

Motor con montaje antivibratorio. Motor eléctrico que utiliza varios materiales para aislar el ruido del bastidor metálico. Este tipo de motor requiere conexión a tierra.

Resistance. The opposition to the flow of an electrical current or a fluid.

Resistencia. Oposición al flujo de una corriente eléctrica o de un fluido.

Resistor. An electrical or electronic component with a specific opposition to electron flow. It is used to create voltage drop or heat.

Resistor. Componente eléctrico o electrónico con una oposición específica al flujo de electrones; se utiliza para producir una caída de tensión o calor.

Restrictor. A device used to create a planned resistance to fluid flow.

Limitador. Dispositivo utilizado para producir una resistencia proyectada al flujo de fluido.

Retrofit guidelines. Guidelines intended to make the transition from a CFC/mineral oil system to a system containing an alternative refrigerant and its appropriate oil.

Directrices de reconversión. Directrices destinadas a facilitar la transición entre un sistema de aceite CFC/mineral y otro con refrigerante alternativo y su aceite correspondiente.

Return well. A well for return water after it has experienced the heat exchanger of the geothermal heat pump.

Pozo de retorno. Pozo donde se acumula el agua de retorno una vez ha pasado por el intercambiador térmico de la bomba de calor geotérmica.

Reverse cycle. The ability to direct the hot gas flow into the indoor or the outdoor coil in a heat pump to control the system for heating or cooling purposes.

Ciclo invertido. Capacidad de dirigir el flujo de gas caliente dentro de la bobina interior o exterior en una bomba de calor a fin de regular el sistema para propósitos de calentamiento o enfriamiento.

Rigid-mount motor. Electric motor that is bolted metal-to-metal to a frame. This type of motor will transmit noise.

Motor con montaje rígido. Motor eléctrico que se encuentra sujeto directamente a un bastidor mediante pernos. Este tipo de motor genera ruido.

Rod and tube. The rod and tube are each made of a different metal. The tube has a high expansion rate and the rod a low expansion rate.

Varilla y tubo. La varilla y el tubo se fabrican de un metal diferente. El tubo tiene una tasa de expansión alta y la varilla una tasa de expansión baja.

Room heater. A gas stove or appliance considered by ANSI to be a heating appliance. This heater will have an efficiency rating.

Calentador de sala. Estufa de gas u otro dispositivo que, según ANSI, es un dispositivo de calefacción. Este tipo de calentador cuenta con una clasificación de eficacia.

Root mean square (RMS) voltage. The alternating current voltage effective value. This is the value measured by most voltmeters. The RMS voltage is 0.707 × the peak voltage.

Voltaje de la raíz del valor medio cuadrado (RMS en inglés). El valor efectivo del voltaje de corriente alterna. Este valor es el que miden la mayoría de los voltímetros. El voltaje RMS es 0,707 por el voltaje pico.

Rotary compressor. A compressor that uses rotary motion to pump fluids. It is a positive displacement pump.

Compresor giratorio. Compresor que utiliza un movimiento giratorio para bombear fluidos. Es una bomba de desplazamiento positivo.

Rotor. The rotating or moving component of a motor, including the shaft.

Rotor. Componente giratorio o en movimiento de un motor, incluyendo el arbol.

Run-load amperage (RLA). The amperage at which a motor can safely operate while under full load, unless it has a service (reserve) factor allowing more amperage.

Amperaje de operación con carga (RLA en inglés). Amperaje bajo el cual el motor puede operar seguramente bajo carga completa, a menos que tenga un factor de servicio (reserva) que permite más amperaje.

Running time. The time a unit operates. Also called the on time.

Período de funcionamiento. El período de tiempo en que funciona una unidad. Conocido también como período de conexión.

Run winding. The electrical winding in a motor that draws current during the entire running cycle.

Devanado de funcionamiento. Devanado eléctrico en un motor que consume corriente durante todo el ciclo de funcionamiento.

Rupture disk. Pressure safety device for a centrifugal low-pressure chiller.

Disco de ruptura. Dispositivo de seguridad para un enfriador centrífugo de baja presión.

Saddle valve. A valve that straddles a fluid line and is fastened by solder or screws. It normally contains a device to puncture the line for pressure readings.

Válvula de silleta. Válvula que está sentada a horcajadas en un conducto de fluido y se fija por medio de la soldadura o tornillos. Por lo general contiene un dispositivo para agujerear el conducto a fin de que se puedan tomar lecturas de presión.

Safety control. An electrical, electronic, mechanical, or electromechanical control to protect the equipment or public from harm.

Regulador de seguridad. Regulador eléctrico, electrónico, mecánico o electromecánico para proteger al equipo de posibles averías o al público de sufrir alguna lesión.

Safety plug. A fusible plug that blows out when high temperature occurs.

Tapón de seguridad. Tapón fusible que se sale cuando se presentan temperaturas altas.

Sail switch. A safety switch with a lightweight, sensitive sail that operates by sensing an airflow.

Conmutador con vela. Conmutador de seguridad con una vela liviana sensible que funciona al advertir el flujo de aire.

Salt solution. Antifreeze solution used in a closed water loop of geothermal heat pumps.

Solución de sal. Líquido anticongelante utilizado en el circuito cerrado de agua de una bomba de calor geotérmica.

Satellite compressor. The compressor on a parallel compressor system that is dedicated to the coldest evaporators.

Compresor auxiliar. Compresor montado en un sistema paralelo y que está dedicado a los evaporadores más fríos.

Saturated vapor. The refrigerant when all of the liquid has just changed to a vapor.

Vapor saturada. El refrigerante cuando todo el líquido acaba de convertirse en vapor.

Saturation. A term used to describe a substance when it contains all of another substance it can hold.

Saturación. Término utilizado para describir una sustancia cuando contiene lo más que puede de otra sustancia.

Scavenger pump. A pump used to remove the fluid from a sump.

Bomba de barrido. Bomba utilizada para remover el fluido de un sumidero.

Scheduled maintenance. The action of performing regularly scheduled maintenance on a unit, including inspection, cleaning, and servicing.

Mantenimiento programado. La acción de dar mantenimiento regularmente programado a una unidad incluyendo inspección, limpieza y servicio.

Schematic wiring diagram. Sometimes called a line or ladder diagram, this type of diagram shows the electrical current path to the various components.

Diagrama de cableado esquematizado. También conocido como de línea o de escalera, este tipo de diagrama muestra la ruta actual que sigue la electricidad para llegar a los diferentes componentes.

Schrader valve. A valve similar to the valve on an auto tire that allows refrigerant to be charged or discharged from the system.

Válvula Schrader. Válvula similar a la válvula del neumático de un automóvil que permite la entrada o la salida de refrigerante del sistema.

Scotch yoke. A mechanism used to create reciprocating motion from the electric motor drive in very small compressors.

Yugo escocés. Mecanismo utilizado para producir movimiento alternativo del accionador del motor eléctrico en compresores bastante pequeños.

Screw compressor. A form of positive displacement compressor that squeezes fluid from a low-pressure area to a high-pressure area, using screw-type mechanisms.

Compresor de tornillo. Forma de compresor de desplazamiento positivo que introduce por fuerza el fluido de un área de baja presión a un área de alta presión, a través de mecanismos de tipo de tornillo.

Scroll compressor. A compressor that uses two scroll-type components, one stationary and one orbiting, to compress vapor.

Compresor espiral. Compresor que utiliza dos componentes de tipo espiral para comprimir el vapor.

Sealed unit. The term used to describe a refrigeration system, including the compressor, that is completely welded closed. The pressures can be accessed by saddle valves.

Unidad sellada. Término utilizado para describir un sistema de refrigeración, incluyendo el compresor, que es soldado completamente cerrado. Las presiones son accesibles por medio de válvulas de dilleta.

Seasonal energy efficiency ratio (SEER). An equipment efficiency rating that takes into account the start-up and shutdown for each cycle.

Relación del rendimiento de energía temporal (SEER en inglés). Clasificación del rendimiento de un equipo que toma en cuenta la puesta en marcha y la parada de cada ciclo.

Seat. The stationary part of a valve that the moving part of the valve presses against for shutoff.

Asiento. Pieza fija de una válvula contra la que la pieza en movimiento de la válvula presiona para cerrarla.

Secondary air. Air that is introduced to a furnace after combustion takes place and that supports combustion.

Aire secundario. Aire que se introduce en un calefactor después que ocurre la combustión y que ayuda la combustión.

Secondary fluid. An antifreeze fluid cooled by a primary (phase-change) refrigerant, which is then circulated by pumps throughout the refrigeration system to absorb heat.
Fluido secundario. Líquido anticongelante refrigerado por otro primario (cambio de fase), que luego es propulsado por las bombas a través del sistema de refrigeración para absorber el calor.
Semiconductor. A component in an electronic system that is considered neither an insulator nor a conductor but a partial conductor. It conducts current in a controlled and predictable manner.
Semiconductor. Componente en un sistema eléctrico que no se considera ni aislante ni conductor, sino conductor parcial. Conduce la corriente de una manera controlada y predecible.
Semi-hermetic compressor. A motor compressor that can be opened or disassembled by removing bolts and flanges. Also known as a serviceable hermetic.
Compresor semihermético. Compresor de un motor que puede abrirse o desmontarse al removerle los pernos y bridas. Conocido también como compresor hermético utilizable.
Sensible heat. Heat that causes a change in temperature.
Calor sensible. Calor que produce un cambio en la temperatura.
Sensor. A component for detection that changes shape, form, or resistance when a condition changes.
Sensor. Componente para la detección que cambia de forma o de resistencia cuando cambia una condición.
Sequencer. A control that causes a staging of events, such as a sequencer between stages of electrical heat.
Regulador de secuencia. Regulador que produce una sucesión de acontecimientos, como por ejemplo etapas sucesivas de calor eléctrico.
Series circuit. An electrical or piping circuit where all of the current or fluid flows through the entire circuit.
Circuito en serie. Circuito eléctrico o de tubería donde toda la corriente o todo el fluido fluye a través de todo el circuito.
Series flow. A flow path in which only one path exists for fluid to flow.
Flujo en serie. Ruta de flujo única para el líquido.
Service valve. A manually operated valve in a refrigeration system used for various service procedures.
Válvula servicio. Válvula de un sistema de refrigeración accionada manualmente que se utiliza en varios procedimientos de servicio.
Serviceable hermetic. See Semi-hermetic compressor.
Compresor hermético utilizable. Véase Compresor semihermético.
Servo pressure regulator. A sensitive pressure regulator located inside a combination gas valve that senses the outlet or working pressure of the gas valve.
Regulador de presión por servomotor. Un regulador de presión sensible que está ubicado dentro de una válvula de gas de combinación que detecta la presión de salida o de trabajo de la válvula de presión.
Set point. The desired control point's magnitude in a control process.
Punto de ajuste. La magnitud deseada de un punto de control en un proceso de control.

Shaded-pole motor. An alternating current motor used for very light loads.
Motor polar en sombra. Motor de corriente alterna utilizado en cargas sumamente livianas.
Shell and coil. A vessel with a coil of tubing inside that is used as a heat exchanger.
Coraza y bobina. Depósito con una bobina de tubería en su interior que se utiliza como intercambiador de calor.
Shell and tube. A heat exchanger with straight tubes in a shell that can normally be mechanically cleaned.
Coraza y tubo. Intercambiador de calor con tubos rectos en una coraza que por lo general puede limpiarse mecánicamente.
Short circuit. A circuit that does not have the correct measurable resistance: too much current flows and will overload the conductors.
Cortocircuito. Corriente que no tiene la resistencia medible correcta: un exceso de corriente fluye a través del circuito provocando una sobrecarga de los conductores.
Short cycle. The term used to describe the running time (on time) of a unit when it is not running long enough.
Ciclo corto. Término utilizado para describir el período de funcionamiento (de encendido) de una unidad cuando no funciona por un período de tiempo suficiente.
Shorted motor winding. Part of an electric motor winding is shorted out because one part of the winding touches another part, where the insulation is worn or in some way defective.
Devanado de motor en cortocircuito. Debido a un aislamiento deficiente u otro defecto, una parte de los elementos del devanado en un motor eléctrico entran en contacto con otra, causando un cortocircuito.
Shroud. A fan housing that ensures maximum airflow through the coil.
Boveda. Alojamiento del abanico que asegura un flujo máximo de aire a través de la bobina.
Sight glass. A clear window in a fluid line.
Mirilla para observación. Ventana clara en un conducto de fluido.
Silica gel. A chemical compound often used in refrigerant driers to remove moisture from the refrigerant.
Gel silíceo. Compuesto químico utilizado a menudo en secadores de refrigerantes para remover la humedad del refrigerante.
Silver brazing. A high-temperature (above 800°F) brazing process for bonding metals.
Soldadura con plata. Soldadura a temperatura alta (sobre los 800°F ó 430°C) para unir metales.
Sine wave. The graph or curve used to describe the characteristics of alternating current and voltage.
Onda sinusoidal. Gráfica o curva utilizada para describir las características de tensión y de corriente alterna.
Single phase. The electrical power supplied to equipment or small motors, normally under $7\frac{1}{2}$ hp.
Monofásico. Potencia eléctrica suministrada a equipos o motores pequeños, por lo general menor de $7\frac{1}{2}$ hp.
Single-phase hermetic motor. A sealed motor, such as with a small compressor, that operates off single-phase power.
Motor de fase sencilla hermético. Un motor sellado, tal como un compresor pequeño, que opera con electricidad de fase sencilla.

Single phasing. The condition in a three-phase motor when one phase of the power supply is open.
Fasaje sencillo. Condición en un motor trifásico cuando una fase de la fuente de alimentación está abierta.
Slinger ring. A ring attached to the blade tips of a condenser fan. This ring throws condensate onto the condenser coil, where it is evaporated.
Anillo tubular. Anillo instalado en los extremos de las palas de un ventilador de condensación. Sirve para lanzar la condensación sobre el serpentín de refrigeración donde es evaporada.
Sling psychrometer. A device with two thermometers, one a wet bulb and one a dry bulb, used for checking air conditions, wet-bulb and dry-bulb.
Sicrómetro con eslinga. Dispositivo con dos termómetros, uno con una bombilla húmeda y otro con una bombilla seca, utilizados para revisar las condiciones del aire, de la temperatura y de la humedad.
Slip. The difference in the rated rpm of a motor and the actual operating rpm when under a load.
Deslizamiento. Diferencia entre las rpm nominales de un motor y las rpm de funcionamiento reales.
Slugging. A term used to describe the condition when large amounts of liquid enter a pumping compressor cylinder.
Relleno. Término utilizado para describir la condición donde grandes cantidades de líquido entran en el cilindro de un compresor de bombeo.
Snap-disc. An application of the bimetal. Two different metals fastened together in the form of a disc that provides a warping condition when heated. This also provides a snap action that is beneficial in controls that start and stop current flow in electrical circuits.
Disco de acción rápida. Aplicación del bimetal. Dos metales diferentes fijados entre sí en forma de un disco que provee un deformación al ser calentado. Esto provee también una acción rápida, ventajosa para reguladores que ponen en marcha y detienen el flujo de corriente en circuitos eléctricos.
Software. Computer programs written to give specific instructions to computers.
Software. Programas de computadoras escritos para darles instrucciones específicas a las computadoras.
Solar collectors. Components of a solar system designed to collect the heat from the sun, using air, a liquid, or refrigerant as the medium.
Colectores solares. Componentes de un sistema solar diseñados para acumular el calor emitido por el sol, utilizando el aire, un líquido o un refrigerante como el medio.
Solar heat. Heat from the sun's rays.
Calor solar. Calor emitido por los rayos del sol.
Solar influence. The heat that the sun imposes on a structure.
Influencia solar. El calor que el sol impone en una estructura.
Solar radiant heat. Solar-heated water or an antifreeze solution is piped through heating coils embedded in concrete in the floor or in plaster in ceilings or walls.
Calor de radiación solar. El agua o líquido anticongelante calentado por energía solar se canaliza a través de serpentines de calefacción instalados en el hormigón del suelo o en el yeso de los techos o paredes.

Soldering. Fastening two base metals together by using a third, filler metal that melts at a temperature below 800°F.
Soldadura. La fijación entre sí de dos metales bases utilizando un tercer metal de relleno que se funde a una temperatura menor de 800°F (430°C).
Solderless terminals. Used to fasten stranded wire to various terminals or to connect two lengths of stranded wire together.
Terminales sin soldadura. Se usan para fijar cable trenzado a varios terminales o para unir dos pedazos de cable.
Solenoid. A coil of wire designed to carry an electrical current producing a magnetic field.
Solenoide. Bobina de alambre diseñada para conducir una corriente eléctrica generando un campo magnético.
Solid. Molecules of a solid are highly attracted to each other, forming a mass that exerts all of its weight downward.
Sólido. Las moléculas de un sólido se atraen entre sí y forman una masa que ejerce todo su peso hacia abajo.
Space cooling and heating thermostat. The device used to control the temperature of a space, such as a home thermostat that controls the temperature in a home or office.
Termostato de enfriamiento o calefacción de espacio. Aparato usado para controlar la temperatura de un espacio, tal como un termostato de hogar que controla la temperatura en un hogar u oficina.
Specific gravity. The weight of a substance compared to the weight of an equal volume of water.
Gravedad específica. El peso de una sustancia comparada con el peso de un volumen igual de agua.
Specific heat. The amount of heat required to raise the temperature of 1 lb of a substance 1°F.
Calor específico. La cantidad de calor requerida para elevar la temperatura de una libra de una sustancia 1°F (−17°C).
Specific volume. The volume occupied by 1 lb of a fluid.
Volumen específico. Volumen que ocupa una libra de fluido.
Splash lubrication system. A system of furnishing lubrication to a compressor by agitating the oil.
Sistema de lubrificación por salpicadura. Método de proveerle lubrificación a un compresor agitando el aceite.
Splash method. A method of water dropping from a higher level in a cooling tower and splashing on slats with air passing through for more efficient evaporation.
Método de salpicaduras. Método dé dejar caer agua desde un nivel más alto en una torre de refrigeración y salpicándola en listones, mientras el aire pasa a través de los mismos con el propósito de lograr una evaporación más eficaz.
Split-phase motor. A motor with run and start windings.
Motor de fase separada. Motor con devandos de funcionamiento y de arranque.
Split suction. When the common suction line of a parallel compressor system has been valved in such a way as to provide for multiple temperature applications in one refrigeration package.
Succión dividida. Cuando la línea común de succión de un sistema de compresor paralelo ha sido dividida de tal manera que provee para aplicaciones de múltiples temperaturas en un empaque de refrigeración.

Split system. A refrigeration or air-conditioning system that has the condensing unit remote from the indoor (evaporator) coil.
Sistema separado. Sistema de refrigeración o de acondicionamiento de aire cuya unidad de condensación se encuentra en un sitio alejado de la bobina interior del evaporador.
Spray pond. A pond with spray heads used for cooling water in water-cooled air-conditioning or refrigeration systems.
Tanque de rociado. Tanque con una cabeza rociadora utilizada para enfriar el agua en sistemas de acondicionamiento de aire o de refrigeración enfriados por agua.
Spring-loaded relief valve. A fluid (refrigerant, air, water, or steam) relief valve that can function more than one time because a spring returns the valve to a seat.
Válvula de alivio de resorte. Una válvula de alivio de fluido (refrigerante, aire, agua o vapor de agua) que puede funcionar más de una vez porque el resorte regresa la válvula a su asiento.
Squirrel cage fan. A cylindrically shaped fan assembly used to move air.
Abanico con jaula de ardilla. Conjunto cilíndrico de abanico utilizado para mover el aire.
Squirrel cage rotor. Describes the construction of a motor rotor.
Rotor de jaula de ardilla. Describe la construcción del rotor de un motor.
Stamped evaporator. An evaporator that has stamped refrigerant passages in sheet steel or aluminum.
Evaporador estampado. Un evaporador que tiene pasajes para el refrigerante estampados en lata o aluminio.
Standard atmosphere or standard conditions. Air at sea level at 70°F when the atmosphere's pressure is 14.696 psia (29.92 in. Hg). Air at this condition has a volume of 13.33 ft^3/lb.
Atmósfera estándar o condiciones estándares. El aire al nivel del mar a una temperatura de 70°F (15°C) cuando la presión de la atmósfera es 14.696 psia (29.92 pulgadas Hg). Bajo esta condición, el aire tiene un volumen de 13.33 ft^3/lb (pies^3/libras).
Standing pilot. Pilot flame that remains burning continuously.
Piloto constante. Llama piloto que se quema de manera continua.
Start capacitor. A capacitor used to help an electric motor start.
Capacitador de arranque. Capacitador utilizado para ayudar en el arranque de un motor eléctrico.
Starting relay. An electrical relay used to disconnect the start capacitor and/or start winding in a hermetic compressor.
Relé de arranque. Relé eléctrico utilizado para desconectar el capacitador y/o el devanado de arranque en un compresor hermético.
Starting winding. The winding in a motor used primarily to give the motor extra starting torque.
Devanado de arranque. Devanado en un motor utilizado principalmente para proveerle al motor mayor para el arranque.
Starved coil. The condition in an evaporator when the metering device is not feeding enough refrigerant to the evaporator.
Bobina estrangulada. Condición que ocurre en un evaporador cuando el dispositivo de medida no le suministra suficiente refrigerante al evaporador.

Static pressure. The bursting pressure or outward force in a duct system.
Presión estática. La presión de estallido o la fuerza hacia fuera en un sistema de conductos.
Stator. The component in a motor that contains the windings: it does not turn.
Estátor. Componente en un motor que contiene los devanados y que no gira.
Steady-state condition. A stabilized condition of a piece of heating or cooling equipment where not much change is taking place.
Condición de régimen estable. Condición estabilizada de un dispositivo de calefacción o de refrigeración en el cual no hay muchos cambios.
Steam. The vapor state of water.
Vapor. Estado de vapor del agua.
Step motor. An electric motor that moves with very small increments or "steps," usually in either direction, and is usually controlled by a microprocessor with input and output controlling devices.
Motor a pasos. Un motor eléctrico que se mueve en incrementos muy pequeños o "pasos", generalmente en cualquier dirección, y generalmente son controlados por un microprocesador de aparatos de control con entradas y salidas.
Strainer. A fine-mesh device that allows fluid flow and holds back solid particles.
Colador. Dispositivo de malla fina que permite el flujo de fluido a través de él y atrapa partículas sólidas.
Stratification. The condition where a fluid appears in layers.
Estratificación. Condición que ocurre cuando un fluido aparece en capas.
Stratosphere. An atmospheric level that is located from 7 to 30 miles above the earth. Good ozone is found in the stratosphere.
Estratosfera. Capa del atmósfera que se encuentra a una altura entre 11 y 48 kilómetros encima de la tierra. Contiene una buena capa de ozono.
Stress crack. A crack in piping or other component caused by age or abnormal conditions such as vibration.
Grieta por tensión. Grieta que aparece en una tubería u otro componente ocasionada por envejecimiento o condiciones anormales, como por ejemplo vibración.
Subbase. The part of a space temperature thermostat that is mounted on the wall and to which the interconnecting wiring is attached.
Subbase. Pieza de un termóstato que mide la temperatura de un espacio que se monta sobre la pared y a la que se fijan los conductores eléctricos interconectados.
Subcooled. The temperature of a liquid when it is cooled below its condensing temperature.
Subenfriado. La temperatura de un líquido cuando se enfría a una temperatura menor que su temperatura de condensación.
Sublimation. When a substance changes from the solid state to the vapor state without going through the liquid state.
Sublimación. Cuando una sustancia cambia de sólido a vapor sin covertirse primero en líquido.

Suction gas. The refrigerant vapor in an operating refrigeration system found in the tubing from the evaporator to the compressor and in the compressor shell.
Gas de aspiración. El vapor del refrigerante en un sistema de refrigeración en funcionamiento presente en la tubería que va del evaporador al compresor y en la coraza del compresor.
Suction line. The pipe that carries the heat-laden refrigerant gas from the evaporator to the compressor.
Conducto de aspiración. Tubo que conduce el gas de refrigerante lleno de calor del evaporador al compresor.
Suction-line accumulator. A reservoir in a refrigeration system suction line that protects the compressor from liquid floodback.
Acumulador de la línea de succión. Un estanque en la línea de succión de un sistema de refrigeración que protege al compresor de una inundación de líquido.
Suction pressure. The pressure created by the boiling refrigerant on the evaporator or low-pressure side of the system.
Presión de succión. La presión creada por el refrigerante hirviendo en el evaporador o en el lado de baja presión del sistema.
Suction service valve. A manually operated valve with front and back seats located at the compressor.
Válvula de aspiración para servicio. Válvula accionada manualmente que tiene asientos delanteros y traseros ubicados en el compresor.
Suction valve. The valve at the compressor cylinder that allows refrigerant from the evaporator to enter the compressor cylinder and prevents it from being pumped back out to the suction line.
Válvula de succión. La válvula en el cilindro de un compresor que permite que el refrigerante del evaporador entre al cilindro del compresor y evita que se bombee nuevamente a la línea de succión.
Suction valve lift unloading. The suction valve in a reciprocating compressor cylinder is lifted, causing that cylinder to stop pumping.
Descarga por levantamiento de la válvula de aspiración. La válvula de aspiración en el cilindro de un compresor alternativo se levanta, provocando que el cilindro deje de bombear.
Sulfur dioxide. A combustion pollutant that causes eye, nose, and respiratory tract irritation and possibly breathing problems.
Dióxido de azufre. Un contaminante por combustión que causa irritación en los ojos, la nariz y las vías respiratorias, y posiblemente problemas respiratorios.
Sump. A reservoir at the bottom of a cooling tower to collect the water that has passed through the tower.
Sumidero. Tanque que se encuentra en el fondo de una torre de refrigeración para acumular el agua que ha pasado a través de la torre.
Superheat. The temperature of vapor refrigerant above its saturation change-of-state temperature.
Sobrecalor. Temperatura del refrigerante de vapor mayor que su temperatura de cambio de estado de saturación.
Surge. When the head pressure becomes too great or the evaporator pressure too low, refrigerant will flow from the high- to the low-pressure side of a centrifugal compressor system, making a loud sound.
Movimiento repentino. Cuando la presión en la cabeza aumenta demasiado o la presión en el evaporador es demasiado baja, el refrigerante fluye del lado de alta presión al lado de baja presión de un sistema de compresor centrífugo. Este movimiento produce un sonido fuerte.
Swaged joint. The joining of two pieces of copper tubing by expanding or stretching the end of one piece of tubing to fit over the other piece.
Junta estampada. La conexión de dos piezas de tubería de cobre dilatando o alargando el extremo de una pieza de tubería para ajustarla sobre otra.
Swaging tool. A tool used to enlarge a piece of tubing for a solder or braze connection.
Herramienta de estampado. Herramienta utilizada para agrandar una pieza de tubería a utilizarse en una conexión soldada o broncesoldada.
Swamp cooler. A slang term used to describe an evaporative cooler.
Nevera pantanosa. Término del argot utilizado para describir una nevera de evaporación.
Sweating. A word used to describe moisture collection on a line or coil that is operating below the dew point temperature of the air.
Exudación. Término utilizado para describir la acumulación de humedad en un conducto o una bobina que está funcionando a una temperatura menor que la del punto de rocío de aire.
System charge. The refrigerant in a system, both liquid and vapor. The correct charge is a balance where the system will give the most efficiency.
Carga del sistema. El refrigerante en un sistema, tanto líquido y vapor. La carga correcta es un balance donde el sistema dará la mayor eficiencia.
System lag. The temperature drop of the controlled space below the set point of the thermostat.
Retardo del sistema. Caída de temperatura de un espacio controlado, por debajo del nivel programado en el termostato.

Tank. A closed vessel used to contain a fluid.
Tanque. Depósito cerrado utilizado para contener un fluido.
Tap. A tool used to cut internal threads in a fastener or fitting.
Macho de roscar. Herramienta utilizada para cortar roscas internas en un aparto fijador o en un accesorio.
Technician. A person who performs maintenance, service, testing, or repair to air-conditioning or refrigeration equipment. *Note:* The EPA defines this person as someone who could reasonably be expected to release CFCs or HCFCs into the atmosphere.
Técnicos. Una persona que lleva a cabo mantenimiento, servicio o reparaciones a equipos de aire acondicionado o refrigeración. *Nota:* Esta persona, según defunido por la EPA, es una persona del cual razonablemente se estaría esparado que libere CFC (clorofluorocarbonos) a la atmósfera.
Temperature. A word used to describe the level of heat or molecular activity, expressed in Fahrenheit, Rankine, Celsius, or Kelvin units.
Temperatura. Término utilizado para describir el nivel de calor o actividad molecular, expresado en unidades Fahrenheit, Rankine, Celsio o Kelvin.
Temperature-measuring instruments. Devices that accurately measure the level of temperature.

Instrumentos que miden temperatura. Aparatos que miden el nivel de la temperatura con precisión.

Temperature reference points. Various points that may be used to calibrate a temperature-measuring device, such as boiling or freezing water.

Puntos de referencia de temperatura. Varios puntos que pueden usarse para calibrar un aparato que mide temperatura, tales como agua congelada o hirvienda.

Temperature-sensing elements. Various devices in a system that are used to detect temperature.

Elementos que detectan temperatura. Varios aparatos en un sistema que se usan para detectar temperatura.

Temperature swing. The temperature difference between the low and high temperatures of the controlled space.

Oscilación de temperatura. Diferencia existente entre las temperaturas altas y bajas de un espacio controlado.

Testing, Adjusting and Balancing Bureau (TABB). A certification bureau for individuals involved in working with ventilation.

Agencia de Prueba, Ajuste y Balanceo (TABB en inglés). Agencia de certificación para los individuos involucrados en el trabajo de ventilación.

Test light. A lightbulb arrangement used to prove the presence of electrical power in a circuit.

Luz de prueba. Arreglo de bombillas utilizado para probar la presencia de fuerza eléctrica en un circuito.

Therm. Quantity of heat, 100,000 BTU.

Therm. Cantidad de calor, mil unidades térmicas inglesas.

Thermistor. A semiconductor electronic device that changes resistance with a change in temperature.

Termistor. Dispositivo eléctrico semiconductor que cambia su resistencia cuando se produce un cambio en temperatura.

Thermocouple. A device made of two unlike metals that generates electricity when there is a difference in temperature from one end to the other. Thermocouples have a hot and cold junction.

Termopar. Dispositivo hecho de dos metales distintos que genera electricidad cuando hay una diferencia en temperatura de un extremo al otro. Los termopares tienen un empalme caliente y uno frío.

Thermometer. An instrument used to detect differences in the level of heat.

Termómetro. Instrumento utilizado para detectar diferencias en el nivel de calor.

Thermopile. A group of thermocouples connected in series to increase voltage output.

Pila termoeléctrica. Grupo de termopares conectados en serie para aumentar la salida de tensión.

Thermostat. A device that senses temperature change and changes some dimension or condition within to control an operating device.

Termostato. Dispositivo que advierte un cambio en temperatura y cambia alguna dimensión o condición dentro de sí para regular un dispositivo en funcionamiento.

Thermostatic expansion valve (TXV). A valve used in refrigeration systems to control the superheat in an evaporator by metering the correct refrigerant flow to the evaporator.

Válvula de gobierno termostático para expansión. Válvula utilizada en sistemas de refrigeración para regular el sobrecalor en un evaporador midiendo el flujo correcto de refrigerante al evaporador.

Three-phase power. A type of power supply usually used for operating heavy loads. It consists of three sine waves that are out of phase by 120° with each other.

Potencia trifásica. Tipo de fuente de alimentación normalmente utilizada en el funcionamiento de cargas pesadas. Consiste de tres ondas sinusoidales que no están en fase la una con la otra por 120°.

Throttling. Creating a planned or regulated restriction in a fluid line for the purpose of controlling fluid flow.

Estrangulamiento. Que ocasiona una restricción intencional o programada en un conducto de fluido, a fin de controlar el flujo del fluido.

Thrust surface. A term that usually applies to bearings that have a pushing pressure to the side and that therefore need an additional surface to absorb the push. Most motor shifts cradle in their bearings because they operate in a horizontal mode, like holding a stick in the palm of your hand. When a shaft is turned to the vertical mode, a thrust surface must support the weight of the shaft along with the load the shaft may impose on the thrust surface. The action of a vertical fan shaft that pushes air up is actually pushing the shaft downward.

Superficie de empuje. Un término que generalmente se aplica a cojinetes que sostienen una presión de empuje a un lado y que necesitan una superficie adicional para absorber este empuje. La mayoría de los ejes de motor están al abrigo en sus cojinetes, porque funcionen en una modalidad horizontal, como cuando uno sostiene una vara en la mano. Cuando un eje está sintonizado a la modalidad vertical, una superficie de empuje debe sostener el peso del eje junto con la carga que el eje puede imponer sobre la superficie de empuje. La moción de un eje vertical de un ventilador que empuja aire hacia arriba está en realidad empujando el eje hacia abajo.

Time delay. A device that prevents a component from starting for a prescribed time. For example, many systems start the fans and use a time-delay relay to start the compressor at a later time to prevent too much in-rush current.

Retraso de tiempo. Un aparato que evita que un componente se encienda por un período prescrito de tiempo. Por ejemplo, muchos sistemas encienden los ventiladores y usando un relé de retraso, encienden el compresor un tiempo después para evitar mucha corriente interna.

Timers. Clock-operated devices used to time various sequences of events in circuits.

Temporizadores. Dispositivos accionados por un reloj utilizados para medir el tiempo de varias secuencias de eventos en circuitos.

Toggle bolt. Provides a secure anchoring in hollow tiles, building block, plaster over lath, and gypsum board. The toggle folds and can be inserted through a hole, where it opens.

Tornillo de fiador. Proveen un anclaje seguro en losetas huecas, bloques de construcción, yeso sobre listón y tablón de yeso. El fiador se dobla y puede insertarse a través de un roto y después se abre.

Ton of refrigeration. The amount of heat required to melt a ton (2,000 lb) of ice at 32°F in 24 hours, 288,000 BTU/24 h, 12,000 BTU/h, or 200 BTU/min.
Tonelada de refrigeración. Cantidad de calor necesario para fundir una tonelada (2.000 libras) de hielo a 32°F (0°C) en 24 horas, 288.000 BTU/24 h, 12.000 BTU/h o 200 BTU/min.
Torque. The twisting force often applied to the starting power of a motor.
Par de torsión. Fuerza de torsión aplicada con frecuencia a la fuerza de arranque de un motor.
Torque wrench. A wrench used to apply a prescribed amount of torque or tightening to a connector.
Llave de torsión. Llave utilizada para aplicar una cantidad específica de torsión o de apriete a un conector.
Total equivalent warming impact (TEWI). A global warming index that takes into account both the direct effects of chemicals emitted into the atmosphere and the indirect effects caused by system inefficiencies.
Impacto de calentamiento equivalente total (TEWI en inglés). Índice de calentamiento de la tierra que tiene en cuenta los efectos directos de los productos químicos emitidos en la atmósfera, y los efectos indirectos causados por la ineficacia de un sistema.
Total heat. The total amount of sensible heat and latent heat contained in a substance from a reference point.
Calor total. Cantidad total de calor sensible o de calor latente presente en una sustancia desde un punto de referencia.
Total pressure. The sum of the velocity and the static pressure in an air duct system.
Presión total. La suma de la velocidad y la presión estática en un sistema de conducto de aire.
Transformer. A coil of wire wrapped around an iron core that induces a current to another coil of wire wrapped around the same iron core. *Note:* A transformer can have an air core.
Transformador. Bobina de alambre devanado alrededor de un núcleo de hierro que induce una corriente a otra bobina de alambre devanado alrededor del mismo núcleo de hierro. *Nota:* un transformador puede tener un núcleo de aire.
Transistor. A semiconductor often used as a switch or amplifier.
Transistor. Semiconductor que suele utilizarse como conmutador o amplificador.
TRIAC. A semiconductor switching device.
TRIAC. Dispositivo de conmutación para semiconductores.
Tube within a tube coil. A coil used for heat transfer that has a pipe in a pipe and is fastened together so that the outer tube becomes one circuit and the inner tube another.
Bobina de tubo dentro de un tubo. Bobina utilizada en la transferencia de calor que tiene un tubo dentro de otro y se sujeta de manera que el tubo exterior se convierte en un circuito y el tubo interior en otro circuito.
Tubing. Pipe with a thin wall used to carry fluids.
Tubería. Tubo que tiene una pared delgada utilizada para conducir fluidos.
Two-speed compressor motor. Can be a four-pole motor that can be connected as a two-pole motor for high speed (3,450 rpm) and connected as a four-pole motor for running at 1,725 rpm for low speed. This is accomplished with relays outside the compressor.
Motor de compresor de dos velocidades. Puede ser un motor de 4 polos que puede conectarse como un motor de 2 polos para velocidades altas (3.450 revoluciones por minuto) y conectarse como un motor de 4 polos para correr a 1.725 revoluciones por minuto para velocidades bajas. Esto se hace con relés fuera del compresor.
Two-temperature valve. A valve used in systems with multiple evaporators to control the evaporator pressures and maintain different temperatures in each evaporator. Sometimes called a hold-back valve.
Válvula de dos temperaturas. Válvula utilizada en sistemas con evaporadores múltiples para regular las presiones de los evaporadores y mantener temperaturas diferentes en cada uno de ellos. Conocida también como válvula de retención.

Ultrasound leak detector. Detectors that use sound from escaping refrigerant to detect leaks.
Detector de escapes ultrasónico. Detectores que usan sonido del refrigerante que está escapando para detectar escapes.
Ultraviolet. Light waves that can only be seen under a special lamp.
Ultravioleta. Ondas de luz que pueden observarse solamente utilizando una lámpara especial.
Ultraviolet light. Light frequency between 200 and 400 nanometers.
Luz ultravioleta. Luz con frecuencia entre 200 y 400 nanómetros.
Upflow furnace. This furnace takes in air from the bottom or from sides near the bottom and discharges hot air out the top.
Horno de flujo ascendente. La entrada del aire en este tipo de horno se hace desde abajo o en los laterales, cerca del suelo. La evacuación se realiza en la parte superior.
Urethane foam. A foam that can be applied between two walls for insulation.
Espuma de uretano. Espuma que puede aplicarse entre dos paredes para crear un aislamiento.
U-Tube mercury manometer. A U-tube containing mercury, which indicates the level of vacuum while evacuating a refrigeration system.
Manómetro de mercurio de tubo en U. Tubo en U que contiene mercurio y que indica el nivel del vacío mientras vacía un sistema de refrigeración.
U-Tube water manometer. Indicates natural gas and propane gas pressures. It is usually calibrated in inches of water.
Manómetro de agua de tubo en U. Indica las presiones del gas natural y del propano. Se calibra normalmente en pulgadas de agua.

Vacuum. The pressure range between the earth's atmospheric pressure and no pressure, normally expressed in inches of mercury (in. Hg) vacuum.
Vacío. Margen de presión entre la presión de la atmósfera de la tierra y cero presión, por lo general expresado en pulgadas de mercurio (pulgadas Hg) en vacío.
Vacuum gauge. An instrument that measures the vacuum when evacuating a refrigeration, air-conditioning, or heat pump system.

Medidor del vacío. Instrumento que se utiliza para medir el vacío al vaciar un sistema de refrigeración, de aire acondicionado, o de bomba de calor.

Vacuum pump. A pump used to remove some fluids such as air and moisture from a system at a pressure below the earth's atmosphere.

Bomba de vacío. Bomba utilizada para remover algunos fluidos, como por ejemplo aire y humedad de un sistema a una presión menor que la de la atmósfera de la tierra.

Valve. A device used to control fluid flow.

Válvula. Dispositivo utilizado para regular el flujo de fluido.

Valve plate. A plate of steel bolted between the head and the body of a compressor that contains the suction and discharge reed or flapper valves.

Placa de válvula. Placa de acero empernado entre la cabeza y el cuerpo de un compresor que contiene la lámina de aspiración y de descarga o las chapaletas.

Valve seat. That part of a valve that is usually stationary. The movable part comes in contact with the valve seat to stop the flow of fluids.

Asiento de la válvula. Pieza de una válvula que es normalmente fija. La pieza móvil entra en contacto con el asiento de la válvula para detener el flujo de fluidos.

Valve stem depressor. A service tool used to access pressure at a Schrader valve connection.

Depresor de vástago de válvula. Una herramienta de servicio que se usa para acceder la presión en una conexión de válvula Schrader.

Vapor. The gaseous state of a substance.

Vapor. Estado gaseoso de una sustancia.

Vapor barrier. A thin film used in construction to keep moisture from migrating through building materials.

Película impermeable. Película delgada utilizada en construcciones para evitar que la humedad penetre a través de los materiales de construcción.

Vapor charge bulb. A charge in a thermostatic expansion valve bulb that boils to a complete vapor. When this point is reached, an increase in temperature will not produce an increase in pressure.

Válvula para la carga de vapor. Carga en la bombilla de una válvula de expansión termostática que hierve a un vapor completo. Al llegar a este punto, un aumento en temperatura no produce un aumento en presión.

Vaporization. The changing of a liquid to a gas or vapor.

Vaporización. Cuando un líquido se convierte en gas o vapor.

Vapor lock. A condition where vapor is trapped in a liquid line and impedes liquid flow.

Bolsa de vapor. Condición que ocurre cuando el vapor queda atrapado en el conducto de líquido e impide el flujo de líquido.

Vapor pressure. The pressure exerted on top of a saturated liquid.

Presión del vapor. Presión que se ejerce en la superficie de un líquido saturado.

Vapor pump. Another term for compressor.

Bomba de vapor. Otro término para compresor.

Vapor refrigerant charging. Adding refrigerant to a system by allowing vapor to move out of the vapor space of a refrigerant cylinder and into the low-pressure side of the refrigeration system.

Carga del refrigerante de vapor. Agregarle refrigerante a un sistema permitiendo que el vapor salga del espacio de vapor de un cilindro de refrigerante y que entre en el lado de baja presión del sistema de refrigeración.

Variable-frequency drive (VFD). An electrical device that varies the frequency (hertz) for the purpose of providing a variable speed.

Propulsión de frecuencia variable (VFD en inglés). Un aparato eléctrico que varía la frecuencia (hertz) con el propósito de proveer velocidad variable.

Variable pitch pulley. A pulley whose diameter can be adjusted.

Polea de paso variable. Polea cuyo diámetro puede ajustarse.

Variable resistor. A type of resistor where the resistance can be varied.

Resistor variable. Tipo de resistor donde la resistencia puede variarse.

Variable-speed motor. A motor that can be controlled, with an electronic system, to operate at more than one speed.

Motor de velocidad variable. Un motor que puede controlarse, con un sistema electrónico, para operar a más de una velocidad.

V belt. A belt that has a V-shaped contact surface and is used to drive compressors, fans, or pumps.

Correa en V. Correa que tiene una superficie de contacto en forma de V y se utiliza para accionar compresores, abanicos o bombas.

Velocity. The speed at which a substance passes a point.

Velocidad. Rapidez a la que una sustancia sobrepasa un punto.

Velocity meter. A meter used to detect the velocity of fluids, air, or water.

Velocímetro. Instrumento utilizado para medir la velocidad de fluidos, aire o agua.

Velometer. An instrument used to measure the air velocity in a duct system.

Velómetro. Instrumento utilizado para medir la velocidad del aire en un conducto.

Vent-free. Certain gas stoves or gas fireplaces are not required to be vented; therefore, they are called "vent-free."

Sin ventilación. Ciertas estufas de gas y chimeneas de gas no requieren una ventilación, por lo que se conocen como estufas "sin ventilación".

Ventilation. The process of supplying and removing air by natural or mechanical means to and from a particular space.

Ventilación. Proceso de suministrar y evacuar el aire de un espacio determinado, utilizando procesos naturales o mecánicos.

Venting products of combustion. Venting flue gases that are generated from the burning process of fossil fuels.

Descargar productos de combustión. Descargar los gases de la chimenea que se generan del proceso de combustión de los combustibles fósiles.

Voltage. The potential electrical difference for electron flow from one line to another in an electrical circuit.

Voltaje. Diferencia de potencial eléctrico del flujo de electrones de un conducto a otro en un circuito eléctrico.

Voltage feedback. Voltage potential that travels through a power-consuming device when it is not energized.

Retroalimentación de voltaje. El voltaje que viaja a través de un aparato de consumo de electricidad cuando no está energizado.

Volt-ohm-milliammeter (VOM). A multimeter that measures voltage, resistance, and current in milliamperes.

Voltio-ohmio-miliamperimetro (VOM en inglés). Multímetro que mide tensión, resistencia y corriente en miliamperios.

Volumetric efficiency. The pumping efficiency of a compressor or vacuum pump that describes the pumping capacity in relationship to the actual volume of the pump.

Rendimiento volumétrico. Rendimiento de bombeo de un compresor o de una bomba de vacío que describe la capacidad de bombeo con relación al volumen real de la bomba.

Vortexing. A whirlpool action in the sump of a cooling tower.

Acción de vórtice. Torbellino en el sumidero de una torre de refrigeración.

Walk-in cooler. A large refrigerated space used for storage of refrigerated products.

Nevera con acceso al interior. Espacio refrigerado grande utilizado para almacenar productos refrigerados.

Water column (WC). The pressure it takes to push a column of water up vertically. One inch of water column is the amount of pressure it would take to push a column of water in a tube up one inch.

Columna de agua (WC en inglés). Presión necesaria para levantar una columna de agua verticalmente. Una pulgada de columna de agua es la cantidad de presión necesaria para levantar una columna de agua a una distancia de una pulgada en un tubo.

Water-cooled condenser. A condenser used to reject heat from a refrigeration system into water.

Condensador enfriado por agua. Condensador utilizado para dirigir el calor de un sistema de refrigeración al agua.

Water-regulating valve. An operating control regulating the flow of water.

Válvula reguladora de agua. Regulador de mando que controla el flujo de agua.

Watt. A unit of power applied to electron flow. One watt equals 3.414 BTU.

Watio. Unidad de potencia eléctrica aplicada al flujo de electrones. Un watio equivale a 3,414 BTU.

Watt-hour. The unit of power that takes into consideration the time of consumption. It is the equivalent of a 1-watt bulb burning for 1 hour.

Watio hora. Unidad de potencia eléctrica que toma en cuenta la duración de consumo. Es el equivalente de una bombilla de 1 watio encendida por espacio de una hora.

Weep holes. Holes that connect each cell in an ice machine's cell-type evaporator that allow air entering from the edges of the ice to travel along the entire ice slab to relieve the suction force and allow the ice to fall off of the evaporator.

Aberturas de exudación. Aberturas que conectan cada célula en el evaporador de tipo celular de una hielera, que permiten que el aire que entra de los bordes del hielo viaja a lo largo de todo el pedazo de hielo para aliviar la fuerza de succión y permitir que el hielo se caiga del evaporador.

Welded hermetic compressor. A compressor that is completely sealed by welding, versus a semi-hermetic compressor that is sealed by bolts and flanges.

Compresor hermético soldado. Un compresor que está completamente sellado por soldadura, contrario a un compresor semihermético que está sellado con tornillos y bridas.

Wet-bulb temperature. A wet-bulb temperature of air is used to evaluate the humidity in the air. It is obtained with a wet thermometer bulb to record the evaporation rate with an airstream passing over the bulb to help in evaporation.

Temperatura de una bombilla húmeda. La temperatura de una bombilla húmeda se utiliza para evaluar la humedad presente en el aire. Se obtiene con la bombilla húmeda de un termómetro para registrar el margen de evaporación con un flujo de aire circulando sobre la bombilla para ayudar en evaporar el agua.

Wet heat. A heating system using steam or hot water as the heating medium.

Calor húmedo. Sistema de calentamiento que utiliza vapor o agua caliente como medio de calentamiento.

Winding thermostat. A safety device used in electric motor windings to detect over-temperature conditions.

Termostato de bobina. Un aparato de seguridad usado en un motor eléctrico para detectar condiciones de exceso de temperatura.

Window unit. An air conditioner installed in a window that rejects the heat outside the structure.

Acondicionador de aire para la ventana. Acondicionador de aire instalado en una ventana que desvía el calor proveniente del exterior de la estructura.

Wire connectors (screw-on). Used to connect two or more wires together.

Conectores de cables (de rosca). Se usan para conectar dos o más cables.

Work. A force moving an object in the direction of the force. Work = Force × Distance.

Trabajo. Fuerza que mueve un objeto en la dirección de la fuerza. Trabajo = Fuerza × Distancia.

WYE transformer connection. Typically furnishes 208 and 115 V to a customer.

Conexión de transformador WYE. Típicamente provee 208 y 115 voltios a los consumidores.

Zeotropic blend. Two or more refrigerants mixed together that will have a range of boiling and/or condensing points for each system pressure. Noticeable fractionation and temperature glide will occur.

Mezcla zeotrópica. Mezcla de dos o más refrigerantes que tiene un rango de ebullición y/o punto de condensación para cada presión en el sistema. Se produce una fraccionación y una variación de temperatura notables.

Zone valve. Zone control valves are thermostatically controlled valves that control water flow in various zones in a hydronic heating system.

Válvula de sector. Se trata de válvulas controladas termostáticamente, que controlan el flujo del agua en varios sectores en un sistema de calefacción hidrónico.

Index

Page numbers in italics refer to figures and tables.

ABS (acrylonitrile butadiene styrene), 224
absolute pressure, 97–99
absolute temperature, 86
absolute zero, 86
AC (armored cable) conduit, *76*
access valves, *116*, 116–117, *117*
accreditation, 267–269
accumulator, 43, *43*
"A" coil, 27
ACR tubing. *See* air conditioning and refrigeration tubing; copper ACR tubing
added value, 257–258
adjustable wrench, 211, *215*
A-frame stepladder, *61*
aftermarket sales, 142
AFUE. *See* annual fuel utilization efficiency
air-acetylene torches, 193, *194*, *195*
air-acetylene torch tips, 198, *198*
air balancing, 224
air conditioners, SEER rating, 141
air conditioning
   history of, 13–14
   troubleshooting case studies, 167–174
air conditioning and refrigeration (ACR) tubing
   described, 32, 34
   guidelines for installing, 34
   identification, 34–35
   wall thicknesses, 34
   See also copper ACR tubing
air conditioning compressors, 23, *24*
Air Conditioning Contractors of America (ACCA), 242, 268
air-cooled condensers, *24*
air ducts
   combustion air requirement, 248
   defined, 28
   fiberglass duct board, 29–30, *30*, *31*, *151*, 222, 224
   flexduct, *29*, *30*, *31*, *151*, 222, *223*
   indoor air quality, 260
   installation problems, 224
   insulation, 222, *223*, 224
   leakage, 246
   maintenance, 151

sheet metal, *28*, 28–29, *29*, *31*, *151*, 222, *223*
   sizing, support, and sealing, 245
   system configurations, 30, 32, *32*
   undersized, 32
air filters
   indoor air quality, 260
   maintenance, 143, *145*, 145–146, *146*
   measuring efficiency, 145–146
   purpose, 143, 145
airflow measuring devices, *208*, *209*
airflow problems, diagnosing, 162
airflow velocity, 245, *245*
air handlers
   installation, 183, *184*, *186*, 186–188, *187*, *188*
   types, 183
air pollution, indoor, 224, 260
air-to-air heat pumps, 43–44
Allen wrenches, 216, *217*
alternating current (AC), 55–56
alternative energy
   defined, 234
   fuel cells, 238–239
   geothermal, 44–46, *45*, 237–238, *238*
   green buildings, 239
   solar, 47–48, 234–237, *235*, *236*
   wind, 237, *237*
American National Standards Institute (ANSI), 243
American Wire Gauge (AWG), 72–73
ammeter, clamp-on, 64–65, *65*
ammonia, 15
ampacity, 72, *73*, 74–75
amperage
   ampacity of wire, 72, *73*, 74–75
   defined, 56
   measuring, 63–64
   measuring in compressors, *149*, 152–153
   Ohm's law, 57–60
   See also current flow
amperes, 56
amps, 56
analog multimeter, 62, *62*
annual fuel utilization efficiency (AFUE), 229

annual maintenance
   benefits, 239
   effects on efficiency, 239–240
   See also scheduled maintenance
apprenticeship programs, 15
arc blast, 53
armored cable (AC) conduit, *76*
atmospheric pressure, 97–99
automatic expansion valve, 134, *135*
auxiliary drain pan, 150
AWG (American Wire Gauge), 72–73
balancing dampers, 224
barometer, 98
batteries, 55
blowers
   gas furnaces, 37
   maintenance, *152*, 152–153, *153*
boilers, clearances, *185*
boiling point, 93
Boyle's law, 102–103, 105
brazing
   overview, 193, 195, *196*, 197–198
   safety, 198–199
breakers, resetting, 168, *173*
British thermal units (BTUs)
   defined, 83
   energy conversions, 230
   overview, 83–84
BTUs. *See* British thermal units
BTUs per hour (BTUHs), 84, *85*, 230
Buffalo Forge Company, 14
burners
   gas furnaces, 37, *39*
   oil furnaces, 40
burrs, removing, *192*

cable ties, 201
capillary action, 197
capillary tube metering device, 131–133, *132*
carbon dioxide, solid, 92, *92*
Carrier, Willis Haviland, 14
CCFs, 230
Celsius temperature scale, 86–87, *87*, *88*
centigrade temperature scale. *See* Celsius temperature scale

centrifugal compressors, 23
certification, 10, 12
CFCs. *See* chlorofluorocarbons
Channel locks, 212
charging devices, 178
Charles's law, 103–104, 105
chillers, 3
chlorofluorocarbons (CFCs), 10. *See also* refrigerants
clamp-on ammeters, 64–65, 65, 66
Clean Air Act, 12
cleaning pads, 216
clearances, 183–184, 185
climate zone map, 244
closed circuits, 57
closed-loop geothermal system, 44, 46
coaxial heat exchanger, 44
codes, 202
coefficient of performance (COP), 231–233
cohesion, molecular, 90
coil cleaners, 146, 147, 147
coils
  "A," 27
  "indoor" and "outdoor," 42
  maintenance, 146–147
  slab, 27
  See also condenser coils
combustion air, 224, 247–248
comfort value, 258–259
comfort zone, 233–234, 234
commercial digital thermostat, 247
commercial ladder, 176
commercial package units, installation components, 183
complete circuits, 57, 57
compound pressure gauges, 99, 100, 119. *See also* manifold gauges
compression stroke, 123
compressors
  cooling capacities, 23–24
  defined, 22–23
  measuring amperage, 149, 152–153
  measuring voltage, 153, 154
  purpose and function, 122
  reciprocating, 123–125, 128–129
  refrigeration cycle, 21, 22–24
  replace versus repair, 261
  rotary, 126–129, 127, 128
  scroll, 125, 126, 127
  terminals, 69, 171
  types, 23, 123
compressor terminals, 66
condensate drain, 223–224
condensate drain line, 222
condenser-coil cleaners, 146, 147

condenser coils
  maintenance, 20, 146–147, 147
  refrigeration cycle, 21
condensers
  defined, 24
  function, 129
  heat pumps, 42
  phases of heat removal, 129
  refrigeration cycle, 21, 24
  types, 24, 129, 130, 131
condensing unit
  clearances, 183, 184
  installation, 189–190
  nameplate, 189
  placement, 184, 186
  replacing, 248
  schematic diagram, 72
  selecting for, 243, 245
conduction, 20, 88, 89
conductors (electrical), 53–54. *See also* wires
conductors (thermal), 88
conduit, 73, 76, 76, 77
consumers, environmentally friendly, 262. *See also* customers
contactors, 66, 67, 68
contact probe, 205
control circuits, 160
control sequence, 160
convection, 89, 89
cooling, forms of, 20–21
cooling and heating diagram, 165
cooling systems
  oversizing, 242
  undersizing, 242
  See also HVAC systems
copper ACR tubing
  bending, 216
  fittings, 220, 220–221, 221
  guidelines for installing, 34
  hard-drawn, 32, 33, 218, 219
  sizes, 219–220, 220
  soft copper, 32, 33, 216, 218, 219
  types, 32, 33, 34, 218, 219
  wall thicknesses, 34
  See also piping; refrigerant piping; tubing
copper-clad wire, 73
copper fittings, 220, 220–221, 221
copper oxide, 198
cordless drills, 205, 206
Core exam, 12
coulomb, 56
crankcase heater, 150
crosspoint screwdriver, 212
cubic feet per minute (CFM), 207

current flow
  defined, 56
  Ohm's law, 57–60
  See also amperage
current rating, 72
customers
  added value to systems, 257–258
  environmentally friendly, 262
  value selling, 258–261
customer satisfaction, 255
customer satisfaction surveys, 255
customer service
  apologizing, 259
  asking questions, 257
  cleaning work sites, 254
  communicating and listening, 158–159
  empathy, 255
  importance, 254
  interactions with customers, 266
  listening, 256–257
  quality and cleanliness, 266–267
customer service checklist, 255
customer surveys, 259

Dalton's law, 104
dampers, 247
delta-T (temperature difference), 27
demand charges, 230
diagnosis, 166
digital humidistat, 248
digital multimeter, 62, 62
digital psychrometer, 205, 206
digital scale, 209, 209
digital thermostats, 4, 247, 247
direct current (DC), 54, 55
discharge line, 21, 34, 35, 111, 111
discharge stroke, 123–124, 124
disconnects, 188–189
distribution systems, 28–32. *See also* air ducts
distributor, 111, 111
distributor line, 35, 35
double-suction riser, 221, 222
drills, cordless, 205, 206
dry bulb thermometer, 143, 144
dry ice, 22, 92, 92
duct board, 29–30, 30, 31, 151, 222, 224
duct fasteners, 223
ducts. *See* air ducts
duct tape, 222, 223, 245
DuPont, 15

EER. *See* Energy Efficiency Ratio
efficiency, in energy transfer, 96. *See also* energy efficiency
electrical fasteners, 201, 202

Index

electrical problems
   diagnosing, 161
   troubleshooting, 177–178
electrical resistance. *See* resistance
electrical shocks, 60
electrical symbols, *67, 68–70, 69*
electric energy, 229–230
electric furnaces, clearances, *185*
electric heating
   coefficiency of performance, 231–233
   power, 60–61
   troubleshooting, 174–176
electricity
   amperes, 56
   complete circuits, 57, *57*
   conductors and wires, 53–54
   incomplete circuits, *60*
   Ohm's law, 57–60
   power, 57, 60–61
   resistance, 56–57
   volts, 54–56 (*see also* voltage)
electric shock, 53, 57
electric strip heat, 42, *42*
electromechanical sequence, 162–164
electromechanical switches, *68*
electromechanical tubing (EMT), 73, 76, *76*
electromotive force. *See* volts
electron flow, 56
electrons, 54, *54*, 56
emergency drain pan, 150
EMT (electromechanical tubing), 73, 76, *76*
energy
   laws of thermodynamics, 102
   transferring, 96
energy bill value, 259–260
energy codes, 240
energy efficiency
   effects of maintenance on, 239–240
   energy codes, 240
   Energy Star program, 241
   government and utility incentives, 240
   overview, 228
   See also alternative energy
Energy Efficiency Ratio (EER), 228, 231
energy-efficient installation
   condensing unit replacement, 248
   cooling and heating load calculations, 242–243
   duct system, 245
   equipment sizing, selection, and efficiency, 243, 245
   miscellaneous components, 247–248
   refrigerant pipe sizing, support, and insulation, 245–247

thermostat installation, 247
energy savings, 259–260, 265
Energy Star program, 241
environmentally friendly consumers, 262
Environmental Protection Agency (EPA), 12
EPA certification, 12
estimators, 5–6
ethics, 176
evaporation, 20–21, 92–93
evaporative condensers, 24, *130*
evaporator-coil cleaners, 147, *147*
evaporator coils, 147, *147, 148*
evaporators
   heat pumps, 42
   overview, 134–136, *135, 136*
   refrigeration cycle, *21*, 26–28, *27*
   types, *27*
exhaust stroke, *124*
extended plenum duct system, *32*
extension ladders, *61, 64*

factory-made duct, 29
Fahrenheit temperature scale, 86, *88*
fans, maintenance, 152–153, *153*
Faraday, Michael, 14
fasteners, 201, *202*
feet per minute (FPM), 207
female connectors, 201, *202*
fiberglass duct board, 29–30, *30, 31, 151, 222, 224*
files, *214*
filter driers, *21*
fire extinguishers, 262, *262*
first-aid kits, 201, *201*
first cost, 228
fish tape, 215–216, *216*
fixed-piston metering device, 131–133, *132*
flaring, 199, *200*, 201, *201*
flaring block, 199, *200*
flaring tool, 199, *200, 201*
flaring wrench, 201, *201*
flaring yoke, 199, *200, 201*
flat blade screwdriver, *212*
flat-plate solar collectors, *236*
flat-rate pricing, 264
flexduct, 29, *30, 31, 151, 222, 224*
float switch, 150
flow control. *See* metering devices
flux, lines of, *57*
flux (used in soldering), 195, *196*, 197, *197*
forced convection, 89
45° fitting, 221, *221*
Freon, 15

frequency, 55
fuel cells, 238–239
fuel oil, 40
fuel oil gun, *41*
fuels
   conversion, 230
   cost comparisons, 230
fuel tanks, 40
furnaces
   clearances, 184, *185*
   multipositional, 183
   oil, *40*, 40–41
   See also gas furnaces
fusion, 91

gas code, 248
gases, 100, *101*. *See also* gas laws; vapor
gas furnaces, *2*
   blower section, 37
   burner and heat exchanger section, 37–38
   components, *39*
   controls section, 37
   efficiency, 228
   overview, 37
   venting section, 38–39
gas heating, 37–39
gas laws
   Boyle's, 102–103, 105
   Charles's, 103–104, 105
   combining, 104–105
   Dalton's, 104
gas valves, *7*
gauge, of wire, 72–73
gauge pressure, 97
gauge repair kit, *205*
gauges. *See* manifold gauges; pressure gauges
geothermal energy, 237–238, *238*
geothermal heat pumps, 44–46, *45*, 237–238, *238*
Gorrie, John, 14
gravity, 101
"green," 228
green buildings, 239
green consumers, 262
greenhouse gases, 241, *241*
grounding, 58, 60, *60*

hacksaw, 211, 215, *215*
HACR circuit breaker, 189
hammers, *214*
hand tools, 210, 211–218
hard-drawn copper, 32, *33, 218, 219*
HCFCs. *See* hydrochlorofluorocarbons

head, 35
heat
 defined, 82
 intensity and quantity, 86
 latent (see latent heat)
 molecular motion, 90–91
 quantity, 87
 sensible, 84–85, 85, 91, 94
 specific, 94–95
 transferring, 20–21, 22, 82, 96
 See also thermodynamics
heat exchangers
 clearances, 185
 gas furnaces, 38–39, 39
 oil furnaces, 40
heating, ventilation, air conditioning, and refrigeration. See HVACR
Heating Season Performance Factor (HSPF), 233
heating systems
 oversizing, 242
 undersizing, 242
 See also HVAC systems
heat of compression, 24
heat pumps
 coefficient of performance, 231–233
 cooling mode, 43, 45
 defrost mode, 44
 geothermal, 44–46, 237–238, 238
 heating mode, 43, 44, 45
 HSPF, 233
 outdoor placement, 184–185
 overview, 42–43
 types, 43
heat strips, 232
heat transfer, 20–21, 22, 82
helical-rotary compressors, 23
hermetic compressors, 123, 124
hertz, 55
hex wrenches, 216, 217
HFCs. See hydrofluorocarbons
household ladder, 176
HSPF. See Heating Season Performance Factor
human comfort zone, 233–234, 234
humidifiers, 234, 247
humidistat, digital, 248
humidity, relative, 234, 235
HVAC Excellence, 269
HVACR (heating, ventilation, air conditioning, and refrigeration)
 importance of, 3–4
 meaning of, 5
HVACR industry
 career advancement, 15–16
 career opportunities, 5–10
 job security, 6
 professional certification and accreditation, 10, 12, 267–269
 work environment, 6
HVACR technicians
 appropriate language, 266
 career advancement, 15–16
 career path, 13
 communication and, 158–159
 customer interactions, 266
 ethics and professional conduct, 176
 goals of, 2
 installation technicians, 5, 7
 professional certification and accreditation, 10, 12, 267–269
 professional image, 21
 record keeping responsibilities, 12–13
 refrigeration technicians, 8–9
 roles of, 10
 as salespeople, 142, 263–264
 service technicians, 7–8, 10
 specializations, 6–7
 training and apprenticeship, 15
 uniforms, 266
 work performed by, 5, 6–10
HVAC systems
 added value, 257–258
 average life, 229
 component matching, 245
 equipment selection and sizing, 243, 245
 performance as perceived by the customer, 270
 replace versus repair, 261–262
 right system, right building, right price, 270
 temperature-related capacity changes, 231, 232
 undersizing or oversizing, 242
 unit and component capacity, 270
 using thermostats in checkouts, 4
 value of investment, 262
hydrochlorofluorocarbons (HCFCs), 10. See also refrigerants
hydrofluorocarbons (HFCs), 10. See also refrigerants
hydrogen fuel cells, 46–47, 47, 238–239
hydronic heating, 36, 36
hydronic piping, 35
hydronics, 35
hydronic water heating, 35

in. Hg (inches of mercury), 98
inches of mercury (in. Hg), 98
inches of vacuum, 99
incomplete circuits, 60
indoor air pollution, 224
indoor air quality (IAQ), 260
"indoor coil," 42
inductive load, 60
industrial ladder, 176
inner duct liner, 223
inside diameter (ID), 32, 33, 219, 220
inspection work. See scheduled maintenance
inspectors, 5
installations
 air handlers, 186–188
 basic methods, 182–183
 basic wiring, 188–189, 189
 clearances, 183–184, 185
 condensing or package unit, 189–190
 condensing unit placement, 184, 186
 energy-efficient, 241–248
 field tools, 182
 first cost, 228
 indoor section, 190–191
 problems and failures, 224
 secondary components, 228–229
 start-up and final checkout, 191
installation technicians, 5, 7
insulation
 condensate drain line, 222
 ductwork, 222, 223, 224
 places used, 222
 refrigerant pipe, 222, 247
 of wire, 77
insulators (electrical), 54, 57
insulators (thermal), 88
intake stroke, 124

Kelvin temperature scale, 87, 88
kilowatt hours (KWH), 229
kilowatts (KW), 229
 conversion to BTUs, 42, 230

ladder diagrams, 71
ladders, types and safety, 60, 61, 61, 64, 168, 176
latent heat
 of boiling, 93
 calculations, 94
 defined, 94
 of fusion, 91, 95–96
 of melting, 91
 overview, 85, 86
 of vaporization, 96
levels, 215, 216
lever-type bender, 216, 218
lifting, correct method, 249
lineman pliers, 212
liquid line, 21, 34–35, 35, 111, 111

liquid-line service valve, *116, 117*
liquids
   boiling point, 93
   evaporation, 92–93
   melting and fusion, 91
   properties, 100, *101*
liquid tight conduit, 77, *77*
liquid-to-air heat pumps, 44. *See also* geothermal heat pumps
liquid-to-liquid heat pumps, 44. *See also* geothermal heat pumps
load, 56
load calculations, 242–243
locked rotor amps (LRA), 189
long radius elbow, 221
low-pressure switch (LPS), 160
lubrication, blowers and fans, 152, *153*
magnetic coils, 69, *69*
magnetic fields, 55, *57*
maintenance, effects on efficiency, 239–240. *See also* scheduled maintenance
maintenance agreements, 142
maintenance calls
   aftermarket sales and, 142
   scheduling, 142
   See also scheduled maintenance
maintenance contracts, 154
maintenance cost, 229
maintenance value, 260
manifold gauges
   checking accuracy, 125
   gauge set, 203, *204, 205*
   repair kit, *205*
   safety tips, 165
   using, 114, 116–117, 118, *118, 119*
Manual D, 245
Manual J, 242, *243*
Manual N, 242
Manual Q, 245, *246*
Manual S, 243
mastics, 245
Material Safety Data Sheet (MSDS), 201, *203*
matter, 100, *101*
MC (metal clad) conduit, 73, *76*
measuring tools, 211, *213*
mechanical codes, 202
mechanical problems, diagnosing, 161
melting, 91
mercury U-tube, 98, *99*
MERV rating. *See* Minimum Efficiency Report Value
metal clad (MC) conduit, 73, *76*
metal files, *214*
metal saws, 211, 215, *215*
metal snap-lock duct, *28*

metering devices
   automatic expansion valve, 134, *135*
   fixed, 131–133
   function, 129–131, *131*
   refrigeration cycle, *21,* 24–26
   thermostatic expansion valve, 9, *133,* 133–134, *134*
   types, 131
meters (electrical)
   clamp-on ammeter, 64–65, *65*
   multimeters, 61–64, 210
   safety tips, 63
   uses, 61
methyl chloride, 15
micron gauge, 211, *211*
Midgley, Thomas, Jr., 15
milliammeter, 210
Minimum Efficiency Report Value (MERV), 145–146
molecular motion, 90–91
molecules, 53–54, *54,* 87
multimeters, 210
   measuring amperage, 63–64
   measuring resistance, 63
   measuring voltage, 62–63
   overview of, 61–64
   safety tips, 63
   selecting, 64
multipositional furnaces, 183
multitap transformers, *169*

nameplates, *189*
National Fuel Gas Code, 248
natural convection, 89
natural gas, 230
needle-nose pliers, *212*
neutrons, 53, 54
nitrogen, 198
NM wire, 73
North American Technician Excellence (NATE), 269
nut drivers, 211, *212*
nylon tie, 201

Ohm, Georg, 57
ohmmeters, 165, 174, 182, 210. *See also* multimeters; VOM
ohms, 56
Ohm's law, 57–60
oil furnaces, *40,* 40–41
oil heating, 40
oil pumps, 40
one-vane rotary compressor, *127*
open circuits, *60*
open-loop geothermal system, 44, *46*

operating costs, 229
"outdoor coil," 42
outside diameter (OD), 32, *33,* 219, 220, *220*
owners/operators, 5
oxyacetylene torch, 195, *196*

package units, *163*
   basic installation components, 183
   functional diagram, *161*
   installation, 189–190
   power supply, 163
   schematic diagram, *71*
   troubleshooting case study, 169–171, *170*
parallel circuits, *70*
passive solar design, 47, 235, *236*
payback, 229
perimeter loop duct system, *32*
permanent split capacitor, *69*
Persia, 13
personal protective equipment (PPE), 52
Phillips screwdriver, *212*
phosphorus, *196*
photovoltaic systems, 47, *48,* 234–235, *235*
piercing vise grips, 211, *213*
pipe-in-a-pipe heat exchanger, 44
pipe insulation, 222
pipe wrench, 211, *215*
piping, 218. *See also* copper ACR tubing; refrigerant piping; tubing
piston metering device, 131–133, *132*
plastic clamps, 201, *202*
pliers, 211, *212, 213*
Plumbing-Heating-Cooling Contractors National Association (PHCC), 268
pocket thermostat, 205
podium ladders, 60
potential difference. *See* volts
pounds per square inch gauge (psig), 97
pounds per square inch (psi), 97
power
   calculating, 60–61
   defined, 57
power supplies, 163, *163*
power tools, 205, *206*
PPE (personal protective equipment), 52
pressure
   absolute, 97–99
   defined, 97
   gas laws, 102–105
   gauge pressure, 97
pressure gauges, 98
   compound, 99, 100, 119, 120
   installing, *118*

pressure gauges (*continued*)
    safety tips, 165
    uses, 182
    See also manifold gauges
pressure-temperature chart (P-T chart), *112–113*, 112–114, *114*, *115*
pricing, flat-rate, 264
problem solving. *See* troubleshooting
professional certification, 10, 12, 267–269
professional conduct, 176
propane, 15
protons, 54, *54*
psig (pounds per square in gauge), 97
psi (pounds per square inch), 97
psychometric chart, *234*
psychrometer, 143, *144*, 205, *206*
P-T chart. *See* pressure-temperature chart
P-traps, 221
pump pliers, *212*
pure resistance, 60
PVC (polyvinyl chloride), 224

R-22 gauge, 165
R-410A gauge, 165
radial duct system, *32*
radiation, 90, *90*
radiators, clearances, *185*
Rankine temperature scale, 87, *88*
reaming tool, *192*
rebates, 240
reciprocating compressors, *23*, 123–125
record keeping, 12–13
reduced trunk duct system, *32*
reexpansion phase, 124–125
refrigerant chart, *11*
refrigerant leaks, *16*
refrigerant-line-piercing pliers, 211, *213*
refrigerant lines
    measuring by outside diameter, 219, *220*
    types by function, 111, *111*
    See also individual lines
refrigerant piping
    energy-efficient installation, 245–247
    installation guidelines, 34
    insulation, 222, 247
    measuring by outside diameter, 220, *220*
    overview, 32–34
    sizing, 245–246
    supporting, 246
    See also copper ACR tubing; piping
refrigerants
    maintenance charge, 149
    professional certification for handling, 10, 12

    properties, 22
    saturation, 112–119
    subcooled, 120–121
    superheated, 120
refrigeration, defined, 21
refrigeration compressors, cooling capacities, 23–24
refrigeration control. *See* metering devices
refrigeration cycle
    components and operation, 21, *21*, *110*, 111, *111*
    compressors, 22–24, 122–129
    condensers, 24, 129
    evaporators, 26–28, 134–136
    metering devices, 24–26, 129–134
    refrigerant, 22
refrigeration ratchet, 210, *210*
Refrigeration Service Engineer Society (RSES), 268
refrigeration system, *21*, 35
refrigeration technicians, 8–9
refrigeration traps, 221
refrigeration wrench, 210, *210*
relative humidity, 234, *235*
relays, *69*
repairs, basic methods, 182–183
residential cooling units, 2
resistance
    defined, 56–57
    measuring, 63
    Ohm's law, 57–60
    pure, 60
    in wire, 77
resistors, 54
restrictive device. *See* metering devices
reversing valve, 43
ring connectors, 201, *202*
Rome, 13
Romex, 73, *73*
room heaters, *185*
rotary compressors, 126–129, *128*
R-values, 29
    ductwork insulation, 222

salespeople, 5, 263–264
sandpaper, 216
saturated mixtures, 130
saturation, 112–119
scale, 198
scheduled maintenance
    air duct leaks, 151
    air filters, 143, 145–146, *146*
    attending to customer complaints, 143
    blowers and fans, 152–153
    cleaning coils, 146–147

    communicating with the customer, 153
    emergency drain pan and float switch, 150
    is not service, 140
    items covered, 10
    maintenance agreements, 142
    maintenance contracts, 154
    measuring amperage and voltage, 152–153
    measuring indoor and outdoor conditions, 143, *143*
    pressure check, 149
    reasons for, 140–141
    reporting recommendations to the customer, 153
    scheduling, 142
    system charge, 149
    thermostat operation, 152
    tools, equipment, and supplies requirements, 155
    wiring, connections, and component inspection, 149–150
scheduled maintenance value, 260
schematic diagrams
    condensing unit, *72*
    defined, 70
    importance, 71–72
    package unit, *71*
    symbols, 69 (*see also* electrical symbols)
    using, 70–72
Schrader valve, 116, *116*
scientific laws, 100–105. *See also* gas laws
Scotch-Brite® pads, 216
screw compressors, *23*
screwdrivers, 211, *212*
scroll compressors, *8*, *9*, *23*, 125, *126*, *127*
sealed-hermetic compressors, 123
Seasonal Energy Efficiency Ratio (SEER), 141, 228, 231
SEC, 76
SEER. *See* Seasonal Energy Efficiency Ratio
semi-hermetic compressors, *124*, 125
sensible heat
    calculations, 94
    defined, 84–85, *85*, 91
    of vapor, 93
service technicians, 7–8, 10
sheet-metal cutters, *213*
sheet-metal duct, 28, *28–29*, *29*, *31*, 151, 222, *223*
shell and tube condenser, *130*
shock. *See* electric shock
silver-bearing solder, 195
simple-series circuits, *70*
sine waves, 55, *55*, *56*

single-phase compressors, 152–153
single-phase voltage, 58
single-pole, double-throw switch, 69
single-pole contractor, 68
slab coil, 28
sling psychrometer, 144, 205
slip joint pliers, 212
"smoke puffers," 151
soft copper tubing, 32, 33, 216, 218, 219
solar energy, 47–48, 48, 234–237, 235, 236
solar panels, 47, 49, 236
solder, 195, 197
soldering
    overview, 193, 194, 195
    purging tubing, 198
    safety, 198–199
solids, physical properties, 100, 101
solid-state circuit board, 7
Solomon, Steve, 62
specific heat, 94–95
split-phase motors, 66, 69
split systems
    basic installation components, 183, 183
    electromechanical sequence, 162–163
    power supplies, 163, 163
spring bender, 216, 218
starting amps, 189
start-up checkout, 191
steam heating, 37
stepladders, 61
subcooling, 120–121
sublimation, 91–92, 92
suction line, 21, 35, 35, 111, 111
suction stroke, 123, 124
superheat, 93, 120
supervisors, 5
surface probe, 205
sustainability, 239
swaging, 192–193, 193, 194
swaging tools, 193
switching relays, 66, 68
symbols, electrical, 66–69, 69
system charge, 149, 171

tax credits, 240
tax deductions, 240
technicians. See HVACR technicians
temperature
    capacity changes in systems, 231, 232
    Energy Efficiency Ratio, 231
    gas laws, 102–105
    scales, 86–87, 87, 88
temperature difference (delta-T), 27
temperature testers, 205

TEV. See thermostatic expansion valves
thermodynamics
    boiling point, 93
    British thermal units, 83–84
    conduction, 88, 89
    convection, 89, 89
    evaporation, 92–93
    heat flow, 87–90, 88
    heat intensity, 86
    heat quantity, 86
    latent heat, 85, 94
    latent heat of fusion, 95–96
    latent heat of vaporization, 96
    laws of, 102
    meaning of, 82
    melting and fusion, 91
    molecular motion, 90–91
    radiation, 90, 90
    sensible heat, 84–85, 85
    sensible heat calculations, 94
    sensible heat of vapor, 93
    specific heat, 94–95
    sublimation, 91–92, 92
    temperature scales, 86–87
    transferring heat, 96
thermostatic expansion valves (TXV or TEV), 9, 133, 133–134, 134
thermostats, 177
    digital, 4
    energy-efficient installation, 247
    inspection, 152
    using in system checkouts, 4
therms, 230
THHN insulation, 77
three-phase compressors, terminals, 171
three-phase voltage, 56, 56
time-delay fuses, 189
tip chart, 198
tools
    airflow measuring devices, 207, 208, 209
    digital psychrometer, 205, 206
    digital scale, 209, 209
    grounding, 58
    hand tools, 210, 211–218
    manifold gauge set, 203, 204, 205
    power tools, 205, 206
    temperature testers, 205
    vacuum pumps, 207, 207
    valve core removal tool, 205, 206
    volt, ohm, and milliammeter, 210 (see also multimeters)
torch tips, 196, 198
transformers, 169
traps, 148, 221
troubleshooting
    case studies, 167–176

communication between technician, dispatcher and customer, 158–159
diagnosing airflow problems, 162
diagnosing electrical problems, 161
diagnosing mechanical problems, 161
electromechanical sequence, 162–164
employee productivity, 166–167
evaluation, 167
finding opens in the control sequence, 160
listing possible causes, 159
making a diagnosis, 166
overview, 177–178
required tools, instrumentation, and supplies, 164–165
sequence of, 167
systematic method, 167
tubing
    brazing, 193, 195, 197–198
    cutting, 191–192, 192
    flaring, 199, 200, 201
    soldering, 193
    swaging, 192–193, 194
    See also copper ACR tubing; piping
tubing benders, 216, 218
tubing cutters, 191–192, 192
TW insulation, 77
two-stage condensing units, 243
TXV. See thermostatic expansion valves
Type DMV ACR tubing, 34
Type I exam, 12
Type II exam, 12
Type III exam, 12
Type K tubing, 34, 219
Type L tubing, 34, 219
Type M tubing, 34, 219

UA STAR, 269
UF, 76
UL 181A duct tape, 245
UL 181B duct tape, 222, 223, 245
ultraviolet (UV) light, 247
Uniform Mechanical Code (UMC), 185
United Association of Journeymen and Apprentices of the Plumbing and Pipefitting Industry of the United States and Canada (UA), 269
United States Department of Energy (DOE), 241
United States Environmental Protection Agency (EPA), 241
United States Green Building Council, 239
upflow heating systems, 183, 184, 187
upflow installation, 183, 184

utilities, electric, 230, 240
utility bills, energy savings, 265
utility knife, *214*
U-tube, 98, *99*

vacuum gauge, micron, 211, *211*
vacuum pumps, 207, *207*
value selling, 258-261
valve core removal tool, 205, *206*
vapor
    boiling point, 93
    evaporation, 93
    sublimation, 91-92
vapor boiling, 93
vaporization, 93
V-belts, 152
venting
    gas furnaces, 38-39
    oil furnaces, 41
vibration isolator, *188*
vibrations, air handler installations, 187, *188*
voltage
    measuring, 62-63
    measuring in compressors, 153, *154*

National Electrical Code levels, 63
    See also volts
voltmeter, 210
    troubleshooting with, 160, 165
    uses, 182
volt-ohm-milliammeter (VOM), 61-64, 210
volts
    alternating current, 55-56
    defined, 54
    direct current, 54, *55*
    Ohm's law, 57-60
    See also voltage
volume, gas laws, 102-105
VOM (volt-ohm-milliammeter), 61-64, 210

water-cooled condensers, *24*, 239
water heaters, clearances, *185*
water-heating, solar, 47
water vapor, 93. *See also* vapor
wattage, 57
watts, 57, 60, 229
wet bulb thermometer, 143, *144*
wind energy, 237, *237*
wind farms, 237, *237*
windmills, 237, *237*

wire brushes, 216, *217*
wire butt, 201
wire cutter pliers, *212*
wire nuts, 201
wires
    allowable ampacities, *74-75*
    current rating, 72
    gauge, 72-73
    identifying markings, 73, 76
    insulation, 77
    physical properties, 53-54
    ratings, 72-73
    resistance, 77
    See also conductors (electrical)
wire stripper, *212*
wiring
    inspection, 149-150, *150*
    installations, 188-189, *189*
wiring diagrams, 71. *See also* schematic diagrams
wood files, *214*
wood saws, 215